Lecture Notes in Artificial

Subseries of Lecture Notes in Computer Science
Edited by J. G. Carbonell and J. Siekmann

Lecture Notes in Computer Science
Edited by G. Goos, J. Hartmanis and J. van Leeuwen

Springer
Berlin
Heidelberg
New York
Barcelona
Hong Kong
London
Milan
Paris
Singapore
Tokyo

Harry S. Delugach   Gerd Stumme (Eds.)

Conceptual Structures: Broadening the Base

9th International Conference
on Conceptual Structures, ICCS 2001
Stanford, CA, USA, July 30-August 3, 2001
Proceedings

Springer

Series Editors

Jaime G. Carbonell,Carnegie Mellon University, Pittsburgh, PA, USA
Jörg Siekmann, University of Saarland, Saarbrücken, Germany

Volume Editors

Harry S. Delugach
University of Alabama in Huntsville
Computer Science Department, TH N-351
Huntsville, AL 35899, USA
E-mail: delugach@cs.uah.edu

Gerd Stumme
Universität Karlsruhe (TH)
Institut für Angewandte Informatik und
Formale Beschreibungsverfahren (AIFB)
76128 Karlsruhe, Germany
E-mail: stumme@aifb.uni-karlsruhe.de

Cataloging-in-Publication Data applied for

Die Deutsche Bibliothek - CIP-Einheitsaufnahme

Conceptual structures : broadening the base ; proceedings / 9th
International Conference on Conceptual Structures, ICCS 2001, Stanford, CA,
USA, July 30 - August 3, 2001. Harry S. Delugach ; Gerd Stumme (ed.). -
Berlin ; Heidelberg ; New York ; Barcelona ; Hong Kong ; London ; Milan ;
Paris ; Singapore ; Tokyo : Springer, 2001
(Lecture notes in computer science ; Vol. 2120 : Lecture notes in
artificial intelligence)
ISBN 3-540-42344-3

CR Subject Classification (1998): I.2, G.2.2, F.4.1, F.2.1

ISBN 3-540-42344-3 Springer-Verlag Berlin Heidelberg New York

Springer-Verlag Berlin Heidelberg New York
a member of BertelsmannSpringer Science+Business Media GmbH

http://www.springer.de

© Springer-Verlag Berlin Heidelberg 2001
Printed in Germany

Typesetting: Camera-ready by author, data conversion by PTP Berlin, Stefan Sossna
Printed on acid-free paper   SPIN 10839891   06/3142   5 4 3 2 1 0

Preface

We are pleased to bring you this collection of papers for the Ninth International Conference on Conceptual Structures (ICCS), representing continued excellence in conceptual structures research. We have adopted the title "Broadening the Base," acknowledging the importance of contributions from scholars in many research areas. The first ICCS meetings focused primarily on Sowa's conceptual graphs; in recent years, however, the ICCS conference series has intentionally widened its scope to stimulate research across domain boundaries. We hope that this stimulation is further enhanced by ICCS 2001 continuing the long tradition of lively conferences about Conceptual Structures.

We wish to express our appreciation to all the authors of submitted papers, to the general chair, to the members of the editorial board and the program committee, and to the additional reviewers for making ICCS 2001 a valuable contribution to the knowledge processing research field. We would also like to acknowledge the leadership of Guy Mineau and Bernhard Ganter in providing a solid framework for an open and effective reviewing process. Very special thanks go to the local organizers for making the conference possible and, furthermore, an enjoyable and inspiring event. We are grateful to the University of Alabama in Huntsville, and the University of Karlsruhe for their generous support.

May 2001

Harry Delugach
Gerd Stumme

Organization

The International Conference on Conceptual Structures (ICCS) is the annual conference and principal research forum in the theory and practice of conceptual structures. Previous ICCS conferences have been held at the Université Laval (Quebec City, 1993), the University of Maryland (1994), the University of California (Santa Cruz, 1995), Sydney (1996), the University of Washington (Seattle, 1997), Montpellier (1998), at Virginia Tech (Blacksburg, 1999), and at Darmstadt University of Technology (2000).

Program Chair

Harry S. Delugach   The University of Alabama in Huntsville, U. S. A.
Gerd Stumme   Universität Karlsruhe, Germany

Local Chair

Robert G. Spillers   Woodside, CA, U. S. A.

Editorial Board

Galia Angelova (Bulgaria)   Deborah McGuinness (U. S. A.)
Michel Chein (France)   Guy Mineau (Canada)
Peter Eklund (Australia)   Bernard Moulin (Canada)
John Esch (U. S. A.)   Marie-Laure Mugnier (France)
Bernhard Ganter (Germany)   Heather Pfeiffer (U. S. A.)
Roger Hartley (U. S. A.)   Uta Priss (U. S. A.)
Mary Keeler (U. S. A.)   John Sowa (U. S. A.)
Lotfi Lakhal (France)   Bill Tepfenhart (U. S. A.)
Wilfried Lex (Germany)   Rudolf Wille (Germany)

Program Committee

Jean-François Baget (France)   Carole Goble (U.K.)
Tru Cao (U.K.)   Robert Godin (Canada)
Claudio Carpineto (Italy)   Michel Habib (France)
Paul Compton (Australia)   Ollivier Haemmerlé (France)
Dan Corbett (Australia)   Adil Kabbaj (Morocco)
Judy Dick (Canada)   Adalbert Kerber (Germany)
David Genest (France)   Sergei Kuznetsov (Germany)
Olivier Gerbé (Canada)   Pavel Kocura (U.K.)

Wolfgang Lenski (Germany)

Graham Mann (Australia)

Philippe Martin (Australia)

João P. Martins (Portugal)

Ralf Möller (Germany)

Aldo de Moor (Netherlands)

Amadeo Napoli (France)

Lhouari Nourine (France)

Peter Øhrstrøm (Denmark)

Silke Pollandt (Germany)

Susanne Prediger (Germany)

Richard Raban (Australia)

Anne-Marie Rassinoux (Switzerland)

Daniel Rochowiak (U. S. A.)

Eric Salvat (France)

Ulrike Sattler (Germany)

Peter Patel-Schneider (U. S. A.)

Stuart Shapiro (U. S. A.)

Finnegan Southey (Canada)

Thanwadee Thanitsukkarn (Thailand)

Petko Valtchev (Canada)

Michel Wermelinger (Portugal)

Karl Erich Wolff (Germany)

Further Reviewers

Alex Borgida (U. S. A.)

Siegfried Handschuh (Germany)

Ralf Molitor (Switzerland)

Table of Contents

Language and Knowledge Structures

Logical and Mathematical Foundations of Conceptual Structures

A Peircean ontology of language

Janos Sarbo and József Farkas

University of Nijmegen, The Netherlands
janos@cs.kun.nl

Abstract. Formal models of natural language often suffer from excessive complexity. A reason for this, we think, may be due to the underlying approach itself. In this paper we introduce a novel, semiotic based model of language which provides us with a simple algorithm for language processing.

1   Introduction

Formal models of natural language often suffer from excessive complexity which, in our opinion, may be due to the underlying approach itself. Their formal character implies that they are doomed to reflect what is 'natural' in language in an ad hoc fashion only.

In this paper we introduce – on the bias of logic – an alternative model of language which is built on the assumptions that (1) language symbols are signs, (2) the meaning of a sign emerges via mediation, and (3) signs arise from a dichotomous relation of perceived qualities. We argue that on the basis of these assumptions and the properties of signs, a simple parsing algorithm can be defined.

2   Sign and perception

In our analysis we follow the principles of Peirce's semiotic ([5], [7]). Accordingly, a sign signifies its object to an agent in some sense, which is called the interpretant of the sign. The inseparable relation of sign, object and interpretant (each of which is a sign, recursively) is called the *triadic relation* of sign. In this paper we start from the observation that the ground for any sign is a *contrast* in the 'real' world. Because sign and object are the primary representation of such contrast, sign and object must be *differ* from each other.

How can we know about signs? We have discussed this problem in [3] and here we will only recapitulate the main results. Following cognition theory ([4]), the recognition of any sign must begin with the sensation of the physical input. Physical stimuli enter the human receiver via the senses which transform the raw data into internal sensation continuously. The output of the senses, a bioelectric signal, is processed by the brain in percepts. The generation of such a percept is triggered by a change in the input, typically, or by the duration of some sampling time, e.g. in the case of visual perception.

H. Delugach and G. Stumme (Eds.): ICCS 2001, LNAI 2120, pp. 1–14, 2001.

The brain compares the current percept with the previous one, and this enables it to distinguish between two sorts of input qualities: one, which was there and remained there, something stable, which we will call a *continuant*; and another, which, though it was not there, is there now (or the other way round), something changing, which we will call an *occurrent*. The collections of continuants and occurrents, which are inherently related to each other, form the basis for our perception of a phenomenon as a sign. We also assume that, by means of *selective attention*, we recognise in these collections coherent sets of qualities: the qualities of the *observed* and those of the *complementary* part of the phenomenon. We will refer to these sets collectively as the *input*.

2.1 The variety of signs

In Peirce's view, the most complete signs are the icon, index, and symbol which represent their object on the basis of, respectively, similarity, causality and arbitrary consensus. Besides this taxonomy, Peirce also distinguishes signs, respectively, according to the categorical status of the sign, and according to the relationship between object and interpretant. From a categorical perspective, signs can be qualisigns, sinsigns or legisigns, which correspond, respectively, to firstness, secondness and thirdness. In other words, a sign can be a quality, an actual event, or a rule. Seen from the perspective of the relationship between object and interpretant, a sign may be a rheme, a dicent or an argument. In other words a sign may signify a qualitative possibility, an actual existence, or a proposition. Thus we obtain nine kinds of sign which may be arranged in a matrix as shown in fig. 1 (the meaning of the horizontal lines and directed edges will be explained later). Although Peirce defined more complex systems of signs, we hold that his 'simple' classification is the most practical.

Here, the expressions 'class of a sign' or 'type of a sign' will be used interchangeably. In our specification of logical and language signs we will make use of Peirce's classes. A comparison between our use of them and his definitions may be found in ([3]).

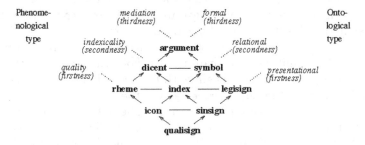

Fig. 1. Peirce's classification of signs

3   Logic and semiosis

We argue that semiosis begins with our experience of the input collections (which are qualities themselves) as signs. Such signs, which are called in Peirce's terminology a qualisign, are special signs for which we have no denotation. Although qualisigns are coherent, by definition, we experience them as independent signs. In order to be able to refer to the qualisigns, we will make use of logical symbols which are the most general of that type of sign. Logical symbols are signs from the logical point of view ([2]). We will represent such signs as *logical functions* on two variables A and B, respectively, for the continuants and the occurrents, over two values 0 and 1, for the complementary and the observed part.

How do complex signs emerge? Again, we refer to [3] where we introduced a semiotic model of signs and elaborated it for logical signs. Briefly, that model is based on a process in which trichotomic relations are generated recursively revealing gradually more accurate and clear *approximations* of the full richness of a sign of an observed phenomenon. Accordingly, in this paper we argue that the proposition of the input as a sign arises from the input qualisigns via a number of signs. By virtue of the fast and continuous nature of cognition we may assume that such signs are not recognised isolatedly, but only as 'temporary' signs. Such signs, which are approximations of the final assertion, are *re*-presentations of the input qualisigns. Their classes are identical to those defined by Peirce. We will argue that language is based on a similar mechanism.

The qualisigns form the ground for our semiosis. Because such signs are perceived as independent signs, but it is their unity that signifies the contrast as a whole, we may assume that there exists a *need* for the representation of the full richness of the relation of the qualisigns, eventually as a proposition. We will argue that this 'representational need' also appears in language in the form of the relational need of symbols.

The sign *mediates* between object and interpretant. In the case of logical signs, the interpretant is defined as the *application* of sign to object, both of which are logical functions. Here, the notion 'application' is used in a broad sense. In the particular case we allow that sign and object 'merge' as a result of such operation. Therefore, this form of semiosis will be called an *interaction*; sign and object will be referred to as its *constituents*.

A derivation of those 'temporary' signs, and, eventually, the proposition of the input as a sign proceeds as follows ([3]). We denote the signs of the collections of the *observed* part by the functions A and B (and those of the complementary part as $\neg A$ and $\neg B$). These collections are similar to the input and appear simultaneously, by definition. Something which is similar to something else can be a sign of it (icon). Such a sign, A *or* B, must refer to an object which includes both A *and* B (sinsign). The interpretant of such a sign and object can refer either to their common origin (via the complementary signs), which is called the *context* (index), or, to their relative difference. The latter provides us with a representation of the input collections independent of each other, the 'abstract' continuants (rheme), and how they co-occur, the 'abstract' occurrents (legisign).

The index can be represented by the Shäffer and Peirce functions, the rheme and the legisign, respectively, by the inhibition and exclusive-or functions.

The application of the index to the rheme and to the legisign yields the complementation of those abstract signs by the context, i.e. the actual constituent 'parts' of the input (dicent), and, the 'property' characterising their co-occurrence (symbol). Dicent and symbol can be represented, respectively, by the implication and the equivalence functions. Finally, by merging this property (as sign) with the sign of those actual parts (as object) we get a proposition of the input as a sign (argument) which can be represented by a syllogism (degenerately).

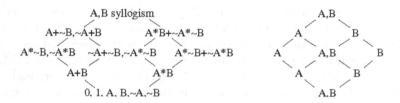

Fig. 2. The classification of logical signs

The logical representation of the qualisign can be completed with two more functions: 0 ('not valid input'), and 1 ('valid input'). As a result, we can conclude that in the semiosis of logical signs *all* Boolean functions (on two variables) can emerge. The resulting classification is depicted in fig. 2 (on the left-hand side).

Notice that in the derivations of the logical signs the interpretant always emerges from *neighbouring* sign and object (in the sense of the triadic relation). Such signs are connected in fig. 1 by a horizontal line. By virtue of the fundamental character of logic it may be conceived that semiosis can always be explained in a similar vein.

4   Language

We will argue that language is logic, *sequentially*. In this section we will show that by means of this single condition a model of language can be derived which is isomorphic to the one of logic. Language signs are symbols which are subject to syntactic and semantic rules. In this paper we will consider only syntactic rules, and restrict language to syntax ([6]).

'Sequential' means that the input signs appear one after the other as qualisigns. Earlier we have pointed out that the qualisigns are the (first) representation of a contrast. In our cognitive model we defined such a contrast between continuant and occurrent. Because language symbols are about 'real' world phenomena (typically), we may assume that an underlying contrast, analogous to the one of cognition, does exist in language, too. The language equivalent of continuant and occurrent is identified, respectively, in the aspects 'thing' and

'change', typically represented by nominals and verbs. Because language signs
are inherently related to memory, a 'change' can also refer to an 'appearing'
new fact, which is a relative change of some thing or event (i.e. an alteration),
typically represented by adjectives and adverbs.

In what follows, we will refer by a sign class to the classification of *syntactic*
symbols. In as much as the logical equivalent of the sign of a collection of 'things'
and 'changes' must be the functions A and B, respectively, we will use the logical
and syntactic names of the qualisigns interchangeably.

4.1   A preliminary classification

We will derive a model of language by transforming our specification of logical
signs to a sequential one. Such a derivation is partly technical. In order not to get
drown in the details, we make a preliminary attempt at a semiotic classification
of the main syntactic concepts.

The argument, which is a proposition, must correspond to the notion of a
sentence in as much as both are expressive of a statement. Hence, the dicent
must be subject, and the symbol predicate. By virtue of its factual meaning (cf.
'modification') the index can be an adjective, an adverb, or a complemented
preposition (in short, prep-compl). A rheme is a possible for the subject, for
example, a noun; a legisign is an actual event in the syntactic sense, that is, a
'structure' event defined as a rule, e.g. a verb(-complement).

4.2   Towards a sequential version of logic

Qualisigns are the representation of a contrast. In the sequential case when
each qualisign consists of a single symbol, also the contrast itself will appear
sequentially. Accordingly, a qualisign will have either the aspect of a 'thing'
(logically A), or a 'change' (logically B). From this it follows that in language
we cannot distinguish between asserted and negated signs (at least not at this
level), and this implies that language phenomena must include their own context.
Hence, a part of the input may have to be devoted to the representation of the
context as a sign. Akin to the general case of signs, we can have access to memory
knowledge, e.g. a lexicon, but such knowledge is *not* related to the perceived
input in any way.

Syntactic qualisigns consist of a single symbol and define a unique universe,
therefore different qualisigns cannot be merged. Because input symbols appear
continuously (i) *'place' has to be created for the appearing next qualisign.*

What can be done with the previous qualisign? The answer is simple, we
have to re-present it by another sign. Following our classification of logic, such
a sign can be an icon or a sinsign.

We mentioned that in the case of language (ii) *a qualisign is either A or B,
but not both.* Because a sinsign is a representation of an event, it must include
the aspect of a 'change'. Accordingly, the re-presentation of a qualisign which
is B can be a sinsign, hence the one which is A must be an icon. Such a representation involves the generation of a new sign, the denotation of which is

identical to the one of the qualisign. From the bottom-up 'nature' of our logical model it follows that the new sign will include the older one degenerately (in the semiotic sense).

Icon and sinsign are different representations of the same qualisigns. Because of (ii) and the uniqueness of signs[1], it follows that icon and sinsign symbols typically will not be adjacent, therefore (iii) *icon and sinsign implement a 'sorting' re-presentation of qualisigns*. In particular, an icon and a sinsign can define a shared universe, for example, in the case of compound (multi-word) symbols, and idiomatic expressions.

By virtue of (i), we have to represent the previous qualisign by a new sign, which is an icon or a sinsign. The appearance of this new sign, in turn, may force us to do the same for the previous icon or sinsign and, eventually, signs may have to be generated in any class. In sum, the conclusion can be drawn that in the sequential model of logic there is sign generation *by need*.

4.3   Sign generation by need

The logical interpretant is defined as the application of sign to object. In language, such application will be called a 'binding'. Due to the sequential nature of language signs, we will also have to consider two degenerate variants of such interaction which are the following.

a) *Accumulation*: in which case an existing sign is combined with another sign of the same type. Such an interaction assigns the same meaning to both constituents thereby rendering them indistinguishable.

b) *Coercion*: in which case a new sign is generated for the denotation of an existing sign (which is said 'coerced'). Coercion applies if the signs, which are to interact, are incapable for accumulation or binding.

Accumulation is possible in any class, except for the qualisign. For example, a series of adjectives can be merged to a single sign via accumulation. Coercion is applicable in any class, except for the argument. In this form of an interaction we refer by the 'constituent' to the sign triggering the interaction.

Coercions play an important role in syntactic sign recognition, more specifically, in what we call as, the *default* scheme. This type of semiosis can be characterised as follows. By virtue of (i), signs are generated by need. The coercion of a qualisign, A or B, yields either an icon or a sinsign. Due to subsequent applications of (i), such an icon or sinsign is coerced, respectively, to a rheme or a legisign, and eventually to a dicent or a symbol. An index cannot emerge this way, because there are no negated signs (so far). By virtue of the appearing different kinds of qualisigns, in the end, we may have a dicent and a symbol sign which are adjacent and generate the argument sign (the sentence as a sign).

Notice that also a dicent or a symbol can be coerced to an argument, but such a sign will be a degenerate one, semiotically, because an argument must

---

[1] Lexical ambiguity of a symbol is treated by introducing a unique denotation for each meaning.

represent the observed phenomenon in its *character* ([5]) and has to include both *A* and *B*. Such signs of the input as a whole, are the dicent and the symbol signs.

In as much as the argument arises from dicent and symbol (subject and predicate), we may conclude that, in the default case, *only* subject and predicate are recognised and their relation represented as a sentence. It will be argued that language sign recognition always follows this scheme and any deviation from it may only occur if otherwise a successful parse cannot be found. Such a case will be described in the next section.

4.4   The genesis of the context

In language we are burdened by the task of the recognition of the entire string of input symbols as a single sign. Although, in some cases, the sign generation operations (coercion, accumulation and binding) may be unsatisfactory, the above goal can yet be achieved if we allow for a sign, which is potentially subject or predicate, to be represented degenerately (in the semiotic sense). Such signs define, what we called as, the *context*. The degenerate representation of symbols also plays an important role in the 'stepwise' construction of signs.

Sign degeneration ($\downarrow$) can be explicated by the phenomenological and ontological types of signs indicated in fig. 1 as follows: dicent$\downarrow$index if subject meaning is not present 'formally'; symbol$\downarrow$index if predicate meaning is not present 'mediationally'; dicent$\downarrow$rheme if subject meaning is unfinished 'indexically'; symbol$\downarrow$legisign if predicate meaning is unfinished 'relationally'.

Sample context signs are, for example, dicent$\downarrow$index in Mary, John likes (the potential subject Mary becomes a context sign for likes); symbol$\downarrow$index in Mary with flowers (the potential property with flowers is represented degenerately as a context sign for Mary).

The existence of degenerate signs is related to our ability of analysing a segment of input symbols (*nested input*) independently from the rest of the input. When such a segment is recognised, its meaning relative to the input as a whole is represented degenerately. Because of this degeneration such a symbol *will not* appear as an isolatedly recognised sign. The semiosis of such *nested signs* is implemented by a recursive application of the sign recognition 'machinery' (for example, in the case of coordination, subordination etc.).

In certain cases we can know in advance if a sign eventually will become a context sign. If a nested sign consists of a single input symbol, the recursive analysis may be replaced by a coercion. For example, an index sign can be directly generated by coercion from such an icon or sinsign. This kind of optimisation is typical for adjective, adverb and prep-compl symbols and it is lexically specified as their syntactic property. The 'stepwise' construction of a sign can be optimised by the immediate generation of the degenerate representation of the interpretant. For example, the interaction of such rheme and index can be directly represented as a rheme or an index sign, without explicitly generating its meaning as a dicent sign.

The representation of logical signs in the sequential case is depicted in fig. 2 (on the right-hand side). Although the same denotations, *A* and *B*, appear many times, each of the occurrences has a different meaning. A language implementation of these signs involves the mapping of the logical functions to their syntactic equivalent. The definition of such a mapping is the subject of the next section.

4.5   Relational need

The requirement that all symbols have to be 'merged' to a single sign and single universe is in conflict with their individual character and unique universe property. How can such symbols be combined?

An answer can be found in the properties of syntactic qualisigns. Such symbols are representations of one 'half' of a contrast. Because we can only reason about a contrast if both of its 'parts' are known, syntactic symbols can be said to be 'longing' for finding their complementary part. This inherent property of syntactic symbols is the language equivalent of the 'representational need' of signs introduced in sect. 3. This interaction 'potential' forms the *ground* of the relational properties of syntactic signs. In as much as the 'parts' of a contrast are different, syntactic sign interactions always arise between symbols of a different type of relational properties.

If, as we argued, logical and language signs are analogous, this must also apply to syntactic symbols. We map continuant and occurrent (via 'thing' and 'change'), respectively, to the *relational types* 'passive' (p) and 'active' (a). Formally, we define the type 'neutral' (n). The logical qualisigns, A and B, are mapped to a syntactic *relational need*, or valency, represented as a pair consisting of a relational type and a set of relational qualities (or, syntactic properties). In this paper it will be assumed that the set of such qualities is *finite*.

Because any sign is a re-presentation of the qualisigns, the relational qualities may contribute to different relational properties in each class. These qualities, which are lexically defined, will be omitted in the specification. Accordingly, we will refer to the relational need of any sign by its type only (and call it an a-, p- and n-need, ambiguously). A sign, which has an n-need, is finished (relationally). Such a sign cannot take part in any interaction, except for a coercion and, vice versa, only such a sign can be subject to a coercion (typically).

The above mapping defines an initial representation of the relational need of signs in the various classes. By virtue of the properties of logical signs, this mapping has to be further developed as follows. Notice that each modification amounts to an adjustment of the definition of the qualisigns.

We introduce an n-need in the qualisign, icon, sinsign and argument classes. Indeed, qualisigns are independent signs which cannot interact; icon and sinsign symbols typically do not establish a relation; and finally, the argument which is the sentence as a sign, must be 'complete' (and neutral, relationally). Because icon, index and symbol do function as sign, in the sense of the triadic relation, we introduce an (optional) p-need for each class which functions as object (e.g. for the legisign class). Finally, when sign and object, respectively, are initially assigned to a p- and an a-need, their relational needs are exchanged in the mapping. The reason for this modification can be explained as follows.

The predicate of the sentence can arise from a legisign (a verb) via the complementation of an index. Traditionally, such complement is lexically specified in the verb's entry. Semiotically, however, it is the complement that points to the verb and selects its actual meaning. This interpretation is conform with the default scheme of sign recognition, according to which, verbs only function as a sign in the predication symbol interaction. Because also adjectives, adverbs etc. function as sign (in the sense of the triadic relation), in our model of language the concept of 'modifier' and 'complement' amalgamate.

In sum, the relational need of a qualisign can be defined as a set, an element of which is a reference to a class in which the qualities of the input symbol can contribute to a relational need (of a sign) via re-presentation. For example, the valency of a transitive verb can be defined as {legisign,symbol} referring to an a-need, respectively, for the complement and the subject. A p-need, which is optional (typically) is omitted (also in the examples).

We demand that a relational need is always satisfied. In particular, a binding satisfies a pair of a- and a p-needs by resolving them; an accumulation merges a pair of needs to a single one; a sign generated by coercion inherits the valency of the sign coerced. In as much as a p-need is optional, when such a need is not present, it is equivalent to an n-need.

Because the index sign does not partake in the default scheme of syntactic sign recognition, the generation of such signs is subject to special conditions. We demand that a symbol can become an index having a p-need, either if any other analysis of that symbol eventually fails, or, if there is an existing a-need of a symbol in the legisign class.

4.6   A Peircean model of language

In this section we will specify language signs and their valency. Instead of using a set representation, we directly refer to the classes of fig. 1. With respect to the lexical denotation of language signs we will refer to the types of speech.

Syntactic qualisigns are defined as follows: A=noun; B=verb, adjective, adverb, prep(-compl), where 'compl' can be a noun, verb, adjective or adverb. Formally, we also define 0='no input' and 1='end of input'. The latter can represent the 'dot' symbol which is considered an A and B sign, having an a-need in any class, but incompatible with any sign except for itself. Hence, the dot symbol is capable of 'forcing' the realisation of pending interactions. In the examples we will assume that the input is closed by a finite number of dots. The specification is depicted in fig. 3 (on the left-hand side). An occurrence of a p-need may also denote an n-need, except for the dicent sign class (cf. subject).

4.7   Morphology

We argue that morphology can be modelled analogously to syntax and logic. Due to space, we can only run through the essential ideas behind such a specification. We characterise morphological symbols as follows (now we refer by a sign class to the Peircean classification of morphological signs).

Dicent and symbol signs are finished morphologically and represent only a *sorting* of signs. This is justified by the different syntactic properties of the signs of these classes. A symbol, for example, an adjective, is a sign which, *syntactically*, requires a complement adjacent to it in the input (on 'surface' level). A dicent, for example, a noun or a verb, does not have such a property.

Rheme, index and legisign, respectively, represent a qualitative possibility, a factuality (e.g. definiteness), and something rule-like (e.g. argument-structure). These signs are involved in the creation of the syntactic relational needs, passive and active (N.B. such a need should not be mixed up with the signs' morphological valency which, akin to syntax, determine the *morphological* sign interactions). Finally, icon and sinsign are either sorting, alike to syntax, or, when they are adjacent, are generating a new syntactic property, e.g. an adjective from a verb (such signs represent the same phenomenon and have a shared universe).

Morphological signs and their *morphological* relational needs are displayed in fig. 3 (on the right-hand side). The dicent sign is defined as a coerced rheme, or a rheme-index interaction, whereas the symbol as a coerced legisign, or an index-legisign interaction. The morphological argument sign coincides with the syntactic qualisign.

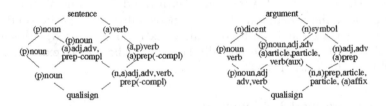

Fig. 3. Classification of syntactic and morphological symbols

Morphological qualisigns are defined as follows: A=noun, verb, adjective, adverb, coordinator; B=preposition, article, particle, affix. Formally, we also define 0='no input' and 1='separation'. The latter can represent the 'space' symbol which is considered an A *and* B sign, incompatible for any interaction, except for accumulation (with any sign triggering the interaction). Hence, the space symbol can 'sweep' out the morphologically finished signs. Dot symbols are treated similarly as in the case of syntax.

4.8   Formal definition

We specify a recogniser for our model as a pushdown automaton. Formally, the automaton is defined as a 8-tuple $M = (K, C, I, \Gamma, \rho, s, F, \Delta)$ where K is a finite set of states, C is a finite set of sign classes, I is a finite set of input symbols, Γ is a finite set of stack symbols, ρ is a function defining the relational need of input symbols, s is the initial state, $F \subseteq K$ is a set of final states, Δ is a transition relation consisting of is a finite set of transition rules.

A transition rule is a mapping $(p, u, \beta) \to (q, \gamma)$ where $p, q \in K$ are, respectively, the states before and after the transition, $u \in I^*$ are the symbols to be read, and $\beta, \gamma \in \Gamma^*$ are the symbols to be popped and pushed. There is a single transition rule for reading the next input symbol onto the stack, all other rules are 'internal' transition rules which operate only on the stack. Such rules will be used for the specification of the syntactic sign interactions as follows.

Because such rules always refer to the set C, in the definitions we will replace p and q by C, but we will specify only those classes of C (by listing them) which are involved in the transition. A further simplification is possible due to the observation that the stack is used in 'frames'. Such a frame (β) has a finite number of locations for each element of C. The first two of these locations will be called the next and the existing sign of a class, all others are temporary locations. A temporary location can be necessary, for example, for the evaluation of a condition. The specification of such computations may require a number of internal rules which we alternatively define as a (logical) expression. Accordingly, the temporary locations are removed from the transition rules.

The value of a next and existing sign can be an input symbol (only in the qualisign class), or a relational need $r = (t, y)$ where $t \subseteq C$ and y is a finite set (of syntactic properties). The logical type of r is defined by the function τ as follows: $\tau((t, y)) = A$ if $t = \emptyset$, and B, otherwise.

Nondeterminism is assumed to be implemented by backtracking ([1]). In the definition of Δ we will allow a reference to the current value of the evaluation mode, forward('f') or backward('b'), via the function $mode$. Finally, we will make use of a graph $G = (C, E)$ where $E = E_d \cup E_h$, and $E_d, E_h \subseteq C \times C$ are, respectively, the set of directed edges and horizontal lines (undirected edges) as shown in fig. 1 (a formal definition is omitted). The successors and neighbours of a class are defined, respectively, by the functions $succ(c) = \{c' | (c, c') \in E_d\}$ and $adj(c) = \{c' | (c, c') \in E_h\}$. An element of $succ(c)$ and $adj(c)$ is denoted, respectively, as c^s and c_a.

In a transition rule, state and stack will be merged. The stack is read non-destructively ($\gamma = \beta'\beta$ where β' and β are, respectively, the next and the current frame). Because input symbols (u) are only related to the qualisign class of C, the treatment of these symbols is specified by a separate rule (as mentioned earlier) in which u is defined to appear in the next sign location of the qualisign class. In sum, we will refer to triples (c, s, s') where c is a class, and s and s' are, respectively, its next and existing signs (any of s and s' may not be specified, in which case it is denoted by a "_" symbol). The triples on the left- and right-hand side of a rule, respectively, refer to the current and next frame on the top of the stack. In the rules below (and also in the examples) the names of the sign classes are abbreviated; ε denotes the empty value.

input :
$(qual, u, \varepsilon) \to (qual, \varepsilon, \rho(u))$.
$(qual, \varepsilon, r), (icon, \varepsilon, \_) \to (qual, \varepsilon, \varepsilon), (icon, r, \_)$ IF $\tau(r) = A$.
$(qual, \varepsilon, r), (sins, \varepsilon, \_) \to (qual, \varepsilon, \varepsilon), (sins, r, \_)$ IF $\tau(r) = B$.

The internal rules will be given by rule schemes for the class variable X ($X \in C\backslash qualisign$). The class of a symbol generated by binding is denoted as X^b. Because of space, the specification of degeneration is omitted and X^b is restricted to X^s (in general, $X^b \in \{X, X_a, X^s\}$). In virtue of the special conditions required by the index class, the triple corresponding to the legisign class is explicitly defined in some of the rule schemes. We will make use of the functions *cmpacc* and *cmpbind* which, respectively, yield true if their arguments can syntactically accumulate and bind in the class specified. Furthermore, we will refer to the functions *acc*, *coerce* and *bind* which, respectively, determine the relational need of the symbols yielded by accumulation, coercion and binding. The function *condix* checks if the special conditions of the index class hold.

accumulation :
$$(X, r, r') \rightarrow (X, \varepsilon, acc(X, r, r')) \quad \text{IF } cmpacc(X, r, r').$$

coercion$_1$:
$$(X, r, r'), (X_a, \varepsilon, \varepsilon), (X^s, \varepsilon, \_), (legi, \_, \lambda) \rightarrow (X, \varepsilon, r), (X^s, r^c, \_)$$
$$\text{IF } \{p, X\} \not\subseteq t' \wedge \neg cmpacc(X, r, r') \wedge condix(X^s, r^c, \lambda)$$
$$\text{WHERE } r^c = coerce(X, r', X^s), \; r' = (t', y').$$

coercion$_2$:
$$(X, \varepsilon, r'), (X_a, r_a, \varepsilon), (X^s, \varepsilon, \_), (legi, \_, \lambda) \rightarrow (X, \varepsilon, \varepsilon), (X_a, \varepsilon, r_a), (X^s, r^c, \_)$$
$$\text{IF } \{p, X\} \not\subseteq t' \wedge \neg cmpbind(X, r', X_a, r'_a) \wedge condix(X^s, r^c, \lambda)$$
$$\text{WHERE } r^c = coerce(X, r', X^s), \; r' = (t', y').$$

binding :
$$(X, r, r'), (X_a, \varepsilon, r'_a), (X^b, \varepsilon, \_), (legi, \_, \lambda) \rightarrow (X, \varepsilon, r), (X_a, \varepsilon, \varepsilon), (X^b, r^b, \_)$$
$$\text{IF } (p \in t' \wedge X_a \in t'_a) \wedge cmpbind(X, r', X_a, r'_a) \wedge condix(X^b, r^b, \lambda)$$
$$\text{WHERE } r^b = bind(X^b, r', r'_a), \; r' = (t', y'), \; r'_a = (t'_a, y'_a).$$

$condix(X, (t, y), (t_l, y_l))$:
$$X = indx \wedge ((p \in t \wedge (mode = \text{`b'} \vee legi \in t_l)) \vee X \in t) \vee \text{TRUE}.$$

A parser can be defined by using temporary locations. Such a location may contain an input symbol, or, one or two constants which are used as pointers to locations of the previous frame.

4.9   Example

In this section we show the analysis of the sentence Mary eats pizza with a fork. The sign matrix will be represented in a tabular form. A column corresponds to a sign class, and a row to the recognition of an input symbol. The treatment of space and dot symbols is omitted. The final step of sign recognition (the generation of an argument sign) is not displayed. The accumulation of symbols is denoted by a "/" sign. Rule names are abbreviated as follows: input(i), accumulation(a), coerce$_1$(c$_1$), coerce$_2$(c$_2$) and binding(b); degeneration is indicated by a subscript 'd'.

The morphological analysis is depicted in table 1. In step 8, the index sign 'a' binds to the rheme 'fork' and complements it with the property of 'definiteness'. The morphological sign 'a fork' is represented degenerately as an index which, then, complements the legisign 'with'. Their interaction yields the sign 'with a fork' (a prep-compl) having adjective- or adverb-like syntactic properties.

Table 1. Morphological analysis

nr.	qual	icon	sins	rhme	indx	legi	dcnt	symb	*rule*
0	Mary(M)								i
1	eat(e)	M							i, c_1
2	-s	e		M					i, c_1, c_1
3	pizza(p)	e	-s	M					i, c_1
4	with(w)	p		e-s			M		i, c_1, b, c_1
5	a(a)		w	p			e-s		i, c_1, c_1, c_1
6	fork(f)		a	p		w	e-s		i, c_1, c_1
7		f			a	w	p		c_1, c_1, c_1, c_1
8				f	a	w	p		b_d
9					a-f	w	p		b, c_1
10								w-a-f	c_1

The final output is (Mary)(eats)(pizza)(with a fork) where an item enclosed in parentheses denotes an argument sign generated.

For the syntactic analysis, the relational needs are defined as follows: 'eats'= {legisign,symbol}, 'with a fork'={index}. The parses are displayed in table 2 and table 3 (in the latter only the steps deviating from the first analysis are given). Notice that, due to the sequentiality of the model, the reference of a sign is fixed when the sign of its object appears.

Table 2. Syntactic analysis (alternative #1)

nr.	qual	icon	sins	rhme	indx	legi	dcnt	symb	*rule*
0	Mary(M)								i
1	eats(e)	M							i, c_1
2	pizza(p)		e	M					i, c_1, c_2
3	with a fork(w_{af})		p	M			e		i, c_1, c_2
4			w_{af}	M	p	e			c_1, b_d
5				M	w_{af}	e-p			c_1
6					w_{af}	e-p	M		b
7							M	e-p-w_{af}	b

Because a prep(-compl) can have an a-need in the legisign class (but only if it is accumulated with a verb), there is a third analysis possible, in which, eats and with a fork combine to a single legisign representing a common meaning, e.g. 'with-a-fork-eating'. This analysis ('with a fork'={legisign}) is shown in table 4.

Table 3. Syntactic analysis (alternative #2)

nr.	qual	icon	sins	rhme	indx	legi	dcnt	symb	rule
3'	with a fork(w_{af})	p		M		e			i, c_2, c_1
4'			w_{af}	p		e	M		c_1
5'				p	w_{af}	e	M		b_d
6'					p-w_{af}	e	M		b
7'							M	e-p-w_{af}	b

Table 4. Syntactic analysis (alternative #3)

nr.	qual	icon	sins	rhme	indx	legi	dcnt	symb	rule
3"	with a fork(w_{af})	p		M		e			i, c_2
4"			w_{af}	M	p	e			c_1, a
5"				M	p	e/w_{af}			c_1
6"					p	e/w_{af}	M		b
7"							M	e/w_{af}-p	b

Further research

The simplicity of the algorithm introduced in the paper allows the generation of efficient parsers. In virtue of the simple 'structure' of its syntactic (and logical) analyses, such parsers can be particularly suitable for text summarisation and for handling incomplete sentences.

References

1. Aho, A.V., Ullman, J.D.: *Parsing*, volume 1 of *The Theory of Parsing, Translation and Compiling*. Prentice-Hall, (1972)
2. Farkas,J.I., Sarbo, J.J.: A Peircean framework of syntactic structure. In *7th International Conference on Conceptual Structures (ICCS'99)*, Lecture Notes in Artificial Intelligence, Vol. 1640. Springer-Verlag, (1999) 112–126
3. Farkas, J.I., Sarbo, J.J.: A Logical Ontology. In: G. Stumme (ed.): *Working with Conceptual Structures: Contributions to ICCS2000*. Shaker Verlag. (2000) 138–151
4. Harnad, S.: *Categorical perception: the groundwork of cognition*. Cambridge University Press, Cambridge (1987)
5. Peirce, C.S.: *Collected Papers of Charles Sanders Peirce*. Harvard University Press, Cambridge (1931)
6. R. Quirk, S. Greenbaum, G. Leech, and J. Svartvik. *A Comprehensive Grammar of the English Language*. Longman, London and New York, 1985.
7. V. Tejera. *Semiotics from Peirce to Barthes*. E.J. Brill, Leiden, 1988.

Word Graphs: The Third Set

C. Hoede and L. Zhang*

Faculty of Mathematical Sciences
University of Twente
P.O. Box 217
7500 AE Enschede, The Netherlands

Abstract. This is the third paper in a series of natural language processing in term of knowledge graphs. A word is a basic unit in natural language processing. This is why we study word graphs. Word graphs were already built for prepositions and adwords (including adjectives, adverbs and Chinese quantity words) in two other papers [2], [3]. In this paper, we propose the concept of the logic word and classify logic words into groups in terms of semantics and the way they are used in describing reasoning processes.

Keywords: Knowledge graphs, word graphs, logic word, classification, ontology.

AMS Subject Classifications: 05C99, 68F99.

1 Introduction

Natural language is a kind of special symbol system that is used to express human ideas and pass on information. Each natural language has evolved into a kind of traditional symbol system which has its own word types and word structures, in which different components have some relationship and these related components play a role as a whole that has some meaning. In science a particularly important part of the special symbol system is the set of words used in logic. It is therefore necessary to study the logic phenomena in natural language and the common rules in different languages in order to reveal and explain pure logical forms, other logical structures and logical rules in sentences of a natural language text.

In particular in knowledge graph theory the meaning, which is considered identical with the graph structure, of a sentence is a function of the meanings, graph structures, of its various parts. Therefore, to understand the meaning of a sentence one first needs to understand the meaning of each word that occurs in the sentence thus gaining the meaning of the whole sentence from these words and their order, one of the syntactical aspects of the sentence. Finally, combining the sentence graphs we can obtain the meaning of the whole text. The goal of this series of papers is to show that words can indeed be represented by graph structures and that in quite different languages, like English and Chinese, the same representational technique can be applied.

---

* On leave from Northwestern University, Xi'an, P.R. China

H. Delugach and G. Stumme (Eds.): ICCS 2001, LNAI 2120, pp. 15–28, 2001.

In case the goal is reached the theory would give firm support to the theory of conceptual graphs, as the words used in that theory can then all be represented by graphs, using a simple, but rather basic, ontology. This may shed light on some of the open problems in the theory of conceptual graphs. Also an alternative theory of linguistic can then be developed, including an attempt to approach the problem of translation completely in terms of graphs.

The paper will study the set of words called *logic words*, which have a logical function in a sentence or even a paragraph. In Section 2 we first discuss some ontological questions and take a stand from the point of view of knowledge graph theory. Logic words are classified in Section 3. Combining the analysis of linguistics with the analysis of logic, we try to reveal common properties of logic words, that are independent of the particular language. In Section 4 we discuss word graphs for logic words.

2   Ontological Aspects

Before starting our particular study, we should say something about the ontology we are using. We refer to the papers of Hoede and Li [2] and of Hoede and Liu [3] for general information on the ontology of knowledge graphs. We recall only the most essential parts for our discussion.

The word graph ontology consists, up till now, of 8 types of binary relationships and 4 types of n-ary relationships, also called *frame* relationships.

The eight binary types describe

Equality	: EQU	Causality	: CAU
Subset relationship	: SUB	Ordering	: ORD
Similarity of sets, alikeness	: ALI	Attribution	: PAR
Disparateness	: DIS	Informational dependency	: SKO.

They are seen as means, available to the mind, to structure the impressions from the outer world, in terms of awarenesses of somethings, that are represented by unlabeled vertices. This structure, a labeled directed graph in mathematical terms, is called *mind graph*. Any part of this graph, can be *framed and named*. Note that here WORDS come into play for the first time, the relationships were considered to be on the sub-language level so to say, on the level of processing of impressions by the brain, using different types of neural networks. The part of the mind graph focused upon is the *word graph* of the chosen word.

Once a subgraph of the mind graph has been framed and named another type of relationship comes in, that between the frame as a unit and its constituent parts. The four n-ary frame-relationships are describing:

Focussing on a situation : FPAR	Possibility of a situation : POSPAR
Negation of a situation : NEGPAR	Necessity of a situation : NECPAR.

The situation is always to be seen as some subgraph of the mind graph. It will already be clear that word graphs for logic words will mainly be constructed using the second set of four n-ary relationships.

Let us compare our ontology with two of the many ontologies proposed in history. The first one is of course that of Aristotle. He distinguished:

- Quantity
- Quality
- Relation
- Location
- Time
- Position
- Substance
- Having
- Doing
- Being affected.

These ten basic notions clearly focus on the physical aspects of the impressions, as do the first eight notions of word graph ontology. The focus there is on the way the world is built.

The second ontology to consider is that of Kant, who distinguished twelve basic notions:

QUANTITY	QUALITY	RELATION	MODALITY
• Unity	• Reality	• Inherence	• Possibility
• Plurality	• Negation	• Causality	• Existence
• Totality	• Limitation	• Commonness	• Necessity.

Note that Kant clearly focuses on the logical aspects, including modal logic concepts like *possibility* and *necessity*. Of course *negation* is included as well. Together with the *and* concept, which is simply two tokens framed together in knowledge graph theory, the negation gives a *functionally complete* set of logical operators, in terms of which first order propositional logic formulae can be expressed. The possibility of describing all known systems of modal logics by means of knowledge graphs was shown by van den Berg [1]. The importance of his results lies in the fact that the formalism of logic is encompassed by knowledge graph theory.

Here some remarks are due concerning the work of C.S. Peirce [5]. Describing logic by graphs, called *existential graphs* by him, was introduced by him before 1900, starting with the idea of simply indicating *and* ($\wedge$) and *negation* ($\neg$) by two different types of frames. The work of van den Berg can be seen as a direct continuation of this set up. It has often been said that Peirce was guided by the ontology of Kant, who presented the twelve basic notions in four triples, see above, when he introduced the notions of *firstness*, *secondness* and *thirdness* of a concept. Peirce's definitions are not very easy to understand. We quote from Sowa [6]

Firstness: The conception of being or existing independent of anything else.

Secondness: The conception of being relative to, the conception of reaction with, something else.

Thirdness: The conception of mediation, whereby a first and a second are brought into relation.

From the point of view of knowledge graph theory the following stand is taken.

For any concept, token or (vertex) of a mind graph, we can distinguish:

- The token itself, which usually has an inner structure, the definition of the concept.
- The token together with its neighbours, including a subgraph of the mind graph, that we call the *foreground knowledge* about the concept.

- The whole mind graph, considered in relation to the concept, also including what we call the *background knowledge* about the concept.

In our view Kant's triples do not correspond precisely to Peirce's notions and we have the idea that the triple of knowledge graph theory: concept, foreground knowledge, background knowledge, is all that Peirce's notions are about. What is extra in our theory is the fact that the mind graph is not a fixed entity but depends on the particular mind (human) and for one human even on the particular circumstances in which the concept word is to be interpreted. Also the *intension* of the concept, its definition, is often not uniquely determined, although it is one of the major goals in science to get at least the definitions straight. The variation in meaning, possible for a word, is an intrinsic and essential part of knowledge graph theory.

3   Logic Words and Their Classification

The words mentioned in traditional logic should be logic words. For example, "and", "or", "not", "if ..... then ....." and "if and only if", which are five words describing connections in proposition logic, are typical logic words. Words like "possible", "necessary", "ought" and "permitted" which are used in modal logic and deontic logic are of course also logic words. But, many other words used in natural language, such as "therefore", "since", "while", "but", "before", etc. are related to logical aspects of utterances as well. So the classification of logic words might include two parts, which are *pure* logic words and *other* logic words. The pure logic words are then, by definition, those words that are mentioned in traditional logic (including proposition logic, predicate logic, tense logic, modal logic, deontic logic and fuzzy logic). This class of words seems quite easy to recognize. The other logic words are words that are somehow related to logical aspects as well. This class turns out not to be as easy to define and to structure. We are therefore in need of more precise classification criteria.

3.1   Classification Criteria

Before discussing possible classification criteria we should mention the objects that we wanted to classify. A corpus of 2000 English words with the property that they occurred most frequently in a set of 15 texts was established by Hofland and Johansson [4]. Our main goal is to develop a system of *structural parsing*, by means of which from a lexicon of word graphs the sentence graph of a given sentence can be constructed. So first the sentence is taken apart, and then from a representation of the parts, the words, a representation of the sentence is constructed.

a) **Subjective classification**

As a start of the general lexicon these 2000 words should be included. For that reason, and to have a natural restriction for the set of words, we considered these 2000 words and tried to classify them on a five point scale:

1. definitely a logic word
2. clearly related to logic, but not basic
3. related to some form of logical reasoning
4. having a vague logical flavor
5. no relation to logic.

As a result the two authors classified the words into classes $C11$, $C12$, $C13$, $C14$, $C15$, $C22$, $C23$, $C24$, $C25$, $C33$, $C34$, $C35$, $C44$, $C45$ and $C55$. The two indices indicate the scale values mentioned by the authors. We give only the first class resulting from this subjective coding process, in order of frequency of the words.

C_{11}: and, if, no, then, must, might, right, however, every, possible, difference, cannot, necessary, therefore, probably, true, thus, nor, everything, else, unless, truth, impossible, neither, doesn't, wouldn't, everyone, ought, isn't, possibly, nevertheless, possibility, existence, maybe, equal, equivalent, necessarily, hence.

b) *Classification by using Kant's ontology*

   Another way of classifying would be to use Kant's ontology and decide whether a word belongs to one of his twelve categories, i.e. expresses something the main feature of which is one of these twelve concepts. As an example, let us consider those words in the class C_{11} that we determined, that would fall in the category of "possibility". We would choose "might", "possible", "cannot", "probably", "impossible", "possibly", "possibility" and "maybe" as elements of this class.

c) *Classification by using knowledge graph theory ontology*

   Looking at these words in b) from the knowledge graph point of view we discover that people, students in our case, making a word graph for those words, use the POSPAR-frame in all cases. This prompts another way of classifying, namely according to the occurrence of FPAR, NEGPAR, POSPAR and NECPAR-frames in the word graphs of the word. Note that these frames correspond to Kant's categories *existence, negation* (seen as a quality by Kant), *possibility* and *necessity*.

Definition 1. A *logic word of the first kind* is a word, the word graph of which contains one of the four types of frames in the knowledge graph ontology.

The existence of two somethings, seen as two components of a frame, puts them in an FPAR-relationship with the frame. That frame can be named "and". Similarly, something in a NEGPAR-frame is put in a NEGPAR-relationship with that frame, that now can be named "not". By functional completeness the other connectives from propositional logic follow from equivalences like $p \vee q \Leftrightarrow \neg(\neg p \wedge \neg q)$, for the "or"connective.

When we consider something, say a situation S (German: Sachverhalt) in the form of a graph a POSPAR-frame may be considered around it. We may describe this by saying "S is a possibility", "It is possible that S" or "Possibly we have S". In Chinese these three utterance are translated as

"ke 3 neng 2 S", literally "possible S"
"S shi 4 ke 3 neng 2 de", literally "S is possible"
"you 3 ke 3 neng 2 S", literally "have possible S",
respectively.

We have chosen to give the Chinese sentences and words in this paper in spelling form, Pin 1 Yin 1, followed by a number 1, 2, 3 or 4, indicating the four intonation forms. The reader, who is not interested in the Chinese way of expressing things, may just skip the corresponding text parts. We think that they support our structural approach to language.

Subtle differences come forward due to the choice of the POSPAR-frame. Suppose S is given by the following graph, in socalled total graph form in which the arc is also described by a vertex,

$$S : \boxed{A} \longrightarrow \!\!\! \text{(CAU)} \longrightarrow \!\!\! \boxed{B}.$$

A POSPAR-frame around the whole of S would describe "(There is) possibility (that) A causes B". A frame around $\boxed{A} \longrightarrow \text{(CAU)}$, which by itself reads "cause A", would lead to a description of "A (is a) possible cause (of) B". A frame around $\text{(CAU)} \longrightarrow \boxed{B}$ would describe "A possibly causes B". In English "possible" is an adjective and "possibly" an adverb. The decision which word to use depends here on "cause" being a noun and "causes" a form of the verb "cause". In Chinese, the three sentences are translated

"A yin 3 qi 3 B shi 4 ke 3 neng 2 de", literally "A cause B is possible of",
"B ke 3 neng 2 you 2 A yin 3 qi 3", literally "B possible have A cause"
"A ke 3 neng 2 yin 3 qi 3 B", literally "A possible cause B".

The reader should remark that the existential and universal quantifiers, "there exists" and "for all", are not falling under Definition 1. In fact, Peirce already pointed out that making the statement S on the paper of assertion (in whatever form), is equivalent to existential quantification (for closed formulae). That is why we put "There exists" between brackets in our example sentences. The universal quantifier in knowledge graph theory is expressed by the SKO-loop on a token, that should be read as "for all".

is to be read as "all dogs bark" or "for all dogs (holds) dog bark(s)". So "all" is not a logic word of the first kind according to Definition 1. We should note here that "all" is falling in the category "totality" of Kant's ontology. "Exist" is a logic word of the first kind as its word graph is the "be"-frame, the empty frame, filled with "something SUB world". In the framing and naming process, a subgraph of the mind graph is framed and in that way the definition of a concept C is given. The description is "C is a ....". Here "is" is a logic word of the first kind too as it describes the FPAR-relationship between C and its frame content. The famous "ISA"-relationship, like in "A dog is a(n) animal", expresses the FPAR-relationship between "animal" (part of the definition of "dog") and the "dog"-frame. So here IS in ISA is a logic word of the first kind too.

For the other logic words the classifying feature in their word graphs is chosen to be one of the other eight types of, binary, relationships. An important example is "causality", basic in words like "cause" or "because", which also is a category of Kant's ontology. In knowledge graph theory we would classify according to the occurrence of a CAU-arc. The ALI-link, for "alike" concepts, corresponds to concepts in the "commonness"-category of Kant's ontology and determines a set of words like "like", "as ..... as", etc. Other logic words, so words with one of the eight binary relationships as dominant link in the word graph, are not really expressing aspects of pure logic, but then pure logic is not the whole of thinking, that we like to describe as "linking of somethings". In expert systems the question "why C?" can be answered if causations are known. If B is a possible cause for C and A a possible cause for B, then the answer may be "Possibly because A". If we analyse this thinking process, we see that we start with C, and then notice that we know that $B \xrightarrow{\text{CAU}} C$ and $A \xrightarrow{\text{CAU}} B$. By linking these data we obtain $A \xrightarrow{\text{CAU}} B \xrightarrow{\text{CAU}} C$, and have found A as possible cause. This process of linking is particularly interesting when the concepts are *expanded*, i.e. they are replaced by the content, of their frame, that is embedded in the rest of the graph considered. This replacement poses problems of its own, but we are not interested in them here. The expansion process plays an important role in thinking. Given some statements, say in mathematics, and the problem to prove that some goal-statement G is true if the given statements are true, then the way to find the proof may be the following. Expand all given statements as far as possible, with available further knowledge, till a graph A is obtained that has the graph G of the goal statement as a subgraph. The basic process of reasoning is namely that whenever a graph is considered to be true, each of its subgraphs must be true (under specific conditions on the structure of these subgraphs).

In trying to find the (answer) graph A one meets the difficulty of expanding the graph in the right direction. From the given statements expansion, combining of "true" graphs, may lead to many answer-graphs A that are all true, but none of which contains the goal graph G as a subgraph.

Both in this general process of reasoning and in the case of expert systems, a "rulebased" version can be given. "If a graph A is true *then* its subgraph G is true" is the rule in the first case, but in natural language we would use the word "so" (which by the way turned up in class C_{12}): "A so G". "If A then B" and "If B then C" are rule-versions for natural language descriptions like "B because of A" and "C because of B". It is for this reason of almost equivalence in description that it makes sense to speak about other logic words. The rule-version has the pure logical setting in which the pure logic words are used. The statements have to be well-formed closed formulae. In natural language the thinking process is often described by not-well formed statements that nevertheless correspond to certain mind graphs that can be used in the description of the expansion process. We will see that this deviation from pure logic allows for dealing with some other linguistic aspects as well.

Definition 2. *Logic words of the second kind* are words the word graph of which contains one of the eight types of binary relationships of the knowledge graph ontology as dominant link.

We have given a restriction here by demanding that the relationship, e.g. the CAU-relationship, is a dominant link as otherwise all words would be logic words. Meant are those words that describe the linking process in its basic form. Word graphs with more than one type of binary relationship are to be excluded, unless one link is clearly dominant. In the first paper in the series the 15 different word graphs for the Chinese word for "in" were given. In them a SUB-link or a PAR-link was clearly dominant. A preposition like "in" can therefore also be seen as a logic word of the second kind, used often in thinking about structuring the world. To determine which link is dominant we need some measure for dominance. In graph theory measures have been developed for concepts like centrality. Similar measures can be chosen for the concept of dominance and thus used to decide whether a word can be called a logic word of the second kind or not.

Finally, some discussion is due on the words "truth" and "true", definitely words often used in logic. There are two ways of looking at these words. First there is the comparison of a statement or proposition p with a model of the situation expressed by p. The outcome of the comparison determines the truth value of p, which in two-valued logic is "true" or "false". For our knowledge graph view, what is happening is comparison of two mind graphs, one for p and one for the model. "Truth" can then be seen as equality of certain frames, one for p and one for (a part of) the model, and hence is a logic word of the first kind according to Definition 3. The truth values "true" and "false" are nothing but instantiations of the outcome of the comparison that may be replaced by, for example, the numbers 1 and 0, as is often done. These are not logic words.

A second way of looking at "truth" is as an attribute of a framed part of the mind graph. That part may represent the content of a contemplation, or, more close to what was said before, the model of a situation, as perceived. Such a perceived situation may be held to be true or false, i.e. truth is attributed. No comparison is made, there is only one frame, that may of course be described by a statement p. One of the arguments in favor of this second way of looking at truth is that both statement and model are parts of the mind graph. The model is also held to be a correct description of the state of affairs, whether this state of affairs is due to a presupposed "outer world", as in physics, or due to ideas in an "inner world", as in mathematics, where e.g. axioms are simply considered to be true. The statements describing axioms are not considered to be in need of comparison.

The knowledge graph theory slogan "the structure is the meaning" is in line with the second way of looking at truth. The statement "it is raining outside", a standard example, does not need comparison with a model. The structure of the part of the mind graph associated by the listener with the statement is all that matters, as far as meaning attribution is concerned. The truth of the statement is depending on the comparison, with the outer world. In socalled *truth conditional* semantics this comparison is stressed. In our *structural* semantics, the outcome of such a comparison is irrelevant. As a major consequence of our stand even statements that are not well-formed also have a well-defined semantics as far as the corresponding mind graph frames are well-defined. A statement like "$x < 5$" is considered to have no well-defined truth conditional semantics even when a model is given, with proper domain and interpretations, because x is free. Any knowledge graph constructed by a mind as corresponding to the statement is the meaning of the statement in structural semantics.

3.2   Classifying Logic Words of the First Kind

Due to our chosen Definition 1 we would have to construct the word graph to decide whether a word is a logic word of the first kind. We already discussed the operators from proposition logic and the quantification operators. Also modal logic operators were discussed. Of all these only the universal quantification did not involve a frame, but was expressed by a SKO-loop and was therefore a logic word of the second kind. The first set of logic words of the first kind corresponds to the four frames themselves and is given in Table I.

Table I

WORD GRAPH	WORD	PARAPHRASE	CHINESE WORD	LITERALLY
Frame	Inherence	be with	Gu 4 you 3 xing 4	primitive have gender
Negframe	Negation	be not	Fei 1	not
Posframe	Possibility	be possible	Ke 3 neng 2 xing 4	possible gender
Necframe	Necessity	be necessary	Bi 4 ran 2 xing 4	necessary gender

Remark first the use of the word xing 4, "gender", which is used to describe the occurrence of an alternative. Literally possibility is circumscribed by "possible male/female", where male/female only functions to express the two values for possibility, possible/impossible. Secondly, the word Fei 1, for "negation", used in the context of logic, literally must be translated as "not".

We have chosen the words inherence and negation as these are two of Kant's categories. Note that the word "negation" has a subjective undertone. Similarly, any subgraph that is framed and named gives a concept with the subgraph as *inherent* property set. The word "inherence" clearly expresses more than just "being".

The second set of logic words of the first kind has graphs containing one of the four frames next to other parts. Let a graph P, corresponding to a proposition p, be contained in a frame, which may be described by "it is so that p", or simply by "being p" or even just "p". In the knowledge graph formalism the graph

$$P : \qquad p \xrightarrow{\text{EQU}} \square \xleftarrow{\text{ALI}} \text{PROPOSITION}$$

would be given, or, simpler, $\boxed{P}$. A frame containing the frames $\boxed{P}$ and $\boxed{Q}$ is the representation of $p \wedge q$, or p AND q in natural language. The word graph for "and" , she 2, respectively "$\wedge$", yu 3, is

without specifying the contents of the two inner frames, or tokens as they are called. By functional completeness other connections in proposition logic are expressable by the "and"-frame and the NEG-frame. Consider a NEG-frame containing a proposition graph P, then it can be described by "it is not so that p", or simply by "negation p" or "not p". Omitting p from the graph the word graph for the word "not" results. In Chinese this is described by, "bu 1 shi 4 p", literally "not be p". Likewise the POS-frame and the NEC-frame allow expressing "possible p" respectively "necessary p". Omitting p again the word graphs for "possible", ke 3 neng 2 de, and "necessary", bi 4 ran 2 de, result as

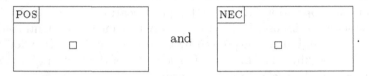

Note that in Chinese the word de is used to express the fact that we are dealing with an adjective.

The frame corresponding to "being" may contain the graph of something considered to exist in the world, so essentially the graph

$$\Box \xrightarrow{\text{FPAR}} \Box \xleftarrow{\text{ALI}} \text{WORLD}$$

describes the word "existence", cun 2 zai 4 xing 4, literally "existence gender". Further information on the "world " considered, which may e.g. be a set of numbers, may be added. We now have word graphs for the logic words of the first kind describing logical operators, see Table II.

Table II

LOGICAL OPERATOR	WORDS	CHINESE WORDS	LITERALLY
Proposition logic	And	Yu 3	and
	Not	Fei 1	not
	Or	Huo 4	or
	If .... then	Ru 2 guo 3 ... ze 2	if ... then
	If and only if	Dang 1 gie 3 jin 3 dang 1	when and only when
Predicate logic	Existence (of)	Cun 2 zai 4	exist
Modal logic	Possible	Ke 3 neng 2 de	possible of
	Necessary	Bi 4 ran 2 de	necessary of

Note that for "if", the word "ru 2 guo 3", is used in "if ... then", whereas in "if and only if", the word "dang", is used, literally meaning "when".

Concluding, the logic words of the first kind are those mentioned in Table I and II. For those words that have word graphs very close to those for these pure logic words we take the stand that although frames are used, the essential meaning is not expressed by the frame. One example should suffice here. The word "both" is used in expressions like "both a and b are numbers", meaning "a is a number and b is a number". "Number $(a) \land$ Number (b)" may be the formulation in predicate logic. The essential meaning of "both" is that the predicate holding for a and b is the same, the "and"-connective has the, different, meaning of combining two propositions that may have no further commonness at all. "Both" is therefore not considered to be a logic word of the first kind. Whether it should be seen as a logic word of the second kind, because the EQU-relationship is clearly present, will be discussed in the next section.

3.3 Classification of Logic Words of the Second Kind

For the logic words of the first kind the four frames in the ontology of knowledge graph theory were chosen as defining entities. For the logic words of the second kind we choose the eight binary relationships. We might call these words *structuring words* as well.

The representation we choose is that of the total graph in which the arc is also represented by a, labeled, vertex. The reason for this is that we can then consider subgraphs of the graph

$$\square \longrightarrow \boxed{\text{TYPE}} \longrightarrow \square \ ,$$

where the label TYPE may be one of the eight types we distinguish. Table III gives the words corresponding to the whole graph of three vertices and the graph in which an encompassing be-frame is considered as well. For the type EQU the graph is considered to represent the word "equal", whereas "being equal" is considered to be a synonym of "equality".

Table III

TYPE	WORD	WORD(+ BE)	CHINESE WORD	CHINESE WORD(+ BE)
EQU	Equal	Equality	Xiang 1 deng 3	Xiang 1 deng 3 xing 4
SUB	In	Containment	Li 3	Bao 1 han 2 xing 4
ALI	Alike	Community	Xiang 1 tiang 4	Gong 4 xing 4
DIS	Distinct	Disparateness	Bu 4 tong 2	Cha 1 yi 4 xing 4
ORD		Ordering	Shun 4 xu 4	Pai 2 xu 4
CAU	Causation	Causality	Qi 3 ying 1	Yin 1 guo 3 xing 4
PAR		Attribution		Shu 3 xing 4
SKO		Dependency	Ying 3 she 4	Yi 3 lai 4 xing 4

Some remarks are due here. First we could not determine an English or Chinese word for the PAR-link. Secondly we notice that for the ORD-link, within a BE-frame, in Chinese the alternative indicating word xing 4, "gender", is not used.

Thirdly the Chinese words tend to extend the description considerably, i.e. a more complicated word graph is expressed. Bao 1 han 2, literally means "around inside", hence "containment". This hints at the fact that also the English word "containment" expresses more than just a SUB-link in a BE-frame. This explains why the two columns for Chinese words given are so different apart from the words for "equal" and "equality". It is simply so that not every knowledge graph has a precisely describing word. We might even have left more places open in the table for "lack of words".

Now let us consider the subgraphs of the form □ ——▸(TYPE) and the corresponding words, given in Table IV, and those of the form (TYPE)——▸ □ , given in Table V.

Table IV

TYPE	WORD	CHINESE WORD	LITERALLY
EQU			
SUB	Of ...	... de yi 1 bu 4 fen	... of one part
ALI			
DIS			
ORD	From	Cong 2	from
CAU	By	You 2 ... ying 3 qi 3	have .... cause
PAR	Of	... de xing 4 zhi 4	... of attribute
SKO	Dependent (on)	Yi 3 lai 4	depend on

Table V

TYPE	WORD	CHINESE WORD	LITERALLY
EQU	Equal (to)	Deng 3 yu 2	equal (to)
SUB	With	Ju 4 you 3 ...	have ...
ALI	Similar (to)	Xiang 1 si 4 ya 2	similar to
DIS	Distinct (from)	Bu 4 tong 2 yu 4	not equal to
ORD	To	Dao 4	to
CAU	With	Yu 4 guo 3 ... xing 4 zhi 4	have ... attribute
PAR			
SKO			

Note, as discussed in [2], that the mereological stand is that SUB, PAR and FPAR are the three mereological relationships, mixed up in the English language by the fact that the words "of" and "with" are used in all three cases, like in the examples:

Part A of B, respectively B with part A,
Attribute A of B, respectively B with attribute A,
Property A of B, respectively B with property A,

or, more concretely:

The tail of the dog,
The beauty of the dog,
The barking of the dog;

if barking is part of the definition of "dog". Some people define a dog as "something that barks and sniffs".

The graphs □ ———→(FPAR) and (FPAR)———→ □ are worded "of" and "with" too, but in Chinese the two graphs are described by "... de te 4 zheng 1" respectively "yu 4 you 3 ... te 4 zheng". We see that in Chinese strict distinction is made for the three mereological relationships. Literally we have

.... de yi 1 bu 4 fen : .... of one part of
... de xing 4 zhi 4   : ... of attribute
... de te 4 zheng 1 : ... of property,

where in English only the word "of" is used, and

ju 4 you 3 ...   : with have ...
ju 4 you 3 ... xing 4 zhi 4 : with have ... attribute
ju 4 you 3 ... te 4 zheng 1 : with have ... property,

where in English only the word "with" is used.

It is quite remarkable that no fourth mereological relationship is used in Chinese, which fact supports our choice of only three mereological relationships. This was not apparent from English.

The EQU, ALI and DIS-relationships are symmetric. This is probably the reason why in Table IV no words are given, which is due to the fact that they do not seem to be present in language.

If y is a function of x,

$x \xrightarrow{\text{ALI}} □ \longrightarrow \text{(SKO)} \longrightarrow □ \xleftarrow{\text{ALI}} y$, we say that y depends (is dependent) on x. Special attention should be paid to the SKO-loop that is considered, by van den Berg and Willems, to represent the words "all",sou 3 you 3, "each",mei 3 ge, "every", mei 3 ge, and "any", ren 4 yi 4. The authors consider the four words to be slightly different so that four different graphs should be presented, although the SKO-loop, indicating something that is informationally dependent only on itself and hence can be anything, clearly stands central. In Chinese three different words are used.

Also note that the CAU-relationship, with chosen word "by" is somewhat out of line with the other relationships that seem more basic as structuring relationships. You 2 ... ying 3 qi 3, literally means "by .... cause".

Of Kant's categories we can, with some difficulty, recognize the following six: Inherence, Negation, Possibility, Necessity, Limitation and Causality. "Existence" we consider as "being in the world", so of less basic nature than "being", "Reality" in our subjectivistic theory is an assumption about correspondence between our image in the mind and a presupposed outer world. We do not discuss the quantity categories of Kant: unity, plurality and totality. Rather, we want to focus on commonness, that we take to be synonym with alikeness, and that puts the ALI-link central.

The ALI-link might be called a primus inter pares as the process of concept formation is seen to result from discovering similarity between a set of perceived objects. A word evokes different subgraphs in different mind graphs, although probably with great similarity.

A further discussion, as well as word graphs for logic words, are available in an extended version of this paper.

References

1. Berg, H. van den, *Knowledge Graphs and Logic: One of Two Kinds*, Dissertation, University of Twente, The Netherlands, ISBN 90–9006360–9 (1993).
2. Hoede, C. and X. Li, Word Graphs: The First Set, in *Conceptual Structures: Knowledge Representation as Interlingua*, Auxiliary Proceedings of the Fourth International Conference on Conceptual Structures, Bondi Beach, Sydney, Australia (P.W. Eklund, G. Ellis and, G. Mann, eds.), ICCS'96 (1996) 81–93.
3. Hoede, C. and X. Liu, Word Graphs: The Second Set, in *Conceptual Structures: Theory, Tools and Applications*, Proceedings of the 6th International Conference on Conceptual Structures, Montpellier, ICCS'98 (M,-L. Mugnier, M. Chein, eds.) Springer Lecture Notes in Artificial Intelligence 1453, (1998) 375–389.
4. Hofland, K. and S. Johansson, *Word Frequencies in British and American English*, The Norwegian Computing Centre for the Humanities, Bergen, Norway, (1982) 43–53.
5. Peirce, C.S., On the algebra of logic, *American Journal of Mathematics*, vol. 7, (1885) 180–202.
6. Sowa, J.F., *Knowledge Representation: Logical, Philosophical, and Computational Foundations*, Draft of a book scheduled to the published by the PWS Publishing Company, Boston Massachusetts, (1994) 99–100.

Aspecto-Temporal Data and Lexical Representations in French within Simple Conceptual Graphs on the Basis of Semantico-Cognitive Schemes

Tassadit Amghar[1], Thierry Charnois[2], and Delphine Battistelli[3]

[1] LERIA, Université d'Angers
2 bd Lavoisier, 49045 Angers Cedex 01, France
Tassadit.Amghar@univ-angers.fr
[2] LIPN - UPRES A 7030, Université Paris 13
Av. J.-B. Clément, 93430 Villetaneuse, France
Thierry.Charnois@lipn.univ-paris13.fr
[3] Equipe LaLic, Université Paris-Sorbonne
96 bd Raspail, 75006 Paris, France
Delphine.Battistelli@paris4.sorbonne.fr

Abstract. This paper deals with the modeling of time, aspect and verbal meanings in natural language processing within Simple Conceptual Graphs (SCG) by way of Semantico-Cognitive Schemes (SCS) and the aspecto-temporal theory. The expression of a semantico-cognitive representation within SCGs is automatically tractable. The SCS allows us to build a representation of a text taking into account the information about time and aspect. It allows us to represent fine subtleties of natural language. On the other hand, the Conceptual Graphs formalism provides a powerful inferential mechanism which makes it possible to reason from texts. Our work bears on French texts. A text is represented by two different structures both represented within the SCG model. The first structure models the semantico-cognitive representation while the second one is the temporal diagram representing the temporal constraints between the situations described in the text. Linking these structures leads us to slightly extend the original SCG model.

1 Introduction

Our interest here is to be able to use *linguistic time* in automatic reasoning processes from representations associated with *narratives* French texts extracted from a real corpus describing traffic accident reports. The following text (A) is extracted from this corpus.

Text (A). *(1)Etant arrêté momentanément sur la file droite du Bd des Italiens, (2)j'avais mis mon clignotant, (3)j'étais à l'arrêt. (4)Le véhicule B arrivant sur ma gauche (5)m'a serré de trop près et (6)m'a abîmé tout le côté avant gauche.*

(1)Stopping momentarily on the right-hand lane of the Boulevard des Italiens, (2)I had put my turn signal on and (3)I was stationary. (4)The other vehicle B [1] arriving from

---

[1] In such reports, implied vehicles are denoted by the letter A or B

H. Delugach and G. Stumme (Eds.): ICCS 2001, LNAI 2120, pp. 29–43, 2001.

my left-hand side (5)squeezed too close up against me and (6)damaged all my vehicle front left side.

Defining automatic reasoning processes from such texts require being able to draw inferences from information bearing on aspectual and temporal information [5]. Two things appear to be particularly important to cope with this kind of information: first, a fine linguistic description of lexical and grammatical knowledge; second a well-suited knowledge representation language.

The *Semantico-Cognitive Schemes* [11] and the *aspectual theory* [10] provide a fine, linguistically well grounded system of semantic primitives and so we choose it to comply with the first requirement. From these works, [4] worked on the construction of the Semantico-Cognitive Representations (SCR) of such *narrative* texts with the aim of building a corresponding pictures sequence.

For the second requirement, it seems relevant to use the Conceptual Graphs formalism [24] [8] which offers a good compromise between inferential facilities and capabilities grounded on formal properties.

Our proposal is a continuation of an earlier work [4] which bears on the construction of Semantico-Cognitive Representations (SCR) of such texts. We propose here a system realizing the transformation of initial semantico-cognitive representations of the texts of the corpus into their conceptual graph representations. Section 2 is devoted to the linguistic background of our work. Two different structures are associated with a text: the first structure is used to model the semantico-cognitive representation of the text and the second one describes the temporal organization of the text through a temporal diagram. Section 3 presents the main principles governing the construction of those structures within Conceptual Graphs (CG) formalism. Before concluding, we dedicate section 4 to a comparative presentation of related works.

2   Linguistic Background

In linguistics and NLP, time and aspectual values are intricately entangled and so are often denoted as aspecto-temporal values. Information helping in their determination comes mainly from two sources : firstly, a lexical one extracted from the verbal lexeme itself (meaning and some other lexical properties of the implied verb)– secondly, a grammatical one which consists of the presence of some specific linguistic markers determining what is usually denoted as the grammatical aspect.

Desclés [10] proposed a general classification of aspecto-temporal values. His study demarcates from the two most famous ones: from Reichenbach's one [22] mainly in that Desclés's temporal entities are of interval type and not of punctual type and from Kamp's one [15] particularly due to the fact that this last one doesn't propose a sufficiently deepened methodology for the computation of fine aspectual values (note also in addition, that [15] distinguishes only two aspectual values and this is certainly not independent of this lack of subtlety). We present here the Semantico-Cognitive Schemes and the aspecto-temporal theories, both introduced by Desclés [10] [11]. They provide the linguistic grounds of this work and make it possible to approach the lexical meanings of verbs in a very narrow adequacy with linguistic intuition. In our application domain, understanding and managing information extracted from car accident reports, the description of movements and temporal information are of prime importance. A taxonomy of aspecto-temporal information of actions in terms of *state event* and *process* has been built[10] and makes it possible to finely represent the situations described in narrative texts.

2.1 Building the Semantico-Cognitive Representation of a Proposition

The *Semantico-Cognitive Schemes* language expresses meanings through abstract structured frames using semantico-cognitive units organized in a system of semantic primitives. The latter are abstract units derived from a linguistic analysis of grammatical and lexical categories occurring in texts but they are not lexicalised anyway. A Scheme describe prototypical relations between two locations or two situations. SCSs are expressed by λ-typed expressions. The French verb *arriver* (*to come*) for example, has the following representation:

$$\lambda x.\lambda y.\ CONTR\ (MOUVT\ or\ (\epsilon_0(ex\ loc)y)(\epsilon_0(inloc)y))y.$$

It can be glossed as follows *an entity y controls (CONTR) an orientated movement (MOUVTor) between two static situations (Sit1 and Sit2). In Sit1, y is located (ϵ_0) to the outside (ex) of the location loc; in Sit2, y is located (ϵ_0) at the inside (in) of this last location.*

An evaluation process where SCS parameters are instanciated with the arguments of corresponding predicative relations transforms SCSs into Semantico-Cognitive Representations (SCR) each of them being linked to a proposition of the involved text. Let us illustrate this process through an example. Consider the following propositions: (1)*la voiture arrivait dans le virage* (*the car was arriving at the bend*) (2)*j'avais mis mon clignotant* (*I had put my turn signal on*). The SCR of (1) is built from the underlying predicative relation (PR) ⟨*arriver_dans, le_virage, la_voiture*⟩ and the linked SCS of *arriver*, through the instanciation of the variables occurring in the SCS by the arguments of PR (i.e. /le virage/ and /la voiture/). This results in the SCR (1').

The building process similar for (2) is glossed by *the entity "je" controls (CONTR) an oriented change (CHANGTor) of the entity le_clignotant between two static situations Sit1 and Sit2.* In Sit1, the entity le_clignotant is depicted as being outside (relationship "ex") the location en_fonctionnement in the context of an activity (ϵ_{act}); whereas in Sit2, le_clignotant satisfies the relation inside (in) for the same location. The associated SCR is (2').

$$(1')CONTR(MOUVTor(\epsilon_0(ex\ le\ \_virage)la\_voiture)$$
$$(\epsilon_0(in\ le\_virage)la\_voiture))la\ \_voiture$$
$$(2')CONTR(CHANGTor\ (\epsilon_{act}(ex\ en\_fonctionnement)le\_clignotant)$$
$$(\epsilon_{act}(in\ en\_fonctionnement)le\_clignotant))je$$

In (1') and (2') the involved entities are typed: *la_voiture, le_clignotant* and *je* belong to the semantico-cognitive type "INDIVIDUAL" and the entities *le_virage* and *en_fonctionnement* to the general semantico-cognitive type LOCALIZATION. As illustrated on Figure 1 SCSs may belong to one of three types: *static* (implying a sole situation Sit without modification), *cinematic* (in the case of a transition between two situations Sit1 and Sit2) or *dynamic* (similar to *cinematic* except that an entity controls the transition). SCSs are

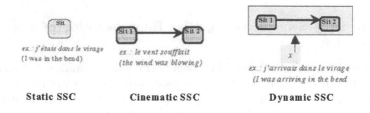

Fig. 1. Static Cinematic and Dynamic SSC

associated with an aspecto-temporal representation taking the form of a set (possibly a singleton) of temporal intervals [3]. For example, a static SCS is associated with an open interval. In order to render relative temporal positions (succession or concomitance) of different propositions of a text, the set of SCRs associated with it has to be organized. So, a text representation relies on SCSs and their corresponding intervals in a temporal diagram. This is obtained by using linguistic indices associated with linguistic temporal knowledge like tenses and grammatical aspectual information occurring in the text. This reflects the discourse structure.

2.2   Aspecto-Temporal Values and Temporal Diagram of a Text

The aspecto-temporal grammatical level is the locus where contextual linguistic indices are taken into account with the aim of associating the relevant aspectual values, morphologically expressed in the tense of a verb. The computation of the aspecto-temporal values is automatically carried out in accordance with a contextual exploration principle [12]. Possible aspectual values are *STATE,EVENT* and *(continous) PROCESS* and this intuitively corresponds to different states or modes of accomplishment of an action which could be *static, accomplished* or *in progress*. For example, in the preceding sentences (1) and (2), actions have *processes* and *resulting state* values, respectively. Figure 3 gives pictorial representations associated with the different kinds of SCR according to their type aspectual values. The temporal information and organization of a text is rendered by temporal links between its associated propositions. These links finally characterize the chronology between states (represented by open intervals), events (represented by closed intervals) and processes (represented by intervals that are semi-opened to the right) corresponding to those propositions.[2] They can then be taken into account in a reasoning process which uses a set of rules

---

[2] refer to [10] for a more complete description of the representation of these aspectual values in terms of topological intervals. Note also that, unlike those used by Reichenbach[22] and Allen[1], they are typed intervals. This last point gives more expressive capabilities to Desclés formalism.

based on temporal connective markers [12] and a set of default rules [4] [3] which renders it possible to state preferences, possibly re-evaluated further in the analysis process. Figure 2 gives the results of the building the semantico-cognitive representation [3] and the temporal diagram [12] of text (A).

Fig. 2. SCR and Temporal Diagram associated with text (A)

3   Computing Aspecto-Temporal Information within SCG Formalism

An SCR represents a proposition of a text. Converting each SCR into an SCG makes it possible to combine the SCGs: this process gives a unique representation of the whole text. All inferences may be expressed in terms of graph operations. This is the case for those based on lexical and grammatical knowledge that are needed to get the SCG associated with the text. In this section, we present the model of the SCGs used and the extension proposed to deal with the aspecto-temporal information. We present the transformation of an SCR to an SCG and how to build a complete SCG associated with a whole text.

3.1   The Basic Model

We restrict ourselves to simple conceptual graphs, or SCG, in the sense of [8] and [20] from which we summarize what is needed here.

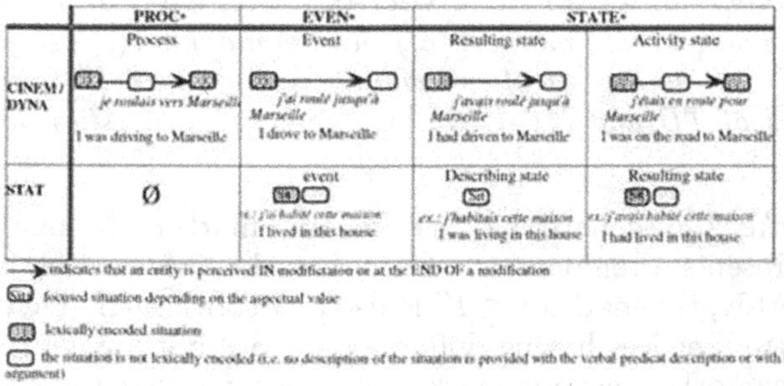

Fig. 3. Different cases of focused situation (Sit2, an intermediate situation, or a posterior situation)

Definition 1. *A conceptual graph* $G = (R, C, E, Lab)$ *is a bipartite labelled graph, not necessarily connected. R and C denote the set of r-vertices (relation vertices) and c-vertices (conceptual vertices) respectively.*
Each vertex $x \in R \cup C$ has a label $Lab(x)$. The label of a c-vertex is a pair $(type(c), marker(c))$ and the set of the labels of C is ordered by a lattice $(T_c, \leq)$. T_c is the set concept types, $marker(c) \in I \cup \{*\}$ where I is the set of individual markers and $*$ the generic one. marker(c) must satisfy the conformity relation τ which assigns a concept type to each individual marker: $type(c) = \tau(marker(c))$. If $r \in R$ then $Lab(r) \in T_R$ where T_R is an ordered set of relation types with a partition: $T_R = T_{R_1} \ldots T_{R_p}$ where T_{R_j} is the set of j-arity relations, $j \neq 0$. Two comparable relations must have the same arity. A signature is associated with each relation type; it specifies the arity and the maximal concept types this relation can link.
E is the set of edges. The set of the edges adjacent to each r-vertex is numbered from 1 to degree(r), and $G_i(r)$ denotes the ith c-vertex adjacent to r.

The SCGs used in our application are in normal form: all c-vertices having the same marker are merged into a unique c-vertex and all r-vertices with the same type and having exactly the same i-th neighbors are replaced by a unique r-vertice. This model is used to represent the temporal diagram (TD) and has been adequately extended to cope with aspecto-temporal information occurring in SCR and in the representation of the whole text.

3.2   Aspectual and Temporal Markers

We introduce a specific marker for temporal information. A concept node is then labelled as a triple ¡type,marker1,marker2¿ where type and *marker1* correspond to type and marker respectively in the original definition above, and *marker2* is an aspecto-temporal marker. Note that marker2 is an individual marker of I'' (the set of markers of TD). In our application, this second marker is a bridge between the temporal interval associated with a textual proposition in the SCG of the temporal diagram and the aspectual values propagated in the SCR corresponding to a textual proposition. Definition 1 is then modified as follows:

Definition 2. New label of a c-vertex. *The label of a c-vertex is a triple $(type(c), marker1(c), marker2(c))$ where $marker1(c) \in I' \cup \{*\}$ and $marker2(c) \in I''$. I' is the set of individual markers and I'' is the set of individual markers of TD and $I' \cap I'' = \{\}$. marker1(c) satisfies the conformity relation.*

Note that $marker2(c)$ cannot be the generic marker. In our application, $marker2(c)$ represents a temporal marker: when this information is not instanciated for a c-vertex, the marker $\varepsilon \in I''$ is used by convention. That is to inhibit the merging of two vertices having different marker2 but having the same type and the same marker1 or generic marker: we consider that the same object at different moments in time must have different representations. The comparison between two labels of two c-vertices is hence defined as follows:

Definition 3. *Let* $e = (t, m_1, m_2)$ *and* $e' = (t', m'_1, m'_2)$ *be two labels of two c-vertices,* $e \leq e'$ *if and only if* $t \leq t'$ *and* $(m'_1 = *$ *or* $m_1 = m'_1)$ *and* $m_2 = m'_2$

Definition 4. Fusion. *Two vertices with labels* e *and* e' *are fusionnable if there exists a vertex different from absurd type with label* e'' *such as* $e'' \leq e$ *and* $e'' \leq e'$. *The fusion of two vertices consists in identifying them in the single vertex which is the lower bound of their labels.*

With definitions 3 and 4, basic operations onto our SCGs (maximal isojoin and projection) are not modified from [8], [20] and [7]. Computing a maximal isojoin between two SCG, G and H, consists in finding a bijection δ between two maximal sub-SCG $G1 \subseteq G$ and $H1 \subseteq H$ such as the underlying graphs of $G1$ and $H1$ are isomorphic and for all $s \in G1$, $\delta(s) \in H1$, and s and $\delta(s)$ are fusionnable. The resulting SCG is obtained by merging each vertex s of $G1$ with $\delta(s)$.

Definition 5. Isojoin. *Let* $G = (R, C, E, Label)$ *and* $G' = (R', C', U', Label')$ *two CG,* α *a bijection from* $A \subseteq R$ *to* $A' \subseteq R'$ *and* β *a bijection from* $B \subseteq C$ *to* $B' \subseteq C'$. *The bijection pair* $\gamma = (\alpha, \beta)$ *is said to be compatible if:*

1. $\forall c \in B$, c *and* $\beta(c)$ *are fusionnable,*
2. $\forall r \in R$, r *and* $\alpha(r)$ *are fusionnable,* $G(r) \subseteq B$, *and* $\forall i, 1 \leq i \leq degree(r)$, $\beta(G_i(r)) = G'_i(r')$ *and if* r'' *is the relation vertex obtained by fusionning* r *and* $\alpha(r)$ *then the vertices arguments of the relation* r'' *must satisfy the conditions given by the signature of the relation* r''

$G[A \cup B]$ *and* $G'[A' \cup B']$ *are said to be isofusionnable*[3]. *The associated fusion is called isojoin between* G *and* G' *according to* γ.

3.3 Representing Intervals

The temporal intervals and relationships between them proposed by Allen [1] are used in our work except that here intervals are typed one's for representing aspectual information (see section 2.2). This leads us to the following CG representations of typed intervals:

```
STATE: ]----[
type STATE(x) is [point]->(])->[interval: *x]->([)->[point]
```

```
EVENT: [----]
type EVENT(x) is [point]->([)->[interval: *x]->(])->[point]
```

```
PROCESS: [----[
type CONTI_PROC(x) is [point]->([)->[interval: *x]->(])->[point]
```

Note that the marker of a type concept *interval* is the temporal marker (*marker2* in definition 2) which is propagated into the corresponding SCR. It links semantic informations and temporal informations of an action (see section 3.4). Allen's relations are defined in terms of conceptual relations, for instance:[4]

---

[3] $G[A \cup B]$ denotes the CG restricted to the vertices of $A \cup B$

[4] the type relation $'|'$ a super type of $'['$ and $']'$

```
relation BEFORE(x,y) is [point:*p1]->(|)->[interval: *x]->(|)
->[point:*p2]->(<)->[point*p3]->(|)->[interval: *y]->(|)->
                                                      [point:*p4]

relation MEETS(x,y) is  [point:*p1]->(|)->[interval: *x]->(|)
->[point*p2]->(|)->[interval: *y]->(|)->[point:*p3]
```

All these primitives allow us to represent the temporal diagram of the text. The
problems of disjunction or propagation constraints are not discussed here. Esch
and Nagle's propositions [13] may be considered to treat these problems.

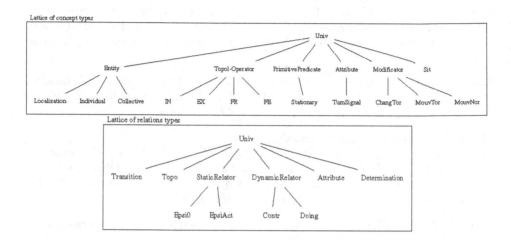

Fig. 4. Part of Lattices of Concepts and Relations

3.4   Our System

As an ontology, we adopt Desclés's one (figure 4) which can be finely tuned to
covers the set of semantics values needed by our application. The final representation of a text is processed thanks to three main modules which operate from
the SCR of each proposition of the text and the temporal diagram to produce
two SCGs representing aspecto-temporal data and the semantic information of
the text respectively.

From SCR to SCG. We have implemented a parser which recognizes an SCR
and generates an SCG. The grammar of SCS we used is given in Figure 5.It is a
context-free grammar. The terminal are given on figure 4. The generation of the
SCG corresponding to the SCR is processed during the recognition of the SCR
(Figure 6). This algorithm is implemented as a DCG rules system in Prolog.
The set of SCG obtained is the set of SCR at a lexical level. It does not contain
any aspecto-temporal information. This information is propagated by the second
component of the system.

```
RSC0 -> RSCDynamic | RSCCinematic | RSCStatic
RSC0 -> &2 RSC0 RSC0
RSCDynamic -> AGTV(MODIF(SIT SIT)) INDIVIDUAL
RSCCinematic->MODIF(SIT SIT)
RSCStatic->SIT
SIT->(StaticRelator Y X)
X -> ENTITY3
Y -> TopologicOperator(ENTITY2) | ENTITY1
ENTITY1-> ATTRIBUTE(INDIVIDUAL)|INDIVIDUAL|PrimitivePredicate
                |PrimitivePredicate(INDIVIDUAL)
ENTITY2->PrimitivePredicate|PrimitivePredicate(INDIVIDUAL)
ENTITY3->ATTRIBUTE(INDIVIDUAL)|INDIVIDUAL
```

Fig. 5. SCR Grammar – RSC0 is the axiom

RECOGNIZED ITEM	ASSOCIATED ACTIONS
AGTV	Create a RELATION node
MODIF	Create a CONCEPT node C1
	And create a ternary RELATION node TRANSITION which links the CONCEPT C1, SIT1 and SIT2
SIT	DEPENDING ON THE CASE :
1)StaticRelator	1)Create a RELATION node R1 (of type 'epsilon')
2)Y=TopologicOperator(ENTITY2)	2)Create a CONCEPT node C2 linked to a relation R2 (of type 'topo') linked dependending on the case (of ENTITY2)
2.1).PrimitivePredicate(INDIVIDUAL)	2.1).to a CONCEPT node PRIMITIVE linked to R1 and to a RELATION node 'DETERMINATION' linked to a CONCEPT node INDIVIDUAL
2.2).PrimitivePredicate	2.2).to a CONCEPT node PRIMITIVE linked to the R1
3)Y=ENTITY1	3)
3.1).Attribute(INDIVIDUAL)	3.1).Create a CONCEPT node ATTRIBUTE C3 linked to R1 and to a RELATION node 'ATTRIBUTION' linked to a CONCEPT node INDIVIDUAL
3.2).INDIVIDUAL	3.2). Create a CONCEPT node INDIVIDUAL linked to R1
3.3).PrimitivePredicate(INDIVIDUAL)	3.3). Create a CONCEPT node PRIMITIVEPREDICATE C4 linked to R1 and to a RELATION node 'DETERMINATION' linked to a CONCEPT INDIVIDUAL
3.4).PrimitivePredicate	f).Create a CONCEPT node PRIMITIVE linked to R1
4)X=ENTITY3	4)
4.1) Attribute(INDIVIDUAL)	4.1).Create a CONCEPT node ATTRIBUT C5 linked to R1 and to a RELATION node 'ATTRIBUTION' linked to a concept node INDIVIDUAL
4.2)INDIVIDUAL	4.2).Create a CONCEPT node INDIVIDUAL linked to R1

Fig. 6. Actions to carry out to build a SCG corresponding to a SCR

Propagating Temporal Markers. This component acts on the input compound of the temporal diagram and the SCGs previously obtained in order to include the temporal information. Remember that the temporal diagram represents the temporal constraints between the situations described in the text. The goal of this module is to propagate the marker of the considered interval according to the aspectual value of the action into the adequate part of the SCR/SCG. For instance, if the aspectual value of an action is EVENT, the set of concepts of the component called *situation 2* is temporally marked by the marker of the c-vertex corresponding to the process of the temporal diagram. Figure 7 gives the necessary rules to take into account each case described in Figure 3. It is the third field of the label of the concept which is inserted. In so doing, we can find the associated interval and its relationships to the other intervals. A strong link between both structures is kept for future inferences.

Type of SSC	Aspectual value	Actions
STATIC	State: 1) Consequent state 2) Description state	1) Not yet treated 2) Mark the set of concepts of SIT
STATIC	Even	Not yet treated
CIN / DYN	State 1) Resulting state 2) Activity state	1) Mark the set of concepts of the situation SIT2 2) Mark the concept MODIF
CIN / DYN	Even	Mark the set of concepts of the situation SIT2
CIN / DYN	Process	Mark the concept MODIF

Fig. 7. Rules to propagate temporal markers

Representing a Text with an Incremental Methodology. Semantic representation of a text is obtained from the compositional principle. The operation used is the maximal isojoin which is more powerful than the directed join proposed by Fargues et al. [14] since the maximal number of concepts to be joined is searched by maximal isojoin. The algorithm corresponding to this operator we use can be found in [7]. After propagating the temporal marker onto adequate

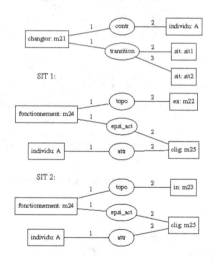

Fig. 8. The representation of *I had put my turn signal on*

c-vertices, and in case of cinematic or dynamic SCR, the SCG associated with the SCR (SCG_{scr}) and the graph corresponding to the concept $Sit1$ (SCG_{sit1}) and the one corresponding to the concept $Sit2$ (SCG_{sit2}) have to be joined. It consists in replacing the c-vertex $Sit1$ from SCG_{scr} by SCG_{sit1} and the concept $Sit2$ by SCG_{sit2}. A c-vertex of SCG_{sit1} must be found to be connected to SCG_{scr}. It is done by searching what has changed between the two situations. So a *set of types difference* is computed between the two SCGs: $Sit1 - Sit2 = c1$ gives the root conceptual vertex of SCG_{sit1} and $Sit2 - Sit1 = c2$ gives the root of SCG_{sit2}. [5] This operation is repeated for each SCG associated with an SCR and then the representation of the text is obtained by the combination of them using maximal isojoins. As an example Figure 8 gives the representation of the second proposition of the text (A) within SCG. Figure 9 gives the full text representation obtained by a succession of maximal isojoins.

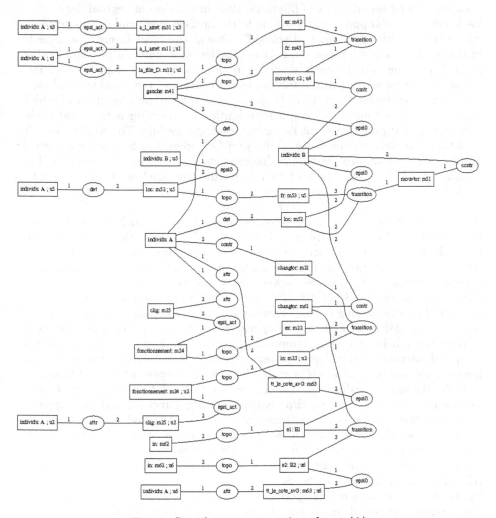

Fig. 9. Complete representation of text (A)

[5] If c1 is not found then an heuristic is applied which searches the c-vertex origin of c2 corresponding to the other graph - and symetrically - (c1 or c2 is always found because there is necessarily a modification between two situations).

4 Related Works

Numerous works tackling the problem of time within the CG framework come from domains related to Knowledge Representation or Natural Language Processing. Depending on the nature of the involved domain, they focus either on some general aspect of representing and managing temporal data, or on more idiosyncratic representations of linguistic time in the case of textual data.

Esch's and Nagle's approach [13] belong to the first category and use an ontology where time entities are "interval shaped". One of the most interesting aspects of this work is that it proposes a solution to the combinatorial problem arising from disjunction, which possibly occurs in the case of the composition of two relations bearing on temporal intervals. In the context of temporal interval logic and interval endpoint logic, Cyre [9] proposes a scheduling algorithm of which the solution is a conceptual graph representation satisfying a temporal problem expressed through an event list corresponding schedule. The whole schedule is obtained mainly on the basis of the join CG operation between conceptual graphs corresponding to the timing diagrams associated with the problem. The complexity of the task is tackled by minimizing the number of temporal relations being taken into account. Pursuing in the general framework of knowledge representation, Sowa [26] studied the modeling of time within the CGs on some different aspects of it such as the representation of process in Natural Language Processing and in programming task modeling. Note here that, although this research has no direct relation with linguistic representation of time, it is nevertheless worth mentioning it here because of its description and computing of relevant temporal relationships between temporal data.

Closest to our works, Sowa [24] [25] gives some bases for a treatment of time in natural language using nested conceptual graphs. His proposition is somehow similar to modal logic with modal operators occurring in CG representations as temporal labels bearing on temporal intervals organized according to Allen's temporal interval relationships [1]. Regardless of the well-known problems arising from the use of some kinds of modal logic for the representation of linguistic time [5], the major weakness of this approach is twofold: first, from a knowledge representation point of view, time entities remain punctual and this appears to be not suitable for linguistic temporal data, second, and from a linguistic point of view, the number of tenses taken into account is too low (since only three tenses, past, present and future) and aspectual values are neglected. For a deeper analysis of the drawbacks of this approach, refer to Moulin and Côté [17] who note, in particular, the disparity of the allowed notations, the ambiguity, the incompleteness of the specification of verbal tenses and the difficulty in harmonizing the verbal tenses and the inter-graph temporal relations.

Closer to our concerns, with the aim of associating temporal conceptual graphs (defined as states) with temporal intervals, [17] proposed a first extension of the conceptual graphs model, for a natural language generation task. To express the semantic value of verbal tenses, they based their approach on Reichenbach's model [22] and represented potential orderings between temporal intervals using Allen's relations. The remarkable points associated with temporal references of an utterance by Reichenbach (i.e. the point of event E, the point of reference R and point of speech S) are embodied in what is called a *period*, and three relations (MRQ-E, MRQ-R, MRQ-S), constitute their respective counterparts. A period is defined as a triplet ¡u, binf, bsup¿ where u characterizes the temporal granularity unit (such as second, minute...) and binf and bsup define respectively the first and the last observation points of a state. In a second step, Moulin [18] remarked that this first proposition was not well suited to express temporal structure of a whole text and extended it by distinguishing three representation levels. The new model drops Reichenbach's theory and is partly based on Desclés's work [10] in particular for the taxonomy of situations (*i.e. states, events and processes*). This allows him to specify the temporal structure of a discourse which is made up of *temporal situations, temporal intervals,* and *temporal relations.* [19] describes an extension of this research with the aim of determining verb tenses in a discourse of which the semantic representation uses the proposed temporal model. Although we work on interpretation and not on generation as Moulin does, our approach is similar and may be considered as a complementary one.

For her part, Zablit [27] also proposed a modeling of time and aspect within CG grounded on Reichenbach's theory for a semantic analyzing task of French texts. Even if an extension of punctual temporal entities is evoked, no hints are given either as to how they could be included in the model or as to how they could establish the links between given temporal relationships between intervals and prepositional content of the involved text. Desclés [10] showed the inadequacy of such type of model to cope with the complexity of all temporal phenomena occurring in natural language.

5   Conclusion

We have presented here an original work which aims at representing the different temporal data which could be extracted from texts. Our representation uses the conceptual graph formalism with some relevant adjuncts to take into account the very fine subtleties of linguistic time with adequacy. As an application, we choose to work from a corpus, made up of accident car reports, in which solving aspecto-temporal problems is of critical importance. Actually, our system is implemented and takes a previously obtained semantico-cognitive representation [3] of a text as its input and translates it into a sole simple conceptual graph. So, this complete treatment from texts to conceptual representations opens the possibility of reasoning from temporal data to a further stage. It is now possible to use the inference capacities of the SCG model to achieve fine deductions bearing on the completion degree of actions described in a text in natural language for example in the context of a questions-answering system. As another perspective, for example, we envisage using temporal information so finely represented with the aim of testing coherence between two accident reports written by those involved. In the context of accident report each party expresses its point of view and by comparing the two resulting texts some incoherences may occur. So following the methodology described in [6] we will be able to detect these incoherences and then clarify the different points of view.

References

1. J.F. Allen. Maintaining knowledge about temporal intervals. *Communication of the ACM*, 26(11): 832-843, 1983.
2. T. Amghar, F. Gayral, B. Levrat. Table 10 left without paying the bill! A good reason to treat metonymy with conceptual graphs. In G. Ellis, Robert Levinson, W. Rich, J. F. Sowa, editors. *Conceptual Structures: Applications, Implementation and Theory.* Lecture Notes in Artificial Intelligence, n 954, Springer Verlag, pp. 129–143, 1995.
3. D. Battistelli. *Passer du texte à une séquence d'images: analyse spatio-temporelle de textes, modélisation et réalisation informatique (système SPAT)* Ph.D. thesis, Université Paris IV - Sorbonne, 2000.
4. D. Battistelli and N. Vazov. Building of semantico-cognitive representation of different types of narratives in French. In *Recent Advances in Natural Language Processing*, pages 373–382, Tzigov Chark, Bulgarie, 1997.
5. H. Bestougeff and G. Ligozat. *Outils logiques pour le traitement du temps, de la linguistique à l'intelligence artificielle*, Masson, 1989.
6. T. Charnois. A natural language processing approach for avoidance of feature interactions. In P. Dini, R. Boutabe and L. Logrippo, editors. *Feature Interactions in Telecommunication Networks IV*, Montréal, IOS Press, pp. 347–363, 1997.
7. T. Charnois. Maximal Isojoin for Representing Software Textual Specifications and Detecting Semantic Anomalies. In *Working with Conceptual Structures*, ICCS'2000, Ed Shaker Verlag, Darmstadt, pp 189-200, 2000.
8. M. Chein and M. L. Mugnier. Conceptual graphs: fundamental notions. *Revue d'Intelligence Artificielle*, 6(4): 365–406, 1992.
9. W. Cyre. Acquiring temporal knowledge from schedules, In Mineau G.W., Moulin B., Sowa J.F. editors. *Conceptual Graphs for knowledge Representation*, Lecture Notes in Artificial, n 699, Springer Verlag, pp. 328–344, 1993.
10. J.P. Desclés. State, Events, Process and Topology. In *General Linguistics* 29(3):159–200, 1989.
11. J.P. Desclés. *Langages applicatifs, Langues naturelles et Cognition*, Hermès, 1990.
12. J.P. Desclés, C. Jouis, H.G. OH, D. Maire–Reppert. Exploration contextuelle et sémantique : un système expert qui trouve les valeurs sémantiques des temps de l'indicatif dans un texte. In *Knowledge modeling and expertise transfer*, Amsterdam, pp. 371–400, 1991.
13. J. W. Esch and T. E. Nagle. Temporal Intervals. In Timothy E. Nagle, Janice A. Nagle, Laurie L. Gehorlz and Peter W. Eklund, editors. *Conceptual Structures and Practice*, Ellis 12. Horwood, 1992.
14. J. Fargues, M.C. Landau, A Dugourd and L. Catach. Conceptual Graphs for Semantics and Knowledge Processing. *IBM Journal of Research and Development*, 30(1) : 70–79, 1986.
15. H. Kamp. Events, Instants, and Temporal Reference. In R. Bauerle, U. Egli, A. von Stechow, editors. *Semantics from different points of view* Springer Verlag, Berlin, pp. 373–417, 1979.
16. D. McDermott. A temporal logic for reasonning about processes and plans. In *Cognitive Sciences 6*, pp. 101–155, 1982.
17. B. Moulin and D. Coté. Extending the Conceptual Graph Model for Differentiating Temporal and Non-Temporal Knowledge. In Timothy E. Nagle, Janice A. Nagle, Laurie L. Gehorlz and Peter W. Eklund, editors. *Conceptual Structures and Practice*, pp. 381–390, Ellis 12. Horwood, 1992.
18. B. Moulin. The representation of linguistic information in an approach used for modelling temporal knowledge in discourses. In Mineau G.W., Moulin B., Sowa J.F. editors. *Conceptual Graphs for knowledge Representation*, Lecture Notes in Artificial, n 699, Springer Verlag, pp. 182–204, 1993.

19. B. Moulin and Stéphanie Dumas. The temporal structure of a discourse and verb tense determination. In W. M. Tepfenhart, J. P. Dick, J. F. Sowa, editors. *Conceptual Structures: Current Practices*, Lecture Notes in Artificial Intelligence, n 835, pp. 45–68, 1994.
20. M.L. Mugnier and M. Chein. Représenter des connaissances et raisonner avec des graphes. *Revue d'Intelligence Artificielle*, 10(1): 7–56, 1996.
21. J.F. Nogier. *Génération automatique de langage et Graphes Conceptuels*, Editions Hermès, Paris, France, 1991
22. H. Reichenbach. *Elements of symbolic logic*, McMillan, New York, 1947.
23. Y. Shoham. *Reasoning about change*, The MIT Press, 1988.
24. J. F. Sowa. *Conceptual Structures: Information Processing in Mind and Machine.* Addison-Wesley, 1984.
25. J. Sowa. Conceptual Graph Notations. *Proceedings of the third annual Workshop on Conceptual Structures*, AAAI Conference, 1988.
26. J.F. Sowa. *Knowledge Representation, Logical, Philosophical, and Computational Foundations*, DUXBURY, 1999.
27. P. Zablit. *Construction de l'interprétation temporelle en langue naturelle: un système fondé sur les graphes conceptuels* Thèse de doctorat, Université Paris Sud-Orsay, 1991.

Learning to Generate CGs from Domain Specific Sentences[1]

Lei Zhang and Yong Yu

Department of Computer Science and Engineering,
Shanghai JiaoTong University,
Shanghai, 200030, P.R.China
{zhanglei,yyu}@cs.sjtu.edu.cn

Abstract. Automatically generating Conceptual Graphs (CGs) [1] from natural language sentences is a difficult task in using CG as a semantic (knowledge) representation language for natural language information source. However, up to now only few approaches have been proposed for this task and most of them either are highly dependent on one domain or use manual rules. In this paper, we propose a machine-learning based approach that can be trained for different domains and requires almost no manual rules. We adopt a dependency grammar – Link Grammar [2] – for this purpose. The link structures of the grammar are very similar to conceptual graphs. Based on the link structure, through the word-conceptualization, concept-folding, link-folding and relationalization operations, we can train the system to generate conceptual graphs from domain specific sentences. An implementation system of the method is currently under development with IBM China Research Lab.

1   Introduction

The first step of using Conceptual Graphs(CGs) in knowledge engineering, in most cases, is to construct conceptual graphs to represent the knowledge contained in the information source. Unfortunately, automating this step for Natural Language(NL) source is a daunting task and often is done using handwritten rules. In this paper, we propose a machine-learning based approach that can be trained for different domains and requires almost no manual rules.

Conceptual graphs, when used to represent knowledge contained in natural language sentences, are used as a semantic representation language. CGs from NL sentences, thus, is a process of semantic analysis. Currently, automating this process in general is infeasible for the set of all possible sentences. The method we proposed in this paper is for domain specific sentences which are sentences that occur only in a specific application domain. Though the sentences are limited in one domain, our method itself is domain independent and the system can be trained for various domains.

The paper is organized as follows. Section 1.1 introduces the concept of "Domain Specific Sentences" used in this paper. Section 1.2 gives a brief introduction for link

[1] This work is supported by IBM China Research Laboratory.

H. Delugach and G. Stumme (Eds.): ICCS 2001, LNAI 2120, pp. 44–57, 2001.

grammar. Section 2 outlines the whole approach by giving an overview. Section 3 presents the detailed operations that generate conceptual graphs from domain specific sentences. Section 4 concludes our work by comparing related work.

1.1 Domain Specific Sentences

The set of sentences that occur only in one given application domain is called domain specific sentences. We assume that domain specific sentences can be characterized as following:

1. vocabulary set is limited
2. word usage patterns exist
3. semantic ambiguities are rare
4. terms and jargon of the domain appear frequently

The notion of sublanguage[3,4] has been well discussed in the last decade. Domain specific sentences actually can be seen as sentences in a domain sublanguage. As previous study has shown, a common vocabulary set and some specific patterns of word usage can be identified in a domain sublanguage [3]. These results provide ground for us to assume the above characteristics about domain specific sentences. In the rest of this paper, we will show how characteristics 1 to 3 are employed in our work. Terms and jargon will be dealt with in the following section by adding them to the link grammar dictionary.

1.2 Link Grammar

Link Grammar is a dependency grammar system we employ in our work. The top diagram in Fig.1 is a link grammar parse result for the example sentence "I go to Shanghai". The labelled arcs between words are called *links*. The labels are the types of the links. For example, the "Sp*I" between "I" and "go" represents the type of links between "I" and a plural verb form. "MVp" connects verb to its modifying prepositional phrases. "Js" connects prepositions to their objects. In link grammar, there is a finite set of such link types.

Each word in link grammar has a *linking requirement* stating what types of links it can attach and how they are attached. The link requirements are stored in a link grammar *dictionary*. The link grammar parse result is called a *linkage* or a *link structure*. The link grammar parser is called a *link parser*. Currently, the link parser from CMU [5] has a dictionary of about 60000 words together with their linking requirements. Although the CMU link parser still has difficulties in parsing complex syntactic structures in real commercial environment, it is now ready for use in relatively large prototypes. Applying link grammar to languages other than English (e.g. Chinese [6]) is also possible.

An important feature of link grammar is that the grammar is distributed among words [2]. There can be a separate grammar definition (linking requirement) for each word. Thus, expressing the grammar of new words or words that have irregular usage is relatively easy. We can just define the grammar for these words and add them to the dictionary. This is how we deal with terms and jargon of a domain in our approach.

Because the vocabulary set of a domain is limited (see section 1.1), we can add all unknown words (including terms and jargon) to the current dictionary of link grammar with affordable amount of work.

The most important reason that makes us adopt link grammar in our work is the structure similarity between link grammar parse result and conceptual graph. Fig.1 shows this similarity by comparing the link grammar parse result, the typical parse tree of constituent grammar and the conceptual graph for the same example sentence. In fact, this similarity comes from the common foundation of both conceptual graph

The link grammar parse result:

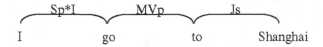

A typical parse tree of constituent grammar:

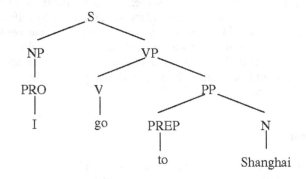

The conceptual graph:

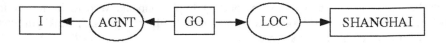

Fig. 1. The similarity between a link structure and a conceptual graph

and link grammar. Conceptual graph consists of concepts and relations. The relations denote the semantic associations between concepts. Similarly, link structure consists of words and links. The links directly connect syntactically and semantically related words [2]. Open words[7] (such as noun, adjective and verb) access concepts from the catalog of conceptual types, while closed words[7] (such as prepositions) and links help clarify the semantic relationships between the concepts [8].

Based on this similarity and restricted to a specific domain, we propose to automatically generate CGs by learning the mapping from link structure to conceptual graph.

2 Overview of the Approach

Our approach of automaticly generating CG is a process consisting of two phases: the training phase and the generating phase. Fig.2 is an overview of the approach.

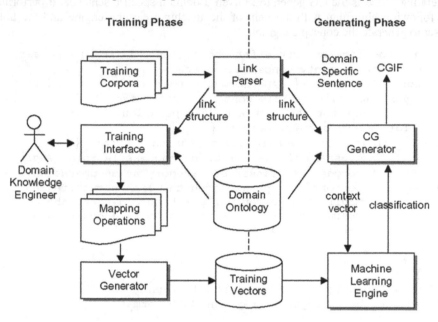

Fig. 2. Overview of the approach

In the training phase, given a link structure generated by the link parser for a sentence in the training corpora, the domain knowledge engineer goes through four steps to manually transform the link structure into a conceptual graph. In each step, one kind of operation is performed. Each kind of operation maps certain part of the syntactic structure into its corresponding semantic representation according to the syntactic and semantic context. The concepts, schemata[2] and relations contained in the semantic representation are selected by the engineer from the domain ontology. Since semantic ambiguities are rare in domain specific sentences (see section 1.1), it is relatively easy for the engineer to perform these mapping operations (the process of semantic analysis).

In order for computer to automatically perform these mapping operations later in the generating phase, we convert the mapping into machine learning area so that it can be learned by machine. The most studied task in machine learning area is inferring a function that classifies a feature vector into one of a finite set of categories [9]. We thus translate the mapping operation into classification operation by encoding the operation as category and encoding the context in which the operation is performed as feature vector. We call the feature vector *context vector* since it encodes the context information of an operation. The vector generator in the left down corner of Fig.2 is

---

[2] Schemata, as defined by Sowa in [1], is a set of conceptual graphs describing background information in a domain.

the component that executes this task. It converts mapping operations into context vectors and corresponding categories.

After the training phase, a set of training vectors and categories can be obtained and the system can enter into the generating phase. The core component in the generating phase is the CG generator. Given a domain specific sentence, it performs the following algorithm with the help of the machine learning engine and the link parser to generate the conceptual graph.

```
1  get the link structure for the sentence from link parser.
2  generate an empty conceptual graph.
3  for (i = 1 to 4) { //perform the four kinds of operations
4    generate all possible context vectors from link
5    structure for the i-th kind of operation.
5    for (every context vector) {
6      if (an operation is needed for the vector) {
7        classify the vector using machine learning engine.
8        decode the classified category as an operation.
9        perform the operation on the link structure and
10       modify the conceptual graph according to the
11       operation result (using concepts, schemata and
12       relations from the domain ontology).
13     }
14   }
15 }
16 do integration on the conceptual graph.
17 output the conceptual graph in CGIF format.
```

Algorithm 1. The algorithm of the CG Generator

This algorithm does not depend on any specific machine learning method. In the current implementation, we are using IBL (Instance Based Learning) in the machine learning engine. IBL makes it easy to determine whether an operation is needed for an arbitrary vector in the above algorithm (line 6). IBL can return a distance value along with the classification result. If the distance value is very large meaning that it can hardly be classified, it can be determined that no operation is needed for the vector because it is far from being similar to the existing training vectors and may be deemed as noise. For other learning method, this determination may not be easily achieved.

In the following sections, we will explain Algorithm.1 and the four operations in detail. We will use the sentence "The polo with an edge is refined enough for work" as an example. The link structure of the sentence is shown in Fig.3. This sentence is excerpted from a corpus of clothes descriptions collected from many clothes shops on the Web. "Polo" is a certain kind of shirt and "edge" actually means collar.

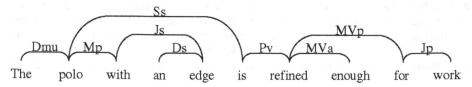

Fig. 3. The link structure for the example sentence

3 Learning to Generate CGs

In this section, the four kinds of mapping operations: word-conceptualization, concept-folding, link-folding and relationalization will be explicated along with Algorithm.1. These four kinds of operations must be performed exactly in the above order in both the training phase and the generating phase because subsequent operations may use information generated in the previous operations. In section 3.5, the integration on conceptual graph (line 16 of Algorithm.1) is explained.

3.1 Word-Conceptualization

- **Function**
Word-conceptualization is the first operation to be performed in Algorithm 1. Its function is to create concept nodes in the initial empty conceptual graph for the open words in the sentence. This operation can be seen as a word sense disambiguation operation.

- **Training**
In the training phase, domain knowledge engineer select all the open words in the link structure one by one. Once an open word is selected, the training interface can provide the engineer a list of possible concepts and schemata retrieved from the domain ontology for the open word. The engineer then selects the appropriate concept or schema from the list.

This operation is then encoded by the vector generator into a context vector and its category. For example, the context vector for the open word "polo" in the example sentence may be <polo, NN, Dmu, Mp>[3] in which "NN" is the POS (part-of-speech) tag[4] and "Dmu" and "Mp" are the innermost left and right link types of "polo" (see Fig.3). All the context information is obtained from the link structure.

The category for the context vector is encoded as the result of the operation – the ID of the selected concept or schema in the domain ontology. In the current implementation, we are using WordNet [10] as an experiment ontology. The encoding is something like "WN16-2-330153" which can be used later as a key to retrieve concept (in WordNet terminology, word sense) from the WordNet database.

- **Generating**
In the generating phase, this operation is indicated in the first loop in Algorithm.1 (line 3). Generating all possible context vectors (line 4 of Algorithm 1.) is actually to generate one context vector for each open word in the link structure of the sentence. The generated context vector is then sent to the machine learning engine to do a classification. The returned category is an encoding of concept or schema ID. In line 9 of Algorithm 1, the CG generator retrieves the concept or schema from the domain

---

[3] The vector is just an example. For brevity, we are not trying to make the vector encoding perfect in this paper. Actually, what context information is encoded into the vector is a separate problem. This problem is isolated into the vector generator component. In the current implementation, we defined a configuration file for the vector generator to address the issue.

[4] We can augment the link parser with a POS tagger so that the accurate POS tag information can be added to the link structure and be obtained from it later.

ontology according to the decoded ID and creates a concept node in the conceptual graph.

Because word usage has patterns in domain specific sentences, we expect that similar context vectors appear for a given open word on a specific word sense. Based on these similar context vectors, we expect that the machine learning engine return correct classification with a high possibility since the semantic ambiguity is also rare in domain specific sentences.

After this step, all the concept nodes of the conceptual graph should be created. For the example sentence, the conceptual graph after this step is shown in Fig.4. For convenience, we use simple concept names in Fig.4. The "S-WORK" is the "SUITABLE-FOR-WORK" schema in domain ontology. We will give more details about it in section 3.5.

Fig. 4. Conceptual graph of the example sentence after word-conceptualization

3.2   Concept-Folding

• Function

Concept-folding operation "folds" concept modifiers into concept nodes. In a standard conceptual graph, generic concepts are modified by an implicit existential quantifier[1, p.86]. It is realized in the indefinite article "a" in English. In the classic example "a cat sits on a mat" in Sowa's book, "a cat" is translated into the generic concept [CAT]. This is basically what this operation does. The indefinite article – "a" – is "folded" into the concept [CAT].

We extended concept modifiers to include generalized quantifiers, modal modifiers and tense modifiers. Several modifiers can be combined together to modify a concept. The modifiers "folded" into a concept node can be removed later in an equivalent standard conceptual graph. This is done in the integration process described in section 3.5.

Generalized quantifiers are used to modify concepts that come from nouns. They include 'a'-quantifier (the original existential quantifier), 'the'-quantifier, 'few/little'-quantifier, 'some'-quantifier, 'any'-quantifier, etc. Modal quantifiers are used to modify concepts that come from verbs. They include 'must'-modifier, 'may'-modifier, 'can'-modifier, etc. Tense modifiers also are used to modify concepts that come from verbs. They include 'past'-modifier, 'future'-modifier, etc. All the types of modifiers can be assigned string Ids to identify them. In our current implementation, about 20 modifiers have been included.

• Training

In the training phase, from the link structure, the domain knowledge engineer can select any link that connect a concept modifier and a concept. Once such a link is selected, the training interface pops up a list of all types of concept modifiers. The engineer then selects the correct one from the list.

For example, in the example sentence, the "Dmu" link between "the" and "polo" is such a link (see Fig.3). The "the" is folded into the [POLO] concept as a 'the'-

quantifier. This operation can be encoded as a context vector <the, Dmu> and its category – the modifier's string ID.

• Generating

In the generating phase, generating all possible context vectors for this operation (line 4 of Algorithm 1.) is actually to generate one context vector for each link that connects a concept modifier with a concept. This needs to consult the concept information generated in the word-conceptualization operations. The generated context vector is then sent to the machine learning engine to do a classification. The returned category is an encoding of the string ID of the modifier's type. In line 9 of Algorithm 1, the CG generator "fold" the modifier into corresponding concept.

In the example sentence, two concept-folding operations are performed. One for "the polo" and one for "an edge". Fig.5 is the conceptual graph of the example sentence after this step.

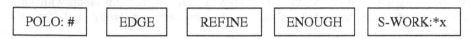

Fig. 5. Conceptual graph of the example sentence after concept-folding

3.3 Link-Folding

• Function

After the concepts have been created in the previous two operations, we then create the semantic relations between the concepts. Closed words (especially prepositions) with their links imply relationships between the concepts of the words they connect. For example, in the example sentence, "… polo --- *with* --- edge …" fragment implies a PART relation between [POLO:#] and [EDGE]. We can "fold" the 'with' and its left and right links and replace them with a PART relation. This is just what the link-folding operation does.

Closed words with their links representing semantic relations can be seen as word usage patterns. In domain specific sentences, we expect such patterns occur frequently. This actually enables the machine to learn the patterns from training corpora. In addition, since semantic ambiguities are rare in domain specific sentences, it can be expected that the result of the learning converge on the correct relation. Similar analysis also applies to the next operation – relationalization in section 3.4.

• Training

In the training phase, the domain engineer can select any closed word that connects two concepts and implies a semantic relation. After the closed word is selected, the engineer can select the implied semantic relation from the domain ontology[5].

The context vector for this operation may encode context information such as the POS tag of the closed word, the left and right link types and the two concepts. For the "… polo --- with --- edge … " case, the context vector may be <with, IN, Mp, Js,

---

[5] It's trivial to handle the direction of a relation. For brevity, we omitted it in this paper.

POLE, EDGE>[6]. The category of the context vector is an encoding of the relation ID in the domain ontology. For the above context vector, the category is the encoding of the ID of the PART relation in the domain ontology.

• **Generating**

In the generating phase, generating all possible context vectors for this operation (line 4 of Algorithm 1.) is actually to generate one context vector for every possible case in which a closed word connects two concepts. This needs consult the concept information generated in the word-conceptualization operations. If an operation is needed for the vector, it is sent to the machine learning engine to do a classification. The returned category is an encoding of the relation ID in domain ontology. In line 9 of Algorithm.1, the CG generator retrieves the relation from domain ontology according to the ID and creates the relation between the two concepts.

For instance, in the example sentence, there are three closed words that need link-folding operation: 'with', 'is' and 'for'. They are shown in Fig.6 along with their links. Among them, the word 'is' is an auxiliary verb and 'with' and 'for' are prepositions.

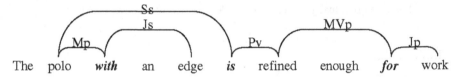

Fig. 6. The three closed words with their links in the example sentence

The relation implied by the auxiliary verb 'is' is THEME and the 'for' between 'refined' and 'work' implies a RESULT relation. The conceptual graph after this step has relations added between concepts. As to our example sentence, the conceptual graph has grown to Fig.7.

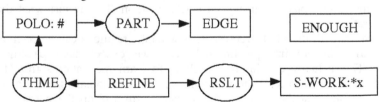

Fig. 7. Conceptual graph of the example sentence after link-folding

3.4 Relationalization

• **Function**

Semantic relations can also be implied by links that connect two concepts in the link structure. For example, the 'MVa' link between 'refined' and 'enough' in the link structure of our example sentence implies a MANNER relation. The relationalization operation directly translates this kind of links into corresponding semantic relations.

---

[6] The POLO and EDGE in the vector are actually the concept IDs in the domain ontology. We will use the same convention in the following vector examples.

- **Training**

In the training phase, domain knowledge engineer can select any link that implies a semantic relation between concepts it connects. The engineer then selects the semantic relation from the domain ontology for the connected two concepts.

The context vector for this operation can include information such as the link type and the concepts. For the "... refined –MVa – enough ... ", the context vector may be <MVa, REFINE, ENOUGH>. The category for the context vector can be encoded as the relation ID in the domain ontology. For the above vector, it is the ID of the MANNER relation.

- **Generating**

In the generating phase, generating all possible context vectors for this operation (line 4 of Algorithm 1.) is actually to generate one context vector for every link that connects two concepts. If an operation is needed for the vector, it is sent to the machine learning engine to do a classification. The returned category is an encoding of the relation ID in domain ontology. In line 9 of Algorithm.1, the CG generator retrieves the relation from domain ontology according to the ID and creates the relation between the two concepts.

After this step, more relations may be created in the conceptual graph. As to the example sentence, the MANNER relation will be created to connect the [REFINE] concept and the [ENOUGH] concept and the whole graph grows to Fig.8.

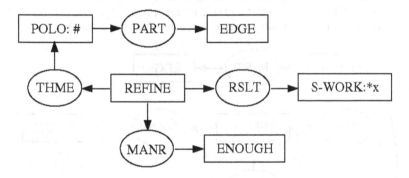

Fig. 8. Conceptual graph of the example sentence after relationalization

3.5 Integration

Integration is the last step(line 16) in Algorithm.1. This step is not a part of the training phase. It only appears in the generating phase and it is the only step that uses manually constructed heuristics. What it does includes simple co-reference detection and nested graph creation.

In the discussion of the previous four operations, we don't involve lambda expressions for brevity. In fact, they may appear when words for concepts are missed in the sentence. They may also be introduced when schema is selected in word-conceptualization phase. In order to complete the conceptual graph, we need to draw co-reference lines between the variables in these lambda expressions.

Although there is machine-learning based approach for co-reference detection [11], in our work we mainly focus on the generation of conceptual graph for a single sentence. Discourse analysis and co-reference detection is left for a separate research work. For different domains, we may construct different heuristics for them. In our current wok we simply make all undetermined references to point to the topic currently under discussion. Nested graph (context) may be introduced by expanding schema definition or removing modal/tense modifiers of a concept. In our example, we have mentioned in section 3.1 that the concept type S-WORK is actually a "SUITALBE-FOR-WORK" schema from the domain ontology. We can do an expansion on it. Fig.9 is the definition for the "SUITALBE-FOR-WORK" schema. SUTB represents the relation SUITABLE.

type SUITABLE-FOR-WORK(x) **is**

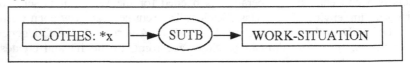

Fig. 9. The definition for SUITABLE-FOR-WORK

After the expansion, we can do a simple co-reference detection that draws a co-reference line between the undetermined variable x and the current topic [POLO:#]. After this step, the final conceptual graph is generated. Fig.10 is the result for our example sentence "The polo with an edge is refined enough for work".

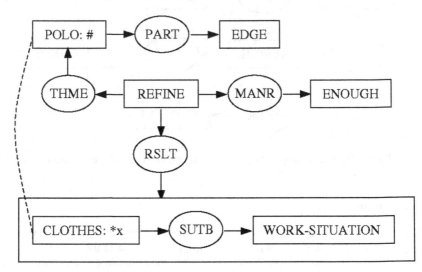

Fig. 10. The final conceptual graph of the example sentence

3.6 Summary

Through the sections from 3.1 to 3.5, we have explained the five operations that map link structure to conceptual graph and show how we translate the mapping into machine learning area. Word-conceptualization and concept-folding build concepts in the conceptual graph. Link-folding and relationalization connect concepts with semantic relations. In the last step, we use manually constructed heuristics to do simple co-reference detection and nested graph creation.

We admit that the whole process is just a learning of mapping although generating CG is surely more than just mapping. This semantic analysis has just made a good beginning. The value of our work lies in the fact that it may be a good start point of a bootstrapping process of semantic analysis.

Currently a system that reads clothes descriptions from clothes shops on the Web and comprehends them in conceptual graphs is under development with IBM China Research Lab. The system is based on the ideas presented in this paper.

4 Related Work

Recently there has been a significant increase in research on learning for natural language by using corpora data [9] and there are growing numbers of successful applications of symbolic machine learning techniques[12,13]. Applying the technique to conceptual graph generation has not yet been seen. Previous work on conceptual graph generation either use manually constructed rules or are highly dependent on one domain.

We roughly divide previous work into slot-filling and structure-mapping categories according to their generating techniques. Slot-filling techniques such as [14,15] fill template graphs with thematic roles identified in the parse tree. Often the conceptual graph of one tree node is constructed using the conceptual graphs of its child nodes according to construction rules on how to fill the slots. This process can be done recursively bottom-up on the parse tree as in [16]. Although this approach has been successfully applied in many applications, it heavily depends on manually created construction rules on the parse tree. These rules are not only hard to create but also difficult to be port to different domains. These rules mix the syntactic knowledge of the grammar used and the semantic knowledge of the domain. They create a tight coupling between the two kinds of knowledge and thus make the independent change of them difficult. In contrast, in our approach, the problem of how to perform mapping operations is translated into machine-learning problems. We need not create rules. We train the machine to learn the rules. The coupling of the syntactic knowledge of link grammar and the semantic knowledge of domain is bound by training. This coupling is loose and can be changed by training in different domains.

Another kind of technique advanced in previous work is to directly map between syntactic structure and semantic structure of CG such as [17] and [18]. We call them structure-mapping. In this respect, they are more similar to our work. To map to more flat structures of conceptual graphs, [17] uses syntactic predicates to represent the grammatical relations in the parse tree. Instead, in our work, link grammar is employed to directly obtain a more flat structure. Syntactic predicates are then translated to all possible semantic interpretations according to a database of translating rules. All the possibilities are finally checked against a LKB. Different from [17]'s approach, our work doesn't uses manual rules. Moreover, we separate the semantic mapping into several steps which greatly reduce the total number of possibilities. In another work in [18], parse tree is first mapped to a "syntactic CG". The "syntactic CG" is then mapped to a real CG. This approach again heavily uses manually constructed mapping rules (such as Parse-To-CG rules and SRTG rules in [18]). Further more, unlike [18]'s two-tier mapping, we do the mapping from syntactic structure in several steps but one-tier.

In [8], a multi-specialists framework for conceptual graph generation is proposed. Our approach can function as a syntax specialist in the framework. Our work presents a preliminary inquest into the use of machine-learning technique to generate conceptual graphs from domain specific sentences. We expect that many improvements are possible and our work may be selectively adopted or enhanced.

References

1. John F. Sowa, Conceptual Structures: Information Processing in Mind and Machine, Addison-Wesley, Reading, MA, 1984.
2. Daniel D.Sleator and Davy Temperley, Parsing English with a Link Grammar, in the *Third International Workshop on Parsing Technologies*, August 1993.
3. Sager Naomi, "Sublanguage: Linguistic Phenomenon, Computational Tool," In R. Grishman and R. Kittredge (eds.), *Analyzing Language in Restricted Domains: Sublanguage Description and Processing*, Lawrence Erlbaum, Hillsdale, NJ, 1986
4. R. Kittredge and J.Lehrberger, "Sublanguage: Study of language in restricted semantic domain", Walter de Gruyter, Berlin and New York, 1982.
5. The information about the link parser from Carnegie Mellon University is available at: http://link.cs.cmu.edu/link/index.html
6. Carol Liu, Towards A Link Grammar for Chinese, Submitted for publication in *Computer Processing of Chinese and Oriental Languages - the Journal of the Chinese Language Computer Society*. Abstract is available at http://bobo.link.cs.cmu.edu/grammar/liu-abstract.html
7. James Allen, "Natural Language Understanding", 2nd edition, pp.24-25, the Benjamin/Cummings Publishing, 1995.
8. Graham A. Mann, Assembly of Conceptual Graphs from Natural Language by Means of Multiple Knowledge Specialists, in *Proc. ICCS'92*, LNAI 754, pp.232-275, 1992.
9. Raymond J.Mooney and Claire Cardie, Symbolic Machine Learning for Natural Language Processing, in the tutorial of *ACL'99*, 1999. Available at http://www.cs.cornell.edu/Info/People/cardie/tutorial/tutorial.html
10. George A.Miller, WordNet: An On-line Lexical Database, in the *International Journal of Lexicography*, Vol.3, No.4, 1990.
11. McCarthy,J., and Lehnert,W., Using Decision Trees for Coreference Resolution. In Mellish, C. (Ed.), *Proceedings of the Fourteenth International Conference on Artificial Intelligence*, pp. 1050-1055. 1995.
12. Claire Cardie and Raymond J.Mooney, Machine learning and natural language (introduction to special issue on natural language learning).   , 34, 5-9, 1999.
13. Brill, E. and Mooney, R.J. An overview of empirical natural language processing, *AI Magazine*, 18(4), 13-24, 1997.
14. Cyre,W.R., Armstrong J.R., and Honcharik,A.J., Generating Simulation Models from Natural Language Specifications, in *Simulation* 65:239-251, 1995.

15. Jeff Hess and Walling R.Cyre, A CG-Based Behavior Extraction System, in *Proc. ICCS'99*, LNAI 1640, pp.127-139, 1999.
16. J.F.Sowa and E.C.Way, Implementing a semantic interpreter using conceptual graphs, in *IBM Journal of Research and Development*, 30(1), pp.57-96, 1986.
17. Paola Velardi, et.,all, Conceptual Graphs for the analysis and generation of sentences, in *IBM Journal of Research and Development*, 32(2), pp.251-267, 1988.
18. Caroline Barrière, From a Children's First Dictionary to a Lexical Knowledge Base of Conceptual Graphs, Ph.D thesis, School of Computing Science, Simon Fraser University, 1997. Available at ftp://www.cs.sfu.ca/pub/cs/nl/BarrierePhD.ps.gz

Solving-Oriented and Domain-Oriented Knowledge Structures: Their Application to Debugging Problem Solving Activity

Slim Masmoudi

University of Tunis I
Faculty of Human and Social sciences of Tunis
Department of Psychology
slim.masmoudi@fshst.rnu.tn

Abstract. This paper describes and analyzes two different sub-systems of knowledge structures – namely "domain-oriented" and "solving-oriented" knowledge structures – in the results of a case study based on interviews and experimental sessions. This case study consisted in the analysis of the activities of an expert programmer in his debugging of a complex computer system. The paper describes the approach adopted to study the conceptual (domain-oriented) structures in relation to the cognitive (solving-oriented) patterns in the context of problem solving activities. The results appear to be consistent with the role of some solving-oriented cognitive patterns in debugging activity and the contextual effects of the problem on the conceptual structures.

1. Introduction

Since the GPS (General Problem Solver) theory developed by [30], several studies have tried to elucidate problem solving strategies and the different stages and operations developed by the human or artificial solver, from an initial state to a final one. These studies, and according to the information processing point of view, have contributed to the better view we now have of the different types of problems and of the role of a set of components, namely problem-space, problem representation, dual search, hypothesis formulation, planning, problem states, and goal-driven solving (for a review, see [18], [39]). Nevertheless, very few studies have addressed the relationships between problem solving, on the one hand, and knowledge structures, and in particular conceptual nets, on the other (see [25]). This may be due to the methodological and empirical difficulties relative to the study of conceptual structures in problem solving contexts, particularly in complex contexts such as debugging. This may also be due to theoretical foundations which do not take into account relationships between data and concepts, and the role of activation processes during problem solving activity.

The present work is based on the assumption that there are two sub-systems of knowledge structures relevant to problem solving: *Solving-oriented* knowledge and *Domain-oriented* knowledge structures. Solving-oriented knowledge structures are cognitive patterns of closely related features of a sub-problem space or a sub-problem

H. Delugach and G. Stumme (Eds.): ICCS 2001, LNAI 2120, pp. 58–71, 2001.

state. Domain-oriented knowledge structures are conceptual nets which represent domain and problem related knowledge. In addition, two activation processes are considered: data-driven and conceptual-driven activation. In the first, concepts are activated by data selected from stimuli and through the instantiation of human cognitive patterns (e.g. the wrong value of the window ordinate on the screen activates some concepts related to the window in question). In the second, some concepts of the conceptual nets are activated by other concepts and thus become available for the cognitive patterns (e.g. a concept related to the window in question activates an other concept related to the general function of the display, which provides information about one or several attributes).

2. Problem Solving and Knowledge Structures

Problem solving activities are often studied in relation to the strategies and heuristics developed by the subject to produce a solution. This point of view, however, does not take into account the role of knowledge structures, their organization and their activation. Neither does it satisfactorily elucidate the dynamic processes between stimuli and knowledge structures on the one hand, and between knowledge structures themselves during problem solving, on the other.

As an alternative, a different problem solving activity model is proposed here. This model consists of three modules: stimulus structures, knowledge structures and dynamics of resolution (Fig. 1).

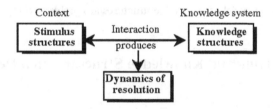

Fig. 1. Model of the problem solving activity. Stimulus structures are the mental representation of the stimulus found in the solving context and essentially in the problem space. The dynamics of resolution is mainly the reflection of the interaction between stimulus structures and knowledge structures.

The key question is to elucidate how one organizes and uses what s/he knows about the problem, and how s/he organizes and uses what s/he knows about his/her domain of expertise during problem solving. In this respect, we start from the assumption that there are essentially two knowledge structure sub-systems which contribute to problem solving. The first, solving-oriented, sub-system is made up of cognitive patterns that represent, manage and process the problem data (data-driven information), activate concepts and aim to solve sub-states of the problem. The second, domain-oriented, sub-system is composed of conceptual nets that represent and manage the concepts of the domain (conceptual-driven information); these concepts are directly activated by the cognitive patterns (Fig. 2).

These two sub-systems are different in the sense that each includes different functional and structural relations. This distinction is based on the nature of information to be processed (either data or concepts), and its relevance to problem-solving. Hence, it is believed to be more appropriate than the one based on the content of information (declarative or procedural).

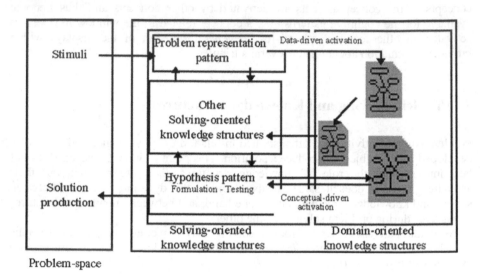

Fig. 2. A general model of knowledge structures in a problem solving activity.

3.   Problem Solving and Knowledge Structures in a Debugging Context

Many attempts have been made to model the debugging activity without considering at the same time the two knowledge structure sub-systems. Instead, scholars have focused solely on either of the two aspects ([22], [32], [33], [40], [41]). In addition, there has been no obvious distinction between the different knowledge structures that can play a crucial role in complex problem solving activities such as debugging. [40], for instance, like [22], proposed a model based on a concatenation of operations through a set of stages, while [32] introduced the role of the abstraction level in program comprehension.

4.   Knowledge Structures

This section consists in the presentation of the basic notions which will be made use of in the subsequent discussion.

4.1 Knowledge Schemas

The notion of schema[1] alone, is not sufficient to describe and explain the cognitive processes underlying knowledge structures. This makes it necessary to consider the different activation and operating processes of the schema, such as the meaning construction processes studied by Kintsch ([23]; see also [29]), and the context dependency aspects of concepts (reviewed by [2]; see also [3]). Thus, the study of these processes in the context of a problem solving activity becomes important. As knowledge schemas, the expert programmer's cognitive patterns elucidated in the debugging activity are contextualized because they process data stemming from the bugs stimuli. They, consequently, activate some concepts from the highly contextualized conceptual structures

4.2 Concepts and Semantic Networks

Basically, a concept is the cognitive meaning of a term and the smallest unit of thought processes. Concepts are used to recognize an object as an instance of the concept, to produce or to understand sentences in which the concept is expressed and to develop constructs or cognitive systems using the concept in question.

First put forward by [6] and reviewed by [5], this notion has since been acquiring more and more importance in cognitive science as an expert knowledge representation technique.

In a semantic/conceptual network, and according to the Spreading Activation Theory (SAT),[2] concepts are represented by nodes, which are interconnected by links or labeled relations. Various types of relationships can define links between concepts The most common representation of conceptual networks, as shown in Fig. 3 (form A), is unhierarchical and can be qualified as static. The alternative suggested by the present work (Fig. 3, form B) represents the knowledge core as a spiral coiling round a central concept; the concept which has the greatest number of direct and indirect links (i.e. embeddedness). This representation, inspired from [15] and [38], is dynamic because the central concept and the general structure of the network can evolve according to the context.

5. Conceptual Structures and Expertise

5.1 Conceptual Chunking and Expertise

Chunking is a mechanism which puts together representation units (such as concepts, schemas, scripts). These units result from the integration and the embeddedness of smaller units that lose their autonomy, thus creating new storage autonomies. Such

---

[1] See [1], [3], [20], [31], [34], [35], [36]; [37]. Concerning program comprehension, see [10], [11], [12], [32], [33]. Concerning planning processes, see [21].

[2] proposed by [5].

representation units emphasize the sophisticated and refined hierarchy in the conceptual organization. Concepts and chunks are used unconsciously by experts, through activation processes, to direct and to interpret perceptual groupings and configurations (see [7], [19]). Conceptual chunking plays a crucial role in the expert problem solving activities ([8], [28]). The following sections will show how concept chunks can be activated to direct the programmer's cognitive patterns, which, in turn, select data from stimuli structures.

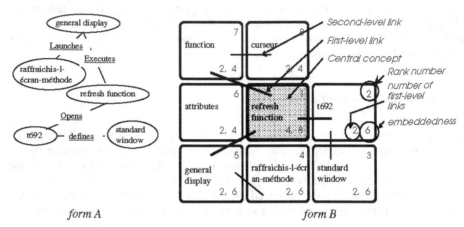

Fig. 3. Semantic network representations: comparison between two forms.

5.2 Conceptual Structures and Contextual Clues

Knowledge nets represent retrieval structures through activation processes that depend both on direct stimulus information and on contextual information. Context is relevant not only to encoding processes, but also to activation processes. [14] presented a review and experimental data showing the importance of contextual information, particularly for the expert (for an extended review, see also [2]). It was argued in [26], [27] that, for an expert programmer, contextual features are more important than direct stimulus information. This is even more the case in problem solving activities such as debugging, where contextual clues help the expert have a dynamic mental representation of the bug, reinforce his hypothesis and activate[3] the relevant concepts.

6. Method

Two different techniques were applied to study both solving-oriented and domain-oriented structures and the relations between them. These are verbal protocol and

---

[3] Activation is unconscious and does not need conscious and logical reasoning.

techniques based on semantic networks. These techniques were used in an extended case study of the activities of an expert programmer in his debugging of complex computer system, through genuine (not artificially created) bugs (see [28]). In addition, a questionnaire was prepared and distributed to 60 expert programmers, in order to verify our assumptions about the knowledge structures identified in the case study.

6.1 Verbal Protocol Tracing and Analysis

6.1.1 Tracing Technique

During the debugging, use has been made of *active tracing*. This corresponds to the real time activity during which the expert programmer describes what he does, both by giving verbal descriptions of his thoughts during the solving process, and by inserting in this description all elements that belong to the solving context, and more particularly, elements that are relevant to his activity. The programmer's activity has also been recorded using a built-in program included in the system to debug.

6.1.2 Types of Corpuses to Analyze

From the raw verbal protocols gathered, and according to the segmenting technique, two types of RTD (Real Time Debugging) corpuses were produced. The "RTD I" corpuses are the result of a "macro-structural" segmenting which is based on type I textual units. These units correspond to phrases constructed around a verbal core with all necessary ingredients to make sense (e.g., subject, object). These corpuses allow the study of the global solving actions, the different cognitive sub-activities implied in the solving process, and the solving-oriented knowledge structures. The "RTD II" corpuses are based on a "micro-structural" segmenting. A type II textual unit corresponds to a fragment of a proposition that either announces something important in the activity or introduces an important concept. The "RTD II" corpuses allow the study of the global solving actions through verbs that the programmer used, as well as the conceptual structures, through what has been called *concept-words* in the present research (e.g. window, list, defun, grep).

6.1.3 Corpus Indexing and Analysis

A *general model of debugging activity indexing* (Table 1) was used to index the "RTD I" corpuses, in order to identify cognitive patterns and cognitive sub-activities. This model contains 12 global units and a total of 30 analysis units. The "RTD I" corpuses are indexed by three indexing and analysis units which are "action", "global action" and "concept-word". The analysis of the indexed corpuses is based on statistical computations.

6.2 Semantic Network Technique

6.2.1 Experimental Procedure to Study Conceptual Structures

Each concept-word identified in the corpus is used as an experimental stimulus manipulated in an experiment. The stimuli are successively presented to the expert programmer who is asked to react by providing as quickly as possible the word which each stimulus triggers in his mind. The result of the experiment is a set of concept/association pairs (Fig. 4).

Table 1. General model of debugging activity indexing.

1. Bug representation	8. Cost management
2. Information search	8.1 management of the cognitive cost
2.1 dynamic search	8.2 management of the material cost
2.2 static search	9. BAC (Bug Appearing Conditions)
2.3 search about state	9.1 BAC generalization
2.4 search about process	9.2 BAC fine-tuning
3. Information search goals	10. Hypothesis formulation
3.1 well-working/badly-working	10.1 problem reformulating
comparison	10.2 new hypothesis
3.2 determining when	10.3 hypothesis generalization
3.3 determining how	10.4 hypothesis fine-tuning
3.4 determining why	11. Hypothesis testing
3.5 determining what	11.1 positive test strategy
4. Interpretation	11.1.2 positive target test
5. Stimulus	11.1.1 positive hypothesis test
5.1 textual stimulus	11.2. negative test strategy
5.2 experimental stimulus	11.2.1 negative hypothesis test
6. Observed phenomena	11.2.2 negative target test
6.1 provoked stimulus	12. Specific domain knowledge evocation
6.2 spontaneous stimulus	
7. Nature of activity cost	
7.1 cognitive cost	
7.2 material cost	

For each debugging session, the list of concept/association pairs resulting from the experiment represents the first material of analysis. The produced semantic network is called "Out-of-Context Semantic Network" (OCSN), since the programmer is not in a problem solving situation (debugging).

The second material of analysis corresponds to other pair lists. The first element of each pair represents a concept-word as stated in the verbal protocol. Each concept-word is associated with the one that follows it. The succession of concept-words as stated in the verbal protocol describes the contextual dynamic links which relate each concept-word to the one that follows it. The resulting semantic network is called "Contextualized Global Semantic Network" (CGSN), since it corresponds to the problem solving situation (debugging activity).

6.2.2 Semantic Network Analysis Technique

Use was made of a computer program called "SemNet"([16], [17]) adapted for the purpose of the present research. This program produces a visual representation of the programmer's knowledge core. For each concept, it calculates two values: n_1 that corresponds to the number of concepts to which it is directly related, and n_2 (embeddedness) which is the sum of n_1 and the number of concepts to which those of the first level are related. The advantages of this analysis are the following:

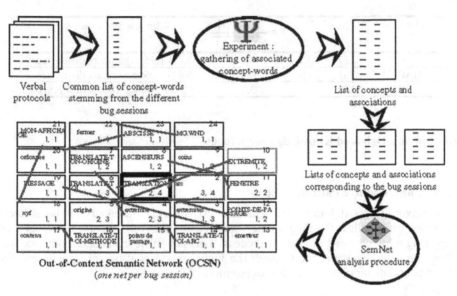

Fig. 4. Procedures used to produce an "Out of Context Semantic Network" (OCSN).

- it provides the possibility to represent the expert's knowledge as a dynamic semantic network;
- it makes it possible to identify differences between two situations (contextualized situation and experimental uncontextualized situation) according to the structure produced, to the network complexity and to the central concept.
- it makes it possible to identify the central concept, to visualize the chunk of a concept and to compare different networks between them.

To sum up, this analysis shows the relationships between the cognitive (solving-oriented) patterns and the conceptual (domain-oriented) structures. Thus, it shows how these conceptual structures evolve through the problem solving activity.

7. Results and Discussion

Domain-oriented and solving-oriented knowledge structures were firstly studied through the "General model of indexing debugging activity". The results of the implementation of this model (Table 1) highlight the importance of essentially three cognitive patterns (Table 2): *Bug Cognitive Pattern* (BCP), *Bug Appearing Conditions Cognitive Pattern* (BACCP) and *Hypothesis Pattern* (HP).[4]

Table 2. This table summarizes the occurrences of textual units indexing the different patterns. Each cognitive pattern has an occurrence rate significantly bigger than the average calculated over the total of textual units (A = 920 textual units / 12 indexing units = 76,7; P_A = 8,3%). The importance of these patterns and their attributes has been confirmed by the questionnaire (Table 3).

Cognitive Pattern	N	%	$\varepsilon = \dfrac{\|N - 76,7\|}{\sqrt{N + 76,7}}$	With α = 5%, ε_{th} = 1,96
BCP	187	20,3	6,8	6,8 > 1,96: significant difference
BACCP	191	20,8	6,4	6,4 > 1,96: significant difference
HP	129	14	3,7	3,7 > 1,96: significant difference

7.1 Solving-Oriented Knowledge Structures

7.1.1 Bug Cognitive Pattern (BCP)

This solving-oriented knowledge structure provides the necessary and sufficient attributes which allow the programmer bug detection and the mental representation triggering a hypothesis formulation process. This pattern is instantiated and updated along the debugging process, through the visualization of different stimuli and the experimental manipulations of the programmer.

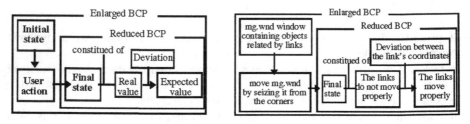

Fig. 5. Schematic representation of the "Bug Cognitive Pattern" and its illustration with the "move-link" bug. The BCP is supposed to have a decoding capacity of all information related to the bug.

Results emerging from the questionnaire show the importance of the instantiation of the BCP attributes. Table 3 summarizes the possible instantiation percentages of the expert programmers who refer to an attribute.

---

[4] For further analysis of the BACCP, see [27]. Because of space constraints, this point cannot be investigated here.

Results show the significant importance of the different BCP attributes identified according to verbal protocols, and their different instantiations.

Table 3. Importance of the different BCP attributes according to the questioned expert programmers. The importance of the pattern attributes as well as their instantiations was confirmed and refined by an open pre-questionnaire distributed to an experimental population of 90 programmers. Other sixty expert programmers have to select one attribute instantiation. Here are the percentages of the expert programmers who define the attribute according to an instantiation.

Pattern attribute	*Attribute instantiation / %*			
1. Defining the program's abnormal behavior	expected value	real value	logical value	specification
	37,5	20,9	20,8	20,8
2. Defining the initial state	data input	initialization variables	before execution	
	42,8	28,6	28,6	
3. Defining the user action	suspected cause	leading to the last expected result		
	75	25		
4. Defining the final state	first computing step incoherent with expected result		program's end	executed code after the bug appearing
	66,7		27,8	5,5
5. Defining the expected value	predicted value	result provided from other computing mean		specified result
	52,6	31,6		15,8
6. Defining the deviation	quantitative / qualitative deviation		gradual deviation	
	55,6		44,4	
7. Defining the values domain of the expected result	interval	type of result	valid / invalid	
	50	38,9	11,1	

7.1.2 Hypothesis Pattern (HP)

The hypothesis is defined, in the present work, as a solving-oriented knowledge structure. The expert programmer is supposed to instantiate, throughout the debugging session, the attributes of this structure in interaction with the bug mental representation (BCP).

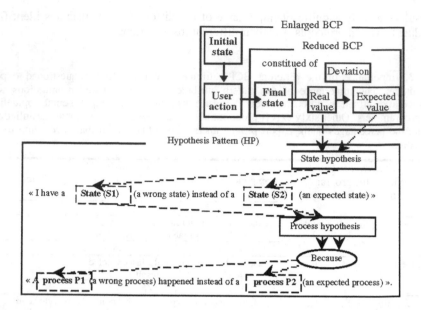

Fig. 6. Relationships between the "Hypothesis Pattern" and "BCP". According to the verbal protocols, the expert programmer formulates two types of hypothesis: a "state hypothesis" and a "process hypothesis". The relationships between the attributes BCP and HP emphasize, on the one hand, the strong relation between the solving-oriented knowledge structures and the stimuli, and, on the other hand, the strong relation between those structures and the domain-oriented ones. The latters are based on the "expected state" and the "expected process". This process corresponds to the way in which the program is executed. In order to have a mental representation of the expected process, the programmer must activate domain-oriented (conceptual) structures.

7.2 Domain-Oriented Knowledge Structures

Conceptual *domain-oriented* structures are activated by the contextual clues (data) through the solving-oriented structures. Four aspects were adopted for the study of the contextual effect: semantic network complexity, general structure, central concept and central micro-network.

CGSN (contextualized) is more complex than OCSN (uncontextualized) since there are more links between its elements. The debugging context has an effect on the programmer's knowledge core of activated concepts, as shown by the greater complexity of the network. Instead, debugging context activates more concepts and provides temporary relations between them. According to the contextual clues, the central concept and the micro-network differ in CGSN and OCSN. The central concept in CGSN, which is the most embedded one, has a strong relationship with the bug solution (the error consists, in this example, in a missing attribute – an unpredicted case – of the "declignot" function).

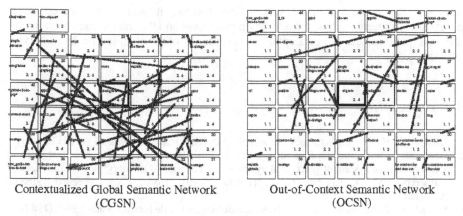

Contextualized Global Semantic Network   Out-of-Context Semantic Network
(CGSN)   (OCSN)

Fig. 7. Comparison between Contextualized Global Semantic Network (CGSN) and Out-of-Context Semantic Network (OCSN).

Thanks to its multiple activations throughout the problem solving activity, and its high degree of embeddedness, a concept becomes central and strongly related to the bug location. The activation processes described above depend highly on the different cognitive pattern instantiations (BCP, BACCP and HP).

With the semantic network technique, the conceptual chunk of a selected concept can be visualized. For each selected concept, this chunk displays the other related concepts, both in CGSN and OCSN. Significant differences were found between the embeddedness of the related concepts of a chunk from CGSN to OCSN. Results showed also strong relationships between the cognitive patterns and the conceptual structures (essentially the different chunks) throughout the debugging activity progress.

8. Conclusion

This paper is based on the assumption that a distinction has to be made between solving-oriented and domain-oriented knowledge structures. It is supposed that the latters correspond to conceptual structures which are activated in interaction with the cognitive patterns of some aspects of the problem space. These patterns play a crucial role in decoding stimulus and contextual information, and in domain-oriented concept activation. The discussion of a part of the presented results emphasizes this distinction which needs to be further explained and emphasized using other experimental techniques, over other problem solving activities. This distinction does not contradict strong relationships between the two knowledge structure sub-classes. It may be a theoretical foundation of future problem solving simulations.

The semantic network representation, that was implemented through the "SemNet" technique, seems to be a powerful means to study conceptual structures and contextual effects. Indeed, it makes it possible to distinguish the stable knowledge core and the contextual-based one.[5] Nevertheless, it would be interesting to investigate, in future studies, the activation processes and the interactions between the cognitive patterns and the conceptual structures, in other complex problem solving activities.

References

1. Alba, J.W., Hasher, L.: Is memory schematic ? Psychological Bulletin. 93 (1983) 203-231.
2. Barsalou, L. W.: Structure, flexibility, and linguistic vagary in concepts: Manifestations of a compositional system of perceptual symbols. In Theories of memory, A. C. Collins, S. E. Gathercole, M. A. Conway (eds.), (1993) 29-101. Hove, UK: Lawrence Erlbaum Associates Ltd.
3. Bastien, C.: Schèmes et stratégies dans l'activité cognitive de l'enfant. Paris : PUF (1987).
4. Chase, W.G., Simon, H.A.: Perception in chess. Cognitive Psychology, 5 (1973) 55-81.
5. Collins, A.M., Loftus, E.F.: A spreading activation theory of semantic processing. Psychological Review, 83 (1975) 402-428.
6. Collins, A.M., Quillian, R.M.: Retrieval time from semantic memory. Journal of Verbal Learning and Verbal Behavior, 8 (1969) 240-248.
7. Cooke, N.J., Atlas, R.S., Lane, D.H., Berger, R.C.: The role of high-level knowledge in memory for chess positions. Unpublished manuscript, Rice University, Houston, TX, (1991).
8. Cooke, N.J.: Modeling human expertise in expert systems. In. R.R. Hoffman (ed.), The Psychology of Expetise. New York : Springer-Verlag (1992) 29-60.
9. de Groot, A.D.: Perception and memory versus thought : Some old ideas and recent findings. In. B. Kleinmuntz (ed.), Problem-solving : Research, method, and theory New York : Wiley (1966) 19-50.
10. Détienne, F.: A schema-based model of program understanding. Eighth interdisciplinary workshop on "Informatics and Psychology". Schaärding (Austria), May (1989) 16-19.
11. Détienne, F.: Expert programming Knowledge : A schema-based Approach. In. J.-M. Hoc, T.R.G. Green, R. Samurçay, D.J. Gilmore (eds.), Psychology of programming. London : Academic Press (1990).
12. Détienne, F.: Une application de la théorie des schémas à la compréhension de programmes. Le travail Humain, numéro spécial : Psychologie ergonomique de la programmation, 51(4) (1988) 335-350.
13. Egan, D.E., Schwartz, B.J.: Chunking in recall of symbolic drawings. Memory and Cognition, 7 (1979) 149-158.
14. Ericsson, K.A., Kintsch, W.: Long-Term Working Memory. Psychological Review, vol. 102, 2 (1995) 211-245.
15. Fisher, K. M., Faletti. J., Patterson, H. A., Thornton, R., Lipson, J., Spring, C.: Computer-based concept mapping: SemNet software - a tool for describing knowledge networks. Journal of College Science Teaching, 19(6) (1990) 347-352.
16. Fisher, K.M.: Semantic networking: New kid on the block. Journal of Research in Science Teaching, 27 (10) (1990) 1001-1018.
17. Fisher, K.M.: SemNet: A tool for personal knowledge construction. In P. Kommers, D. Jonassen, & T. Mayes (eds.), Cognitive tools for learning. Berlin: Springer-Verlag, 63-76 (1992).
18. Frensch, P. A., Funke, J.: Complex problem solving: The European Perspective. Hillsdale, NJ: Lawrence Erlbaum Associates (1995).

---

[5] Conceptual domains, conceptual progress during a problem solving activity and solving-oriented knowledge structures are further investigated in [28].

19. Glaser, R., Chi, M. T. H.: Overview. In. M. T. H. Chi, R. Glaser, M. J. Farr (eds.), The nature of expertise (p. xv-xxviii) Hillsdale, NJ : Erlbaum (1988).
20. Graesser, A.C., Nakamura, G.V.: The impact of a schema on comprehension and memory. In. G.H. Bower (Ed.), The Psychology of learning and motivation New York : Academic Press, Vol. 16 (1982) 60-109.
21. Hoc, J.M.: Psychologie cognitive de la planification. PUG, Grenoble (1987).
22. Katz, I.R., Anderson, J.R.: Debugging : An analysis of bug location strategies. Human-Computer Interaction, 3 (1988) 351-400.
23. Kintsch, W., & Welsch, D. M.: The construction-integration model: A framework for studying memory for text. In W. E. Hockley & S. Lewandowsky (eds.), Relating theory and data: Essays on human memory in honor of Bennett B. Murdock. Hillsdale, NJ.: Erlbaum (1991) 367-385.
24. Kintsch, W.: Comprehension: A paradigm for cognition. Cambridge: Cambridge University Press (1995).
25. Marshall, S. P.: Schemas in problem solving. New York: Cambridge University Press (1995).
26. Masmoudi, S., Martin, R.: Cognitive Model of Debugging Expertise. Neuvième Conférence Européenne en Ergonomie Cognitive : ECCE-9, University of Limerick, Ireland, September (1998).
27. Masmoudi, S.: A Psychological Cognitive Approach to Debugging. In. B. Blumental, J. Gornostaev, C. Unger (eds.), Proceedings of EWHCI'95 (The 5th East-West International Conference of Human-Computer Interaction) Moscow, Russia, July 4-7 (1995).
28. Masmoudi, S.: Modélisation cognitive de l'expertise de débogage de systèmes informatiques complexes. Lille : Presses Universitaires du Septentrion (2000).
29. McKoon, G., & Ratcliff, R.: Contextually relevant aspects of meaning. Journal of Experimental Psychology: Learning, Memory, and Cognition, 14 (1988) 331-343.
30. Newell, A., Simon, H. A.: Human problem-solving. Englewood Cliffs, NJ : Prentice Hall (1972).
31. Norman, D.A.: Learning and memory. San Francisco, CA : W.H. Freeman (1982).
32. Pennington, N.: Cognitive components of expertise in computer programming : A review of litterature. Psychological Documents, 15 (1985) N° 2702.
33. Pennington, N.: Stimulus structures and mental representations in expert computer programs. Cognitive Psychology, 19 (1987) 295-341.
34. Richard, J-F., Kekenbosch, C.: Les structures de connaissances. In. R. Ghiglione, J.-F. Richard (eds.), Cours de Psychologie. Paris: Dunod (1995) 208-226.
35. Richard, J-F.: Les activités mentales. Paris : Armand Colin (1990).
36. Schank, R.C.: Dynamic memory, a theory of reminding and learning in computers and people. Cambridge : Cambridge University Press (1982).
37. Smith, E. E., Medin, D. L.: Concepts and Categories. Cambridge, Mass.: Harvard University Press (1981).
38. Sowa, J. F.: Conceptual Structures : Information Processing in Mind and Machine. Menlo Park, CA: Addison-Wesley (1983).
39. Sternberg, R. J., Frensch, P. A.: Complex problem solving: Principles and mechanisms. Hillsdale, NJ: Lawrence Erlbaum Associates (1991).
40. Vessey, I.: Expertise in debugging computer programs : A process analysis. International Journal of Man-Machine Studies, 23 (1985) 459-494.
41. Vessey, I.: Toward a theory of computer program bugs : An empirical test. International Journal of Man-Machine Studies, 30 (1989) 23-46.

Concept Graphs and Predicate Logic

Frithjof Dau

Technische Universität Darmstadt, Fachbereich Mathematik
Schloßgartenstr. 7, D-64289 Darmstadt, `dau@mathematik.tu-darmstadt.de`

Abstract. In the ICCS 2000 proceedings we introduced negation to simple concept graphs without generic markers by adding *cuts* to their definition. The aim of this paper is to extend this approach of cuts to simple concept graphs *with* generic markers. For these graphs, a set-theoretical semantics is presented. After this a modification of Peirce's beta-calculus is provided, and definitions for mappings Φ and Ψ between concept graps and first order logic are given. If we consider both concept graphs and first order logic formulas, together with their particular derivability relations, as quasiorders, Φ and Ψ are mutually inverse quasiorder isomorphisms between them. The meaning of this fact is elaborated. Finally we provide a result that links the semantics of concept graphs and the semantics of first order logic. This result can be used to show that the calculus for concept graphs is sound and complete.

1 Motivation and Overview

In [Da00], we introduced negation to simple concept graphs without generic markers by adding *cuts* to their definition. These concept graphs are closely related to the α-part of the existential graphs of Charles Sanders Peirce, which consist only of cuts and propositional variables. The aim of this work is to extend the approach of [Da00] to simple concept graphs with generic markers. These graphs correspond to the β-part of existential graphs, namely to existential graphs which are built up of cuts, relation names, and lines of identity. It is accepted that these graphs are equivalent to first order logic. An argumentation to support this (but, in our view, not a strict mathematical proof) can be found in [Ro73]. Therefore it seems to be evident that a class of simple concept graphs with generic markers and negations (i.e. cuts) are equivalent to first order logic, too. Indeed, this equivalence can be described and proven in a mathematically precise way, which will be in [Da01]. In this paper, we want to elaborate some aspects of this work.

To start, we provide the necessary definitions for concept graphs with cuts. A mathematical semantics for these graphs which is based on power context families is presented. After this sematical part, we provide a calculus which is based on the β-calculus for existential graphs, but which captures the specific properties of concept graphs. A mathematical definition for the version of the well-known Φ-operator which maps (in this case) simple concept graphs with cuts to first order logic is given, as well as the definition for a mapping Ψ in the inverse

H. Delugach and G. Stumme (Eds.): ICCS 2001, LNAI 2120, pp. 72–86, 2001.

direction. It turns out that Φ and Ψ are mutually inverse isomorphismns between the quasiordered sets of concept graphs and first order logic. The proof for this is very extensive (and will be given in [Da01]). In this paper, we investigate only the meaning of these isomorphismns. Finally, we give a result that links the semantics for concept graphs to the usual relational semantics for first order logic. This result can be used to show that the calculus for concept graphs is sound and complete.

2   Basic Definitions for Simple Concept Graphs

First, we start with an underlying set of variables and names which are needed in concept graphs as well as in first order logic.

Definition 1.

1. *Let Var be a countably infinite set. The elements of Var are called* variables. *In concept graphs, we need a sign* $*$*, the* generic marker. *Further we assign a new sign* $*_\alpha$ *to each variable* $\alpha \in Var$.
2. *An alphabet is a triple* $\mathcal{A} := (\mathcal{G}, \mathcal{C}, \mathcal{R})$ *such that*
   - $\mathcal{G}$ *is a finite set whose elements are called* object names.
   - $(\mathcal{C}, \leq_{\mathcal{C}})$ *is a finite ordered set with a greatest element* $\top$. *The elements of this set are called* concept names.
   - $(\mathcal{R}, \leq_{\mathcal{R}})$ *is a familiy of finite ordered sets* $(\mathcal{R}_k, \leq_{\mathcal{R}_k})$, $k = 1, \ldots, n$ *(for an* $n \in \mathbb{N}$ *with* $n \geq 1$*) whose elements are called* relation names. *Let* $id \in \mathcal{R}_2$ *be a special name which is called* identity.

Now we can define the underlying structures of concept graphs with cuts. There are two slight changes compared to Definition 2 in [Da00]: First, for purely technical reasons, we add the sheet of assertion to the definition. Second, we change the definition of the mapping *area* so that the area of a cut c contains vertices, edges, and other cuts which are enclosed by c, *but not* if they are nested deeper inside other cuts. The only reason for this is that this definition reflects the meaning of area in existential graphs better than the definition in [Da00].

Definition 2. *A relational graph with cuts is a structure* $(V, E, \nu, \top, Cut, area)$ *such that*

- V, E *and* Cut *are pairwise disjoint, finite sets whose elements are called* vertices, edges *and* cuts, *respectively,*
- $\nu : E \to \bigcup_{k=1}^{n} V^k$ *(for a* $n \in \mathbb{N}, n \geq 1$*) is a mapping,*
- $\top$ *is a single element, the* sheet of assertion, *and*
- $area : Cut \cup \{\top\} \to \mathfrak{P}(V \cup E \cup Cut)$ *is a mapping such that*
   - a) $c_1 \neq c_2 \Rightarrow area(c_1) \cap area(c_2) = \emptyset$,
   - b) $V \cup E \cup Cut = \bigcup_{k \in Cut \cup \{\top\}} area(k)$,
   - c) $c \notin area^n(c)$ *for each* $c \in Cut$ *and* $n \in \mathbb{N}$ *(with* $area^0(c) := \{c\}$ *and* $area^{n+1}(c) := area^n(c) \cup \{area(c') \mid c' \in area^n(c)\}$*).*

For an edge $e \in E$ with $\nu(e) = (v_1, \ldots, v_k)$ we define $|e| := k$ and $\nu(e)|_i := v_i$. For each $v \in V$, let $E_v := \{e \in E \mid \exists i\colon \nu(e)|_i = v\}$. Analogously, for each $e \in E$, let $V_e := \{v \in V \mid \exists i\colon \nu(e)|_i = v\}$. If it cannot be misunderstood, we write $e|_i$ instead of $\nu(e)|_i$.

The empty graph has the form $\mathfrak{G}_\emptyset := (\emptyset, \emptyset, \emptyset, \top, \emptyset, \emptyset)$.

We say that all edges, vertices and cuts in the area of c and all items which are deeper nested are *enclosed by* c:

Definition 3. *For a relational graph with cuts $(V, E, \nu, \top, Cut, area)$ we define $\overline{area} : Cut \cup \{\top\} \to \mathfrak{P}(V \cup E \cup Cut)$, $\overline{area}(c) := \bigcup_{n \in \mathbb{N}} area^n(c)$. Every element k of $\overline{area}(c)$ is said to be* enclosed by c, *and vice versa: c is said to* enclose k. *For every element of $area(c)$, we say more specificly that it is* directly enclosed by c. *Because every $k \in V \cup E \cup Cut$ is directly enclosed by exact one $c \in Cut \cup \{\top\}$, for every $k \in area(c)$ we can write $c = area^{-1}(k)$, or even more simply and suggestive: $c = cut(k)$.*

By $c_1 \le c_2 :\Longleftrightarrow c_1 \in \overline{area}(c_2)$ a canonical tree ordering on $Cut \cup \{\top\}$ with $\top$ as greatest element is defined.

In this work, we will only consider graphs in which vertices must not be deeper nested than any edge they are incident with. This is captured by the following definition:

Definition 4. *If $cut(e) \le cut(v)$ holds for every $e \in E$ and $v \in V$ such that e is incident with v, then $\mathfrak{G}$ is said to have* dominating nodes.

Now simple concept graphs with cuts are derived from relational graphs with cuts by additionally labeling the vertices and edges with concept names and relation names, respectively, and by assigning a reference to each vertex.

Definition 5. *A simple concept graph with cuts and variables over the alphabet $\mathcal{A}$ is a structure $\mathfrak{G} := (V, E, \nu, \top, Cut, area, \kappa, \rho)$ where*

- *$(V, E, \nu, \top, Cut, area)$ is a relational graph with cuts*
- *$\kappa : V \cup E \to \mathcal{C} \cup \mathcal{R}$ is a mapping such that $\kappa(V) \subseteq \mathcal{C}$, $\kappa(E) \subseteq \mathcal{R}$, and all $e \in E$ with $\nu(e) = (v_1, \ldots, v_k)$ satisfy $\kappa(e) \in \mathcal{R}_k$*
- *$\rho : V \to \mathcal{G} \cup \{*\} \cup \{*_\alpha \mid \alpha \in Var\}$ is a mapping.*

If additionally $\rho : V \to \mathcal{G} \cup \{*\}$ holds, then $\mathfrak{G}$ is called simple concept graph with cuts over the alphabet $\mathcal{A}$. *If even $\rho : V \to \mathcal{G}$ holds, then $\mathfrak{G}$ is called* nonexistential simple concept graph with cuts over the alphabet $\mathcal{A}$. *For the set E of edges, let $E_{id} := \{e \in E \mid \kappa(e) = id\}$ and $E_{nonid} := \{e \in E \mid \kappa(e) \ne id\}$. The elements of E_{id} are called* identity-links.

In the rest of this work, we will mainly talk about (existential) simple concept graphs with cuts and dominating nodes over the alphabet $\mathcal{A}$, and will call them 'concept graphs' for short. This set of concept graphs is denoted by CG.

The mathematical definitions make up an exact and solid foundation for concept graphs which can serve as a precise reference and a basis for mathematical proofs on concept graphs. But in order to work with concept graphs, their mathematical representations are too clumsy and too difficult to handle. Hence one may prefer the well known graphical representations of conceptual graphs. Because we added cuts as new syntactical elements to the graphs, we have to explain how concept graphs with cuts are drawn. This shall be done now.

Vertices are usually drawn as small rectangles. Inside the rectangle for a vertex v, we write first the concept name $\kappa(v)$ and then the reference $\rho(v)$, seperated by a colon. These rectangels are called *concept boxes*. An edge e is drawn as a small oval with its relation name $\kappa(e)$ in it. The name 'id' for the identity is often replaced by the symbol '$=$'. These ovals are called *relation ovals*. For an edge $e = (v_1, \ldots, v_n)$, each concept box of the incident vertices $v_1, \ldots, v_n$ is connected by a line to the relation oval of e. These lines are numbered $1, \ldots, n$. If it cannot be misunderstood, this numbering is often omitted. There may be graphs such that its lines cannot be drawn without their crossing one another. To distinguish such lines from each other, Peirce introduced a device he called a 'bridge' (see [Ro73], Page 55). But, except for bridges between lines, all the boxes, ovals, and lines of a graph must not intersect. Nearly all graphs which occur in applications do not need bridges. Finally, a cut is drawn as a bold curve (usually an oval) which exactly contains in its inner space all the concept boxes, ovals, and curves of the vertices, edges, and other cuts, resp., which the cut encloses (not necessarily directly). The curve of a cut may not intersect any other curves, ovals or concept boxes, but it may intersect lines which connect relation ovals and concept boxes (this is usually inevitable).

To illustrate these agreements, consider the following graph over the alphabet $\mathcal{A} := (\emptyset, \{\text{CAT}, \text{ANIMAL}, \top\}, \{\text{cute, id}\})$ in its mathematical form:

$$\mathfrak{G} := (\{v_1, v_2\}, \{e_1, e_2\}, \{(e_1, (v_1, v_2)), (e_2, v_2)\}, \top, \{c_1, c_2\},$$
$$\{(\top, \emptyset), (c_1, \{v_1\}), (c_2, \{v_2, e_1, e_2\})\},$$
$$\{(v_1, \text{CAT}), (v_2, \text{ANIMAL}), (e_1, \text{id}), (e_2, \text{cute})\}, \{(v_1, *), (v_2, *)\})$$

In Figure 1, we give one (possible) diagram for this graph. The indices v_1, v_2, e_1, e_2, c_1, c_2 do not belong to the diagram. They are added to make the translation from $\mathfrak{G}$ to the diagram more transparent. Now one can see that the intuitive meaning of the graph is 'it is not true that there is a cat which is not a cute animal', i.e. 'every cat is a cute animal'. This will be worked out in Section 3.

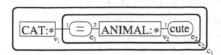

Fig. 1. One example for a simple concept graph with cuts and dominating nodes

For our further work, especially for the calculus, the notion of a *subgraph* is needed. We distinguish beween *subgraphs* and *closed subgraphs*. Informally spoken, a *subgraph* is a part of a graph such that

- if the subgraph contains a cut c, then it contains whatever is scribed inside c, i.e. $\overline{area}(c)$, and
- if the subgraph contains an edge, then it contains all vertices which are incident with the edge.

If it holds furthermore that for each vertex of subgraph all incident edges are part of the subgraph, too, then the subgraph is called a *closed subgraph*. Instead giving a formal definition for this (which will be in [Da01]), we only provide some examples.

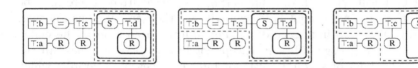

The marked area in the left example is no subgraph. The marked area in the middle example is a subgraph which is not closed. The marked area in the right example is a closed subgraph in the outermost cut.

We use the following teminology: For each vertex, edge, cut, or subgraph, we say that it is *evenly enclosed* iff the number of cuts which enclose it is even, otherwise it is *oddly enclosed*. Evenly enclosed cuts are also called *positive cuts*, and oddly enclosed cuts are called *negative cuts*. The formal definitions can be found in [Da00].

3   Semantics for Simple Concept Graphs

As in the papers of Prediger (cf. [Pr98a], [Pr98b]) or as in [Da00], we use power context families as model structures for concept graphs. Their definition can be found in [Wi97], [Pr98b] or in [Da00]. But we want to repeat the definition of a $\vec{\mathbb{K}}$-interpretation which we presented in [Da00], because the identity-relation $id \in \mathcal{R}_2$ caused a slight change to Prediger's definition of a $\vec{\mathbb{K}}$-interpretation.

Definition 6. *For an alphabet $\mathcal{A} := (\mathcal{G}, \mathcal{C}, \mathcal{R})$ and a power context family $\vec{\mathbb{K}}$, we call the union $\lambda := \lambda_\mathcal{G} \dot{\cup} \lambda_\mathcal{C} \dot{\cup} \lambda_\mathcal{R}$ of the mappings $\lambda_\mathcal{G} : \mathcal{G} \to G_0$, $\lambda_\mathcal{C} : \mathcal{C} \to \underline{\mathfrak{B}}(\mathbb{K}_0)$ and $\lambda_\mathcal{R} : \mathcal{R} \to \mathfrak{R}_{\vec{\mathbb{K}}}$ a $\vec{\mathbb{K}}$-interpretation of $\mathcal{A}$ if $\lambda_\mathcal{C}$ and $\lambda_\mathcal{R}$ are order-preserving, $\lambda_\mathcal{C}(\top) = \top$, $\lambda_\mathcal{R}(\mathcal{R}_k) \subseteq \underline{\mathfrak{B}}(\mathbb{K}_k)$ for all $k = 1, \ldots, n$, and $(g_1, g_2) \in Ext(\lambda_\mathcal{R}(id)) \Leftrightarrow g_1 = g_2$ hold for all $g_1, g_2 \in \mathcal{G}$. The pair $(\vec{\mathbb{K}}, \lambda)$ is called* context-interpretation *of $\mathcal{A}$ or, according to classical logic, $\mathcal{A}$-structure.*

Now we can define whether a concept graph is valid in an $\mathcal{A}$-structure. To do this, we read and evaluate the graph from the outside. Hence we start with the sheet of assertion $\top$, and proceed with the inner cuts. This method (reading a graph from the outside and proceeding inwardly) was called 'endoporeutic method' by Peirce (see [Ro73]). Before we give a precise definition, we exemplify this method on the graph $\mathfrak{G}$ from Figure 1.

We start the evaluation of this graph on the sheet of assertion $\top$. Because only the cut c_1 is directly enclosed by $\top$, the graph $\mathfrak{G}$ is true if the part of $\mathfrak{G}$ which is enclosed by c_1 is false. Because c_1 contains the vertex v_1 and the cut c_2, we come to the following conclusion: $\mathfrak{G}$ is true iff it is not true that the following two conditions hold: There exists an object such that o_1 is a cat (lets call it o_1) and the information which is enclosed by c_2 is false. Now we have to evaluate the area of c_2. Intuitively spoken, the area of c_2 is true iff there is an object that is a cute animal and identical to o_1. Note that during this evaluation, we refer to the object o_1. That is why the endoporeutic method goes from the outside to the inside: We cannot evaluate the inner cut c_2 unless we know which object is assigned to the generic marker in the vertex v_1. Finally $\mathfrak{G}$ is true if there is no cat such that there is no other object which is identical to the cat (hence the cat itself) and which is a cute animal. In simpler words: $\mathfrak{G}$ is true if there is no cat which is not a cute animal, i.e. if every cat is a cute animal.

Hopefully this example helps the reader to understand the following definitions. The assignment of objects to vertices by valuations:

Definition 7. *Let* $(\vec{\mathbb{K}}, \lambda)$ *be a context-interpretation of the alphabet* $\mathcal{A}$ *and* $\mathfrak{G} = (V, E, \nu, \top, Cut, area, \kappa, \rho)$ *be a graph. A* partial valuation *of* $\mathfrak{G}$ *is a mapping* $ref : V' \subseteq V \to \mathbb{K}_0$ *such that* $V' \supseteq \{v \in V \mid \rho(v) \in \mathcal{G}\} =: V_{\mathcal{G}}$ *and* $ref(v) = \lambda_{\mathcal{G}}(\rho(v))$ *holds for all* $v \in V_{\mathcal{G}}$. *If* $V' = V$ *holds, then* ref *is called* (total) valuation *of* $\mathfrak{G}$.

With the endoporeutic method, we can evaluate the area of a cut c in a concept graph $\mathfrak{G}$ if we already have a partial valuation ref which assigns objects to the vertices which are placed outside of c. This is written down $(\vec{\mathbb{K}}, \lambda) \models \mathfrak{G}[c, ref]$ and defined as follows:

Definition 8. *Let* $(\vec{\mathbb{K}}, \lambda)$ *be a context-interpretation of an alphabet* $\mathcal{A}$ *and let* $\mathfrak{G} := (V, E, \nu, \top, Cut, area, \kappa, \rho)$ *be a concept graph. Inductively on the tree* $Cut \cup \{\top\}$, *we define* $(\vec{\mathbb{K}}, \lambda) \models \mathfrak{G}[c, ref]$ *for every cut* $c \in Cut \cup \{\top\}$ *and every partial valuation* $ref : V' \to \mathbb{K}_0$ *with* $V' \supseteq \bigcup\{area(d) \mid d \in Cut \cup \{\top\}, d > c\}$ *and* $V' \cap \overline{area}(c) = \emptyset$:

$(\mathbb{K}, \lambda) \models \mathfrak{G}[c, ref] :\Longleftrightarrow$

ref *can be extended to a partial valuation* $\widetilde{ref} : V' \cup (V \cap area(c)) \to \mathbb{K}_0$ *(i.e.* $\widetilde{ref}(v) = ref(v)$ *for all* $v \in V'$*) such that the following conditions hold:*

- $\widetilde{ref}(v) \in Ext(\lambda_C(\kappa(v)))$ *for each* $v \in V \cap area(c)$ *(vertex condition)*
- $\widetilde{ref}(e) \in Ext(\lambda_{\mathcal{R}}(\kappa(e)))$ *for each* $e \in E \cap area(c)$ *(edge condition)*
- $(\mathbb{K}, \lambda) \not\models \mathfrak{G}[c', \widetilde{ref}]$ *for each* $c' \in Cut \cap area(c)$ *(iteration over* $Cut \cup \{\top\}$*)*

For $(\vec{\mathbb{K}}, \lambda) \models \mathfrak{G}[\top, \emptyset]$ *we write* $(\vec{\mathbb{K}}, \lambda) \models \mathfrak{G}$. *If we have two concept graphs* $\mathfrak{G}_1$, $\mathfrak{G}_2$ *such that* $(\vec{\mathbb{K}}, \lambda) \models \mathfrak{G}_2$ *for each* $\mathcal{A}$-*structure with* $(\vec{\mathbb{K}}, \lambda) \models \mathfrak{G}_1$, *we write* $\mathfrak{G}_1 \models \mathfrak{G}_2$.

Note that this definition (in particular the edge condition) relies on the condition that we consider concept graphs with dominating nodes only.

Now we are prepared to present the calculus for concept graphs.

4   The Calculus for Simple Concept Graphs with Cuts

The following calculus is based on the β-calculus of Peirce for existential graphs with lines of identity. More precisely: The first five rules of the calculus are a concept graph version of Peirce's β-calculus. The rules 'generalization' and 'specialization' encompass the orders on the concept- and relation names. The rules '$\top$- and id-Insertion' and '$\top$- and id-Erasure' are needed to comprehend the specific properties of the concept name $\top$ and the relation name id. For the sake of intelligibility, the whole calculus is described using common language. An appropriate mathematical definition as in [Da00] will be given in [Da01].

Definition 9. *The calculus for simple concept graphs with cuts over the alphabet* $\mathcal{A}$ *consists of the following rules:*

- **erasure**
  In positive cuts any directly enclosed edge, isolated vertex and closed subgraph may be erased.
- **insertion**
  In negative cuts any directly enclosed edge, isolated vertex and closed subgraph may be inserted.
- **iteration**
  Let $\mathfrak{G}_0$ *be a subgraph of* $\mathfrak{G}$ *and let* $\bar{c} \leq cut(\mathfrak{G}_0)$ *be a cut that does not belong to* $\mathfrak{G}_0$. *Then a copy of* $\mathfrak{G}_0$ *may be inserted into* $\bar{c}$. *For every vertex* v *with* $cut(v) = cut(\mathfrak{G}_0)$, *an identity-link from* v *to its copy may be inserted.*
- **deiteration**
  If $\mathfrak{G}_0$ *is a subgraph of* $\mathfrak{G}$ *which could have been inserted by rule of iteration, then it may be erased.*
- **double cuts**
  Double cuts (two cuts c_1, c_2 *with* $cut^{-1}(c_2) = \{c_1\}$*) may be inserted or erased.*
- **generalization (unrestriction)**
  For evenly enclosed vertices and edges, their concept names or object names resp. their relation names may be generalized.
- **specialization (restriction)**
  For oddly enclosed vertices and edges, their concept names or object names resp. their relation names may be specialized.
- **isomorphism**
  A graph may be substituted by an isomorphic copy of itself.
- $\top$- **and** id-**Insertion**

1. **⊤-rule**
   For $g \in \mathcal{G} \cup \{*\}$, an isolated vertex $\boxed{\top : g}$ may be inserted in arbitrary cuts.

2. **identity**
   Let $g \in \mathcal{G}$, let $\boxed{P_1 : g}$, $\boxed{P_2 : g}$ be two vertices in cuts c_1, c_2, resp., and let $c \leq c_1, c_2$ be a cut. Then an identity-link between $\boxed{P_1 : g}$ and $\boxed{P_2 : g}$ may be inserted in c.

3. **splitting a vertex**
   Let $g \in \mathcal{G} \cup \{*\}$. Let $v = \boxed{P : g}$ be a vertex in cut c_0 and incident with relation edges $R_1, \ldots, R_n$, placed in cuts $c_1, \ldots, c_n$, resp.. Let c be a cut such that $c_1, \ldots, c_n \leq c \leq c_0$. Then the following may be done: In c, a new vertex $v' = \boxed{\top : g}$ and a new identity-link between v and v' is inserted. On $R_1, \ldots, R_n$, arbitrary instances of v are substituted by v'.

4. **congruence**
   Let $g \in \mathcal{G}$ and let $v = \boxed{P : g}$ be a vertex in cut c. Then the following may be done: In c, a new vertex $v' = \boxed{P : *}$ and a new identity link between v and v' is inserted. On every edge, every instance of v is substituted by v'.

– **⊤- and id-Erasure**
   The ⊤- and id-Insertion-rule may be reversed.

Example 1. Here is an example for the iteration-rule. Note that for one of the two possible vertices an identity link is inserted to its copy.

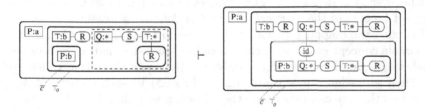

Below please find two examples for the ⊤- and *id*-Insertion rule 2 (splitting a vertex).

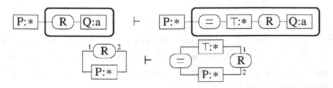

Here is an example for the ⊤- and *id*-Insertion rule 4 (congruence).

$$\boxed{\text{CAT:Yoyo}} - (\text{on}) - \boxed{\text{MAT:*}} \quad \vdash \quad \boxed{\top\text{:Yoyo}} - (=) - \boxed{\text{CAT:*}} - (\text{on}) - \boxed{\text{MAT:*}}$$

A very common operation for concept graphs is the *juxtaposition*. We ommit the mathematical definition here and describe only its graphical notation. The juxtaposition of a given set $\{\mathfrak{G}_i \mid 1 \le i \le n\}$ of graphs is simply writing them side by side: $\mathfrak{G}_1\ \mathfrak{G}_2\ \ldots\ \mathfrak{G}_n$. The juxtaposition of a empty set of graphs is the empty graph. The juxtaposition is needed to define the syntactical entailment relation:

Definition 10. *Let $\mathfrak{G}_a$, $\mathfrak{G}_b$ be two nonexistential concept graphs. Then $\mathfrak{G}_b$ can be derived from $\mathfrak{G}_a$ (which is written $\mathfrak{G}_a \vdash \mathfrak{G}_b$) if there is a finite sequence $(\mathfrak{G}_1, \mathfrak{G}_2, \ldots, \mathfrak{G}_n)$ with $\mathfrak{G}_1 = \mathfrak{G}_a$ and $\mathfrak{G}_b = \mathfrak{G}_n$ such that each $\mathfrak{G}_{i+1}$ is derived from $\mathfrak{G}_i$ by applying one of the rules of the calculus. The sequence is called a proof for $\mathfrak{G}_a \vdash \mathfrak{G}_b$).*

If $\{\mathfrak{G}_i \mid i \in I\}$ is a (possibly empty) set of nonexistential concept graphs, then a graph $\mathfrak{G}$ can be derived from $\{\mathfrak{G}_i \mid i \in I\}$ if there is a finite subset $\{\mathfrak{G}_1, \ldots, \mathfrak{G}_n\} \subseteq \{\mathfrak{G}_i \mid i \in I\}$ with $\mathfrak{G}_1 \ldots \mathfrak{G}_n \vdash \mathfrak{G}$.

5   The Syntactical Equivalence between CG and FOL

Another way to understand concept graphs if to translate them to formulas of first order logic. Mappings which translate conceptual graphs to formulas of predicate logic (first or higher order) are usually denoted by Φ. In this work, we give a mathematical definition for an operator Φ which translates concept graphs to formulas of first order logic with equality and with relation names (i.e. the concept and relation names of $\mathcal{A}$) but without function names. The set of these formulas over $\mathcal{A}$ is denoted by $\mathrm{FOL}^{\mathcal{A}}$ or, even simpler, by FOL. For a formula f, the set $\mathrm{Free}(f)$ of the *free variables in f* is defined as usual.

To define the mapping $\Phi : \mathrm{CG} \to \mathrm{FOL}$, let $\mathfrak{G} := (V, E, \nu, \top, Cut, area, \kappa, \rho)$ be a simple concept graph with cuts, variables, and dominating nodes over the alphabet $(\mathcal{G}, \mathcal{C}, \mathcal{R})$. First we define $\mathrm{Free}(\mathfrak{G}) := \{\alpha \in \mathrm{Var} \mid \text{there is a } v \text{ with } \rho(v) = *_\alpha\}$. Then we assign a new variable $\alpha_v \notin \mathrm{Free}(\mathfrak{G})$ to each vertex $v \in V$ with $\rho(v) = *$ so that we can now define the following mapping Φ_t on V:

$$\Phi_t(v) := \begin{cases} \alpha_v \text{ for } \rho(v) = * \\ \alpha \text{ for } \rho(v) = *_\alpha \text{ and } \alpha \in \mathrm{Var} \\ a \text{ for } \rho(v) = a \text{ for } a \in \mathcal{G} \end{cases}$$

Finally we can define the mapping $\Phi : \mathrm{CG} \to \mathrm{FOL}$ inductively on the tree $Cut \cup \{\top\}$. So let $c \in Cut \cup \{\top\}$ be an arbitrary cut. First we define a formula f which encodes all edges and vertices which are directly enclosed by c. If c does not directly enclose any edges or vertices, simply set $f := (\exists x.x = x)$. Otherwise let f be the conjunction of the atomic formulae

$$\kappa(w)(\underline{\Phi_t(w)}) \text{ with } w \in V \cap area(c),$$
$$\Phi_t(w_1) \underline{=} \Phi_t(w_2) \text{ with } k \in E_{id} \cap area(c) \text{ und } \nu(k) = (w_1, w_2), \quad \text{and}$$
$$\kappa(e)(\underline{\Phi_t(w_1), \ldots, \Phi_t(w_j)}) \text{ with } e \in E_{nonid} \cap area(c) \text{ and } \nu(e) = (w_1, \ldots, w_j).$$

(The signs which have to be understood literally are underlined. For example, the first formula is the sequence of signs which consists of the result of the evaluation of $\kappa(w)$, a left bracket, the result of the evaluation of $\Phi_t(w)$ and a right bracket.)

Let $v_1, \ldots, v_n$ be the vertices $\mathfrak{G}$ which are enclosed by c and which fulfill $\rho(v_i) = *$, and let $area(c) \cap Cut = \{c_1, \ldots, c_l\}$ (by induction we already assigned formulas to these cuts). If $l=0$, set $\Phi(c) := \exists \alpha_{v_1} \ldots \exists \alpha_{v_n}.f$, otherwise set

$$\Phi(c) := \exists \alpha_{v_1} \ldots \exists \alpha_{v_n}.\underline{(}f\underline{\wedge\neg}\Phi(c_1)\underline{\wedge}\ldots\underline{\wedge\neg}\Phi(c_l)\underline{)} \quad,$$

and the definition of $\Phi : \mathrm{CG} \to \mathrm{FOL}$ is finished.

Now we want to give a definition for a mapping $\Psi : \mathrm{FOL} \to \mathrm{CG}$ in the opposite direction. The formulas of FOL are built up inductively, hence the definition of Ψ is done inductively too. To start we set the images of the terms of FOL by a mapping Ψ_t: We set $\Psi_t(C) \equiv C$ for concept names $C \in \mathcal{C}$ and $\Psi_t(\alpha) \equiv *_\alpha$ for variables $\alpha \in \mathrm{Var}$. Now we can define Ψ inductively on the composition of formulas (but we only give graphical descriptions, because the mathematical definition of Ψ is rather technical and therefore omitted here).

- $g \equiv c(t)$ for a term t and a concept name $C \in \mathcal{C}$: $\Psi(C(t)) \equiv \boxed{C : \Psi_t(t)}$
- $R(t_1, \ldots, t_n)$ for a n-ary relation name R and terms $t_1, \ldots, t_n$:

$$\Psi(R(t_1, \ldots, t_n)) \equiv \boxed{\top : \Psi_t(t_1)} \overset{1}{-} \left(\ R\ \right) \overset{n}{-} \boxed{\top : \Psi_t(t_n)}$$

with $2 \ldots n{-}1$ above R.

- $t_1 = t_2$ for two terms t_1, t_2:

$$\Psi(t_1 = t_2) \equiv \boxed{\top : \Psi_t(t_1)} \overset{1}{-} \left(\ =\ \right) \overset{2}{-} \boxed{\top : \Psi_t(t_2)}$$

- $f_1 \wedge f_2$ for two formulas f_1 and f_2: $\Psi(f_1 \wedge f_2) \equiv \Psi(f_1)\ \Psi(f_2)$
  (i.e. the juxtaposition of $\Psi(f_1)$ and $\Psi(f_2)$).
- $\neg f$ for a formula f: $\Psi(\neg f) \equiv \left(\ \Psi(f)\ \right)$

- $\exists \alpha(f)$ for a formula f and a variable α:
  If $\alpha \notin \mathrm{Free}(f)$, let $\Psi(\exists \alpha f) \equiv \Psi(f)$
  For $\alpha \in \mathrm{Free}(f)$, the following steps have to be taken:
  1. A concept box $v := \boxed{\top : *}$ is drawn besides $\Psi(f)$: $\boxed{\top : *}\quad \Psi(f)$
  2. On every edge every instance of a concept box $\boxed{\top : *_\alpha}$ is substituted by the new concept box v (note that the concept boxes which are incident with an edge can only bear the concept name $\top$).
  3. Every (isolated) concept box $\boxed{P : *_\alpha}$ is substituted by a concept box $\boxed{P : *}$, which is linked to the new concept box v with an identity link:
  $$v \ \underline{-}\left(\ =\ \right)\underline{-}\ \boxed{P : *}$$
  4. All concept boxes $\boxed{\top : *_\alpha}$ are erased.

Note that Ψ translates formulas without free variables to simple concept graphs with negations and dominating nodes.

To illustrate the procedures in Φ and Ψ, we give a little example. A well-known example for translating first order logic to existential graphs is the formula which expresses that a binary relation F is a (total) function. This formula is written down only by using the $\exists$-quantor and the junctors $\land$ and $\neg$. Please see below:

$$f :\equiv \neg\exists x.\neg\exists y.(xFy \land \neg\exists z(xFz \land \neg(y = z)))$$

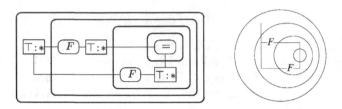

Fig. 2. The concept graph $\Psi(f)$ and the appropriate existential graph for f

The translations of this formula into a concept graph (by Φ) and into an existential graph are shown in Figure 2. Note the structural similarities of these two different kinds of graphs. Now the concept graph in Figure 2 can be translated back to a first order logic formula by the mapping Φ. One possible result (depending on the chosen variables and to the order of the subformulas) is:

$$\Phi(\Psi(f)) \equiv \neg\exists x.(\top(x) \land \neg\exists y.(\top(y) \land xFy \land \neg\exists.z(\top(z) \land xFz \land \neg(\land y = z))))$$

If we erase all subformulas $\top(\dots)$ of this formula, we get f again. In particular, we have $f \vdash \Phi(\Psi(f))$ and $\Phi(\Psi(f)) \vdash f$ (see Theorem 1).

Assume that we have a sound and complete calculus $\vdash$ on FOL. To encompass the orders on the concept- and relation names and the specific properties of the concept name $\top$, we add the following axioms to the calculus:

- $\forall x.\top(x)$
- $\forall x.(C_1(x) \to C_2(x))$ for two concept names $C_1, C_2 \in \mathcal{C}$ with $C_1 \leq_{\mathcal{C}} C_2$
- $\forall x_1 \dots \forall x_n.(R_1(x_1, \dots, x_n) \to R_2(x_1, \dots, x_n))$ for two n-ary relation names R_1, R_2 with $R_1 \leq_{\mathcal{R}} R_2$

Now we have reached the following situation. We have two logical systems with a calculus $\vdash$, and two mappings between them:

$$(\mathrm{CG}^{\mathcal{A}}, \vdash) \underset{\Psi}{\overset{\Phi}{\underset{\longleftarrow}{\longrightarrow}}} (\mathrm{FOL}^{\mathcal{A}}, \vdash)$$

It is well accepted that existential graphs and hence simple conceptual graphs with negations are equivalent to first order logic. The meaning of this equivalence shall be elaborated now.

Baader, Molitor and Tobies gave mathematical definitions for Φ and Ψ and a proof for $\Phi(\Psi(f)) \equiv f$ for each formula $f \in$ FOL (see [BMT98]). This is an important part in the equivalence between FOL and CG, but it is in our view not sufficient. To comprehend this, note that *every* injective mapping $\tilde{\Psi} :$ FOL $\to$ CG and its inverse mapping $\tilde{\Phi} := \tilde{\Psi}^{-1} : \mathrm{CG} \to \mathrm{FOL}$ fulfill $\tilde{\Phi}(\tilde{\Psi}(f)) \equiv f$. So the condition $\Phi(\Psi(f)) \equiv f$ seems not to capture the whole meaning of the statement 'simple conceptual graphs with cuts are equivalent to first order logic'. The crucial point is that FOL and CG are not only sets of graphs resp. formulas but sets which are quasiordered by their particular syntactical entailment relations $\vdash$ (i.e. each relation $\vdash$ is reflexive and transitive). Of course, one expects that Φ and Ψ respect these entailment relations. In fact Φ and Ψ are quasiorder-isomorphisms between $(\mathrm{CG}, \vdash)$ and $(\mathrm{FOL}, \vdash)$ which are mutually inverse. This is captured by the following crucial theorem:

Theorem 1 (Main Syntactical Theorem for the Mappings Φ and Ψ).

Let $\mathfrak{G}$, $\mathfrak{G}_1$ and $\mathfrak{G}_2$ be concept graphs over $\mathcal{A}$ and let f, f_1, f_2 be FOL-formulas over $\mathcal{A}$. Then the following implications hold:

1) $\mathfrak{G}_1 \vdash \mathfrak{G}_2 \quad \Rightarrow \quad \Phi(\mathfrak{G}_1) \vdash \Phi(\mathfrak{G}_2)$
2) $f_1 \vdash f_2 \quad \Rightarrow \quad \Psi(f_1) \vdash \Psi(f_2)$
3) $\mathfrak{G} \vdash \Psi(\Phi(\mathfrak{G}))$ *and* $\Psi(\Phi(\mathfrak{G})) \vdash \mathfrak{G}$
4) $f \vdash \Phi(\Psi(f))$ *and* $\Phi(\Psi(f)) \vdash f$

Instead of giving the proof for this theorem here (the proof for it will be in [Da01]), we want to emphasize that four conditions of the theorem are logically independent, e.g. none of the conditions can be derived from the remaining three (hence all conditions have to be proven seperately). To illustrate this, we provide two extremely simple examples. Each of these examples consists of two very small ordered sets (which represent for the quasiordered sets $(\mathrm{CG}^{\mathcal{A}}, \vdash)$ and $(\mathrm{FOL}^{\mathcal{A}}, \vdash)$) and two mappings between them (which represent the mappings Φ and Ψ).

To see that 1) cannot be derived from 2)-4) have a look at the two ordered sets in the left example of Figure 3, each consiting of two elements and the two mappings between them. Note that the mappings are mutually invers, therefore the conditions 3) and 4) are fulfilled. But only one mapping is order-preserving, hence 2) is fulfilled but 1) is not. Because we have one example which satisfies 2)-4), but not 1), we are done. Analogously it can be shown that 2) cannot be derived from 1), 3) and 4).

To see that 3) cannot be derived from 1),2) and 4), consider the two ordered sets with their mappings between them in the right example of Figure 3. It is easy to see that this example satisfies the conditions 1), 2) and 4), but does not satisfy 3), hence we are done again. In turn again it can be shown analogously that 4) cannot be derived from 1)–3).

Fig. 3. Two examples for Theorem 1

6   The Semantical Equivalence between FOL and CG

In Section 3 we have introduced a semantics for concept graphs. Now we proceed with the well-known semantics for first order logic. Usually the models for FOL-formulas are not power context families, but relational structures. Relational structures are pairs $\mathcal{M} = (M, I)$, consisting of a *universe* M and a function I with $I : \mathcal{G} \to M$, $I : \mathcal{C} \to \mathfrak{P}(M)$ and $I : \mathcal{R}_k \to \mathfrak{P}(M^k)$ for each k. Relational structures are obviously closely related to power context families. Roughly spoken: If we remove all intensional information from a power context family, we get a relational structure. To put it formally:

Definition 11. *If $(\mathbb{K}, \lambda)$ is a $\mathbb{K}$-interpretation, define $\mathcal{M}(\mathbb{K}, \lambda) := (M, I)$ as follows: $M := \underline{G}_0$, $I(G) := \lambda_{\mathcal{G}}(G)$ for all $G \in \mathcal{G}$, $I(C) := Ext(\lambda_{\mathcal{G}}(C))$ for all $C \in \mathcal{C}$ and $I(R) := Ext(\lambda_{\mathcal{G}}(R))$ for all $R \in \mathcal{R}$. The relational structure $\mathcal{M}(\vec{\mathbb{K}}, \lambda)$ is called the* relational structure of $(\vec{\mathbb{K}}, \lambda)$.

If we look back to the definition of the relation $\models$ between power context families and concept graphs, in particular to the vertex condition and edge condition of Definition 8, we realize that only the extensions of formal concepts were checked. This yields the following lemma:

Lemma 1. *For a $\mathcal{A}$-structure $(\vec{\mathbb{K}}, \lambda)$ and a simple concept graph with cuts $\mathfrak{G}$, we have $(\vec{\mathbb{K}}, \lambda) \models \mathfrak{G} \iff \mathcal{M}(\vec{\mathbb{K}}, \lambda) \models \Phi(\mathfrak{G})$*

The proof for this lemma will be given in [Da01]. Now we are ready to connect syntax and semantics both for FOL and CG. Remember that we assumed to have a sound and complete calculus $\vdash$ on FOL, i.e. we have

5) $f_1 \vdash f_2 \iff f_1 \models f_2$.

Furthermore Lemma 1 yields an equivalence between the two $\models$-relations on FOL and CG. If we resume these facts, we get the following theorem:

Theorem 2 (Main Semantical Theorem for the Mappings Φ and Ψ).
Let $\mathfrak{G}_1$ and $\mathfrak{G}_2$ be concept graphs over $\mathcal{A}$ and let f_1 and f_2 be FOL-formulas over $\mathcal{A}$. Then the following equivalences hold:

6) $\mathfrak{G}_1 \models \mathfrak{G}_2 \iff \Phi(\mathfrak{G}_1) \models \Phi(\mathfrak{G}_2)$

The proof for this theorem is by far not as extensive as the proof for the main syntactical theorem. But the proof has to be done, since conditions 5) and 6) are logically independent from the conditions 1)-4) of Theorem 1. This seems to be evident, since in condititions 1)-4) only the relations $\vdash$ on FOL and CG appear. But to show exactly that all conditions 1)-6) are logically independent, we need three little considerations.

1. To see that none of the conditions 1)-4) can be derived from the remaining conditions, simply *define* the relations $\models$ on FOL and CG as follows: For two formulas f_1 and f_2 define $f_1 \models f_2 :\iff f_1 \vdash f_2$. For two concept graphs

$\mathfrak{G}_1$ and $\mathfrak{G}_2$, define $\mathfrak{G}_1 \models \mathfrak{G}_2 :\Longleftrightarrow \Phi(\mathfrak{G}_1) \vdash (\mathfrak{G}_2)$. Obviously, 5) and 6) hold for arbitrary relations $\vdash$ on FOL and CG. Hence the indepence results for 1)-4) after Theorem 1 yield appropriate independence results for 1)-6). E.g. 1) cannot be derived from 2)-6).

2. To see that 5) cannot be derived from 1)-4) and 6), define $\models$ on FOL and CG as follows: For two formulas f_1 and f_2 always set $f_1 \models f_2$, and for two concept graphs $\mathfrak{G}_1$ and $\mathfrak{G}_2$ always set $\mathfrak{G}_1 \models \mathfrak{G}_2$ too. Since $\models$ does not appear in the condititions 1)-4), these conditions still hold, and it is easy to see that 6) is fulfilled, too, but 5) is not satisfied.

3. To see that 6) cannot be derived from 1)-5), define $\models$ on FOL and CG as follows: For two formulas f_1 and f_2 define $f_1 \models f_2 :\Longleftrightarrow f_1 \vdash f_2$. For two concept graphs $\mathfrak{G}_1$ and $\mathfrak{G}_2$ always set $\mathfrak{G}_1 \not\models \mathfrak{G}_2$. Again it is easy to see that 1)-5) hold, but 6) does not.

To summarize the above argument: If we want a full syntactical and semantical equivalence, we have to prove all six conditions 1)-6) from Theorems 2 and 1. But once we have done this, we immediately get the soundness and completeness for the calculus on concept graphs:

Theorem 3 (soundness and completeness for concept graphs).
   Let $\mathfrak{G}_1$ and $\mathfrak{G}_2$ be concept graphs over $\mathcal{A}$. Then it holds:

$$\mathfrak{G}_1 \vdash \mathfrak{G}_2 \quad \Longleftrightarrow \quad \mathfrak{G}_1 \models \mathfrak{G}_2$$

Proof: $\mathfrak{G}_1 \vdash \mathfrak{G}_2 \overset{1)-4)}{\Longleftrightarrow} \Phi(\mathfrak{G}_1) \vdash \Phi(\mathfrak{G}_2) \overset{5)}{\Longleftrightarrow} \Phi(\mathfrak{G}_1) \models \Phi(\mathfrak{G}_2) \overset{6)}{\Longleftrightarrow} \mathfrak{G}_1 \models \mathfrak{G}_2$ $\square$

References

[BMT98]   F. Baader, R. Molitor, S. Tobies: The Guarded Fragment of Conceptual Graphs. RWTH LTCS-Report.
http://www-lti.informatik.rwth-aachen.de/Forschung/Papers.html

[Da00]   F. Dau: Negations in Simple Concept Graphs, in: B. Ganter, G. W. Mineau (Eds.): Conceptual Structures: Logical, Linguistic, and Computational Issues. Lectures Notes in Artificial Intelligence 1867, Springer Verlag, Berlin–New York 1997, 263–276.

[Da01]   F. Dau, Negations in Concept Graphs. PhD-Thesis. To appear.

[GW99a]   B. Ganter, R. Wille: Formal Concept Analysis: Mathematical Foundations. Springer, Berlin-Heidelberg-New York 1999.

[Pe98]   C. S. Peirce: Reasoning and the Logic of Things. The Cambridge Conferences Lectures of 1898. Ed. by K. L. Kremer, Harvard Univ. Press, Cambridge 1992.

[Pr98a]   S. Prediger: Kontextuelle Urteilslogik mit Begriffsgraphen. Ein Beitrag zur Restrukturierung der mathematischen Logik, Shaker Verlag 1998.

[Pr98b]   S. Prediger: Simple Concept Graphs: A Logic Approach, in: M, -L. Mugnier, M. Chein (Eds.): Conceptual Structures: Theory, Tools and Applications, Springer Verlag, Berlin–New York 1998, 225–239.

[Ro73]   D. D. Roberts: The Existential Graphs of Charles Sanders Peirce, Mouton The Hague – Paris 1973.

[So84]   J. F. Sowa: Conceptual Structures: Information Processing in Mind and Machine. Addison Wesley Publishing Company Reading, 1984.

[So99]   J. F. Sowa: Conceptual Graphs: Draft Proposed American National Standard, in: W. Tepfenhart, W. Cyre (Eds.): Conceptual Structures: Standards and Practices, Springer Verlag, Berlin–New York 1999, 1-65.

[So00]   J. F. Sowa: Knowledge Representation: Logical, Philosophical, and Computational Foundations. Brooks Cole Publishing Co., Pacific Grove, CA, 2000.

[We95]   M. Wermelinger: Conceptual Graphs and First-Order Logic, in: G. Ellis et al. (Eds.): Conceptual Structures: Applications, Implementations and Theory, Springer Verlag, Berlin–New York 1995, 323–337.

[Wi97]   R. Wille: Conceptual Graphs and Formal Concept Analysis, in: D. Lukose et al. (Hrsg.): Conceptual Structures: Fulfilling Peirce's Dream, Lectures Notes in Artificial Intelligence 1257, Springer Verlag, Berlin–New York 1997, 290–303.

Generalized Quantifiers and Conceptual Graphs

Tru H. Cao

Artificial Intelligence Group
Department of Engineering Mathematics, University of Bristol
United Kingdom BS8 1TR
Tru.Cao@bristol.ac.uk

Abstract. Conceptual graphs have been shown to be a logic that has a smooth mapping to and from natural language, in particular generally quantified statements, which is one of its advantages over predicate logic. However, classical semantics of conceptual graphs cannot deal with intrinsically vague generalized quantifiers like *few*, *many*, or *most*, which represent imprecise quantities that go beyond the capability of classical arithmetic. In this paper, we apply the fuzzy set-theoretic semantics of generalized quantifiers and formally define the semantics of generally quantified fuzzy conceptual graphs as probabilistic logic rules comprising only simple fuzzy conceptual graphs. Then we derive inference rules performed directly on fuzzy conceptual graphs with either relative or absolute quantifiers.

1 Introduction

Natural language is a principal and important means of human communication. It is used to express information as inputs to be processed by human brains then, very often, outputs are also expressed in natural language. How humans process information represented in natural language is still a challenge to science, in general, and to Artificial Intelligence, in particular. However, it is clear that, for a computer with the conventional processing paradigm to process natural language, a formalism is required. For reasoning, it is desirable that such a formalism be a logical one.

The object of natural language (specifically English) that our attention is focused on in this paper is one of *generalized quantifiers*, which can be classified into *absolute quantifiers* and *relative quantifiers*, where the quantities expressed by the latter are relative to the cardinality of a set. Examples of absolute quantifiers are *only one*, *few*, or *several*, while ones of relative quantifiers are *about 9%*, *half*, or *most*. In practice, there are quantifying words, e.g. *few* and *many*, that may be used with either meaning depending on the context. For instance, *few* in "*Few* people in this conference are from Asia" may mean a small number of people, while *few* in "*Few* people in the United Kingdom are from Asia" may mean a small percentage of population.

Classical predicate logic with only the existential quantifier, equivalent to the absolute quantifier *at least 1*, and the universal quantifier, equivalent to the relative quantifier *all* or *every*, cannot deal with general quantification in natural language. In fact, the logic of generalized quantifiers has been the quest and focus of significant research effort.

In [14], for instance, the semantics of generally quantified statements were defined relatively to each other, e.g., "*Few A's are B's*" if and only if "*Not many A's are B's*". As such, that work does not define the primary meaning of generalized quantifiers, which is the quantity that they express, and thus is not adequate for

H. Delugach and G. Stumme (Eds.): ICCS 2001, LNAI 2120, pp. 87–100, 2001.

quantitative reasoning with them. A path-breaking work in formalizing generally quantified statements was [2], where a generally quantified set was interpreted as a set of sets of individuals. For example, "*only N A*'s" was interpreted as the set $\{X \subseteq U \mid |X \cap A| = N\}$, where U is a universe of discourse and $|S|$ denotes the cardinality of S, whence "*Only N A*'s are *B*'s" means $B \in \{X \subseteq U \mid |X \cap A| = N\}$. Still, no definition was given for words expressing imprecise quantities, like *several* or *most*; only some semantic postulates stating their relations to each other were suggested as in [14].

Meanwhile, fuzzy logic ([23]) based on fuzzy set theory ([22]) has been developed for approximate representation of, and reasoning with, imprecise information often encountered in the real world as reflected in natural language. While there are still many unresolved theoretical issues regarding the uncertainty management problem in general, and fuzzy logic in particular, fuzzy logic has been successfully applied to several areas, such as expert systems, knowledge acquisition and fusion, decision making, and information retrieval, among others.

In particular, fuzzy arithmetic has provided a basis for defining and computing with generalized quantifiers, e.g. [11], [12], [24], where they are treated in a unified way as fuzzy numbers, which are fuzzy sets on the set of real numbers. A relative quantifier Q in a statement "Q A's are B's" can be interpreted as the proportion of objects of type A that belong to type B, i.e., $Q = |A \cap B|/|A|$, which is a fuzzy number. Equivalently, as discussed recently in [11], it can also be interpreted as the fuzzy conditional probability, which is a fuzzy number on [0, 1], of $B(x)$ being **true** given $A(x)$ being **true** for an object x picked at random uniformly.

Although the interpretation of relative quantifiers as fuzzy probabilities was also mentioned in [25], the cardinality-based interpretation was used instead for deriving inference rules on relatively quantified statements. Meanwhile, [11] has shown that the probability-based interpretation lends a new perspective of reasoning with generalized quantifiers as probabilistic logic programming. For instance, on the basis of Jeffrey's rule ([9]), it allows one to have an answer as a fuzzy probability to a question like "How likely is it that John is *not fat*?" given that "*Most* people who are *tall* are *not fat*" and "John is *fairly tall*".

However, a logic for handling natural language should have not only a capability to deal with the semantics of vague linguistic terms, but also a structure of formulas close to that of natural language sentences. Significantly, while Zadeh has regarded a methodology for computing with words as the main contribution of fuzzy logic ([24], [26]), Sowa has regarded a smooth mapping between logic and natural language as the main motivation of conceptual graphs (CGs) ([16], [17]). It shows that fuzzy logic and conceptual graphs are two logical formalisms that emphasize the same target of natural language, each of which is focused on one of the two mentioned desired features of a logic for handling natural language.

In [15] and [16], Sowa showed that natural language expressions with generalized quantifiers could be smoothly represented in conceptual graphs but not in predicate logic, yet did not adequately formalize them. A substantial effort that has been made in formalizing generally quantified CGs was [19] and [20]. Therein, the authors extended predicate logic with equality by adding to it numerical quantifiers, then used it to define semantics and inference rules for conceptual graphs with numerical quantifiers. However, limited by classical arithmetic, vague quantifiers like *few*, *many*, or *most* were not considered in that work.

In this paper, we aim to combine the advantages of both conceptual graphs and fuzzy logic into a formalism for representing and reasoning with generally quantified statements in particular, and linguistic information in general. In [21], while extending fuzzy conceptual graphs (FCGs) ([13]) with fuzzy conceptual relations, the authors also introduced generalized quantifiers into them. However, that work actually just adapted inference rules in [25] to the conceptual graph notation, without formally defining the semantics of fuzzy conceptual graphs with generalized quantifiers. In contrast, here we use the probability-based interpretation to expand generally quantified FCGs into probabilistic FCG rules, formally defining their semantics in terms of simple FCGs without generalized quantifiers. Then, on the basis of this semantics, we derive inference rules performed directly on FCGs with either relative quantifiers or absolute quantifiers; reasoning with the latter was not considered in [25] or [11].

Firstly, in Section 2, we summarize the basic notions of fuzzy set theory and fuzzy arithmetic, more details of which can be found in [7] and [10]. Section 3 presents the expansion rule that formally defines the semantics of generally quantified FCGs in terms of simple FCGs. Then, Section 4 derives inference rules for relative quantifiers and absolute quantifiers, and Jeffrey's rule on generally quantified FCGs. Due to space limitation we omit the proofs for the soundness of these inference rules, which were presented in the submitted version of this paper. Finally, Section 5 concludes the paper and suggests future research.

2   Fuzzy Arithmetic

2.1   Fuzzy Sets and Fuzzy Numbers

For a classical set, an element is to be or not to be in the set or, in other words, the membership grade of an element in the set is binary. Fuzzy sets are to represent classes of objects where the boundary for an object to be or not to be in a class is not clear-cut due to the vagueness of the concept associated with the class. Such vague concepts are frequently encountered in the real world as reflected in natural language, like *young* or *old*, *small* or *large*. So the membership grade of an element in a fuzzy set is expressed by a real number in the interval [0, 1].

Definition 2.1 A *fuzzy set* A on a domain U is defined by a membership function μ_A from U to [0, 1]. It is said to be a *fuzzy subset* of a fuzzy set B also on U, denoted by $A \subseteq B$, if and only if $\forall u \in U: \mu_A(u) \leq \mu_B(u)$.

In this work we apply the voting model interpretation of fuzzy sets ([1], [8]) whereby, given a fuzzy set A on a domain U, each voter has a subset of U as his/her own crisp definition of the concept that A represents. For example, a voter may have the interval [0, 35] representing human ages from 0 to 35 years as his/her definition of the concept *young*, while another voter may have [0, 25] instead. The membership function value $\mu_A(u)$ is then the proportion of voters whose crisp definitions include u. As such, A defines a probability distribution on the power set of U across the voters, and thus a fuzzy proposition "x is A" defines a family of probability distributions of the variable x on U.

Operations on fuzzy sets also generalize those on classical sets, computing with real numbers in [0, 1] instead.

Definition 2.2 Let A and B be two fuzzy sets on a domain U. Then the *fuzzy intersection* of A and B is a fuzzy set denoted by $A \cap B$ and defined by $\forall u \in U$: $\mu_{A \cap B}(u) = min\{\mu_A(u), \mu_B(u)\}$. The *fuzzy union* of A and B is a fuzzy set denoted by $A \cup B$ and defined by $\forall u \in U$: $\mu_{A \cup B}(u) = max\{\mu_A(u), \mu_B(u)\}$. The *fuzzy complement* of A is a fuzzy set denoted by $\tilde{A}$ and defined by $\forall u \in U$: $\mu_{\tilde{A}}(u) = 1 - \mu_A(u)$.

As a special category of fuzzy sets, fuzzy numbers defined as fuzzy sets on the set $\mathbf{R}$ of real numbers are to represent imprecise numeric values. Using fuzzy numbers is one step further than intervals in representing imprecise numeric values, whereby intervals can be considered as special fuzzy numbers. For example, an interval $[a, b]$ can be represented as the fuzzy number I defined by $\mu_I(x) = 1$ if $x \in [a, b]$, or $\mu_I(x) = 0$ otherwise.

2.2 Operations on Fuzzy Numbers

One method for defining operations on fuzzy numbers is based on the *extension principle*. According to this principle, any function f: $U_1 \times U_2 \times ... \times U_n \to V$ induces a function g: $X_1 \times X_2 \times ... \times X_n \to Y$ where X_1, X_2, ..., X_n and Y are respectively sets of fuzzy sets on U_1, U_2, ..., U_n and V and, for every $(A_1, A_2, ..., A_n) \in X_1 \times X_2 \times ... \times X_n$, $g(A_1, A_2, ..., A_n) \in Y$ is defined by:

$$\forall v \in V: \mu_{g(A_1, A_2, ..., A_n)}(v) = sup\{min\{\mu_{A_1}(u_1), \mu_{A_2}(u_2), ..., \mu_{A_n}(u_n)\} \mid$$
$$(u_1, u_2, ..., u_n) \in U_1 \times U_2 \times ... \times U_n \text{ and } v = f(u_1, u_2, ..., u_n)\}.$$

From now on, for simplicity, we use the same notation for a real number function (e.g. f above) and its fuzzy extension (e.g. g above).

For example, using this method, the product of two fuzzy numbers A and B is the fuzzy number $A.B$ defined as follows:

$$\forall z \in \mathbf{R}: \mu_{A.B}(z) = sup\{min\{\mu_A(x), \mu_B(y)\} \mid (x, y) \in \mathbf{R} \times \mathbf{R} \text{ and } z = x.y\}.$$

Figure 2.1 illustrates this multiplication operation on fuzzy numbers.

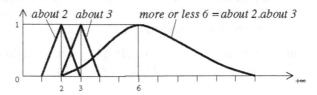

Fig. 2.1. Fuzzy multiplication

Similarly, the method can be applied to extend *min* and *max*, which are functions from $\mathbf{R} \times \mathbf{R}$ to $\mathbf{R}$, for fuzzy numbers as defined below:

$$\forall z \in \mathbf{R}: \mu_{min\{A, B\}}(z) = sup\{min\{\mu_A(x), \mu_B(y)\} \mid (x, y) \in \mathbf{R} \times \mathbf{R} \text{ and } z = min\{x, y\}\}$$
$$\forall z \in \mathbf{R}: \mu_{max\{A, B\}}(z) = sup\{min\{\mu_A(x), \mu_B(y)\} \mid (x, y) \in \mathbf{R} \times \mathbf{R} \text{ and } z = max\{x, y\}\}.$$

A fuzzy version of the less-than-or-equal-to relation $\leq$ on real numbers can then be defined for fuzzy numbers such that $A \leq B$ if and only if $min\{A, B\} = A$ or, equivalently, $max\{A, B\} = B$. However, we note that, while $\leq$ is a total order on real numbers, its fuzzy version is just a partial order on fuzzy numbers, because $min\{A, B\}$

and $max\{A, B\}$ may equal to neither A nor B. For illustration, Figure 2.2 shows two comparable fuzzy numbers and two incomparable ones with respect to $\leq$.

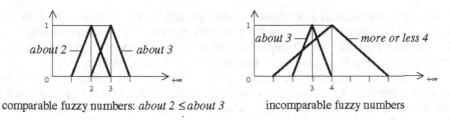

comparable fuzzy numbers: *about 2* $\leq$ *about 3*   incomparable fuzzy numbers

Fig. 2.2. Comparable and incomparable fuzzy numbers

2.3 Fuzzy Lower and Upper Bounds

For a real number x, the notion "at least x" is represented by the interval $[x, +\infty]$, which includes all the real numbers y such that $x \leq y$. Similarly, "at most x" is represented by $[-\infty, x]$. For a fuzzy number A, we denote the least specific fuzzy number that is greater than or equal to A by *at least A*. That is, $A \leq$ *at least A* and $B \subseteq$ *at least A* for every B such that $A \leq B$, whence the membership function of *at least A* can be specified as follows:

$\forall x \in \mathbf{R}$: $\mu_{at\ least\ A}(x) = sup\{\mu_A(y) \mid y \in \mathbf{R}$ and $y \leq x\}$.

Similarly, the fuzzy number *at most A*, such that *at most A* $\leq A$ and $B \subseteq$ *at most A* for every B where $B \leq A$, is defined by:

$\forall x \in \mathbf{R}$: $\mu_{at\ most\ A}(x) = sup\{\mu_A(y) \mid y \in \mathbf{R}$ and $x \leq y\}$.

Figure 2.3 illustrates such lower bound and upper bound fuzzy numbers.

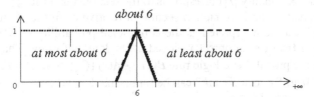

Fig. 2.3. Fuzzy lower bound and upper bound

We call a fuzzy number whose membership function is non-decreasing a *non-decreasing fuzzy number*, and a fuzzy number whose membership function is non-increasing a *non-increasing fuzzy number*. According to the above definition of fuzzy lower bounds, if A is a non-decreasing fuzzy number, then *at least A = A*. Similarly, if A is a non-increasing fuzzy number, then *at most A = A*. The following property also holds, where A/B denotes the fuzzy division of A by B, and B is said to be *positive* if and only if $\mu_B(x) = 0$ for every $x \leq 0$.

Proposition 2.1 For any fuzzy number A and positive fuzzy number B:
1. *at least (A/B) = at least A / B*.
2. *at most (A/B) = at most A / B*.

3   Representation of Generalized Quantifiers

3.1   Quantification as Conditional Probability

Firstly, in reasoning with quantifiers, *absolute quantifiers* and *relative quantifiers* on a set have to be distinguished, where the quantities expressed by the latter are relative to the cardinality of the set. Examples of absolute quantifiers are *only one*, *few*, or *several*, while ones of relative quantifiers are *about 9%*, *half*, or *most*.

In the crisp case, absolute quantifiers can be defined by natural numbers, and relative quantifiers by non-negative rational numbers that are not greater than 1 measuring a proportion of a set, where 0 means 0% and 1 means 100%. Correspondingly, in the fuzzy case, absolute quantifiers can be defined by fuzzy sets on the set $\mathbf{N}$ of natural numbers, i.e., fuzzy numbers whose domain is restricted to $\mathbf{N}$, and relative quantifiers by fuzzy numbers whose domain is restricted to the set of rational numbers in [0, 1]. For simplicity without rounding of real numbers, however, we assume absolute quantifiers to be defined by fuzzy numbers on [0, +∞] and relative quantifiers by fuzzy numbers on [0, 1].

The existential quantifier in classical logic corresponds to *at least 1* in natural language, which is an absolute quantifier whose membership function is defined by $\mu_{at\ least\ 1}(x) = 1$ if $x \geq 1$, or $\mu_{at\ least\ 1}(x) = 0$ otherwise. Meanwhile, the universal quantifier, which corresponds to *all* or *every* in natural language, is a relative quantifier and its membership function is defined by $\mu_{all}(1) = 1$ and $\mu_{all}(x) = 0$ for every $0 \leq x < 1$.

Arithmetic operations for fuzzy numbers as presented in Section 2 are also applicable to absolute quantifiers with [0, +∞] being used in place of $\mathbf{R}$, and to relative quantifiers with [0, 1] being used instead. Also, on the basis of the extension principle, each absolute quantifier Q on a type T whose denotation set in a universe of discourse has the cardinality $|T|$ corresponds to the relative quantifier $Q_T = Q/|T|$.

As mentioned in the introduction section, a relative quantifier Q in a statement "Q A's are B's" can be interpreted as the conditional probability of $B(x)$ being **true** given $A(x)$ being **true** for an object x picked at random uniformly. That is, it can be represented by the probabilistic logic rule $B(x) \leftarrow A(x)$ [Q] where $Q = Pr(B(x) \mid A(x))$. For example, "*Most* Swedes are *tall*" can be represented by $tall(x) \leftarrow$ Swede(x) [*most*] where *most* $= Pr(tall(x) \mid$ Swede$(x))$.

In general, $A(x)$ or $B(x)$ can also be represented in conceptual graphs or any other logical formalism. In [3], for dealing with vagueness and imprecision in the real world as reflected in natural language, fuzzy conceptual graphs were developed with fuzzy concept or relation types ([6]) and fuzzy attribute values defined by fuzzy sets. Here, focusing on representing and reasoning with generalized quantifiers, we consider *simple FCGs* with only fuzzy attribute values. For example, the simple FCG in Figure 3.1 expresses "John is *fairly tall*", where *fairly tall* is a linguistic label of a fuzzy attribute value, and [PERSON: John] is called an entity concept whereas [HEIGHT: *@*fairly tall*] an attribute concept.

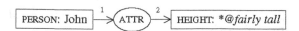

Fig. 3.1. A simple FCG

We now formulate generally quantified FCGs as logical expressions that smoothly match with generally quantified statements in natural language.

3.2 Generally Quantified FCGs

In [15], set referents and numeric quantifiers were introduced into concepts to represent plural noun phrases and quantification in natural language, where the semantics of such an extended CG was defined by its expansion into a CG without numeric quantifiers. For example, Figure 3.2 shows an extended CG G and its defining expansion E, which literally says "There exists a set of two persons who see John".

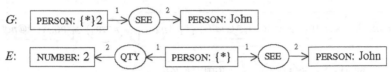

Fig. 3.2. A CG with a numeric quantifier and its defining expansion

However, that defining expansion does not capture the actual meaning of generalized quantifiers, because a quantifier on a type does not simply mean that there exists a set of objects of that type that have some property, and the cardinality of the set is defined by the quantifier. Rather, a quantifier on a type constrains the number of objects of that type that can have some property. For example, "*Only one* person is the President of the United States" means not only "There exists a set of *only one* person who is the President of the United States", but also "There is no more than one person who is the President of the United States", as the former alone does not exclude "There are two different sets each of which has *only one* person who is the President of the United States".

Here we apply the conditional probability interpretation of relative quantifiers presented above to define an FCG with a relative quantifier as a probabilistic FCG rule, where the quantifier is the conditional probability of the head given the body of the rule, which are both represented by simple FCGs. Meanwhile, an FCG with an absolute quantifier is semantically equivalent to the one with the relative quantifier converted from that absolute quantifier as noted above.

In this paper, we consider only FCGs that contain only one generally quantified concept, excluding ones with the generic referent *, whose quantifiers are implicitly the existential quantifier. For example, Figure 3.3 shows a generally quantified FCG G and its defining expansion E, expressing "*Most* Swedes are *tall*", where *most* and *tall* are linguistic labels of fuzzy sets. We note that this defining expansion rule can be seen as a generalization of the one for universally quantified CGs defined in terms of CG rules as introduced in [15] and studied in [5], as the universal quantifier is a special relative quantifier.

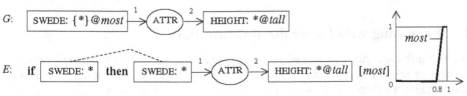

Fig. 3.3. A generally quantified FCG and its defining expansion

Furthermore, the type in a generally quantified concept can be represented by a simple FCG as a lambda expression defining that type as for CGs ([15], [18]). We call such a simple FCG a *lambda FCG*, which is like a simple FCG except that it has one concept, which we call a *lambda concept*, whose referent is denoted by λ to be distinguished from the generic and individual referents. For example, Figure 3.4 illustrates a generally quantified FCG *G* and its defining expansion *E*, expressing "*Most* people who are *tall* are *not fat*". As such, a lambda FCG corresponds to a relative clause in natural language.

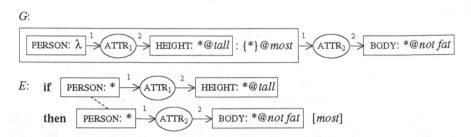

Fig. 3.4. Quantification on a type defined by a lambda FCG

We now formally define generally quantified FCGs and the expansion rule for their semantics exemplified above. Since a generally quantified concept [T:{*}*Q*] is semantically equivalent to [[T: λ]:{*}*Q*], where [T: λ] is a special lambda FCG with only the lambda concept, one can always assume the type of a generally quantified concept to be represented by a lambda FCG.

Definition 3.1 A *generally quantified concept* is defined by a triple of a concept type represented by a lambda FCG, the set referent {*}, and a generalized quantifier. A *generally quantified FCG* is a simple FCG with one generally quantified concept. It is said to be a *relatively quantified* or an *absolutely quantified* FCG if the generalized quantifier in it is a relative quantifier or an absolute quantifier, respectively.

Definition 3.2 The defining expansion of a relatively quantified FCG *G* is the probabilistic FCG rule **if** *F* **then** *H* [*Q*] where:
1. *F* is obtained from the lambda FCG in *G* by replacing its lambda concept [T: λ] with [T: *], *H* is obtained from *G* by replacing its generally quantified concept with [T: *], and there is a coreference link between these two concepts [T: *] of *F* and *H*.
2. *Q* is the relative quantifier in *G* and $Pr(H \mid F) = Q$.

4   Reasoning with Generalized Quantifiers

For the inference rules presented below, generalizing an FCG means replacing its concept or relation types and concept referents with less specific ones, as for a CG, and its fuzzy attribute values with ones whose defining fuzzy sets are fuzzy supersets of those defining the former. It is dually similar for specializing.

4.1 Inference Rules for Relative Quantifiers

Proposition 4.1 A generally quantified FCG *G* with a relative quantifier *Q* entails a generally quantified FCG *G** obtained from *G* by generalizing it except for its generally quantified concept, and replacing *Q* with *at least Q*.

For example, in Figure 4.1, *G* expressing *"Most* people who are *tall* are *not fat"* entails *G** expressing *"Most* people who are *tall* are *not very fat"*, provided that *not fat ⊆ not very fat*, and *most* is a non-decreasing fuzzy number whereby *at least most = most* as noted in Section 2.

G:

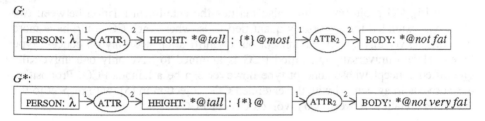

Fig. 4.1. Generalization inference rule for relative quantifiers

Proposition 4.2 A generally quantified FCG *G* with a relative quantifier *Q* entails a generally quantified FCG *G** obtained from *G* by specializing it except for its generally quantified concept, and replacing *Q* with *at most Q*.

For example, in Figure 4.2, *G* expressing *"About 9%* people who are *tall* are *fat"* entails *G** expressing *"At most about 9%* people who are *tall* are *very fat"*, provided that *very fat ⊆ fat*.

G:

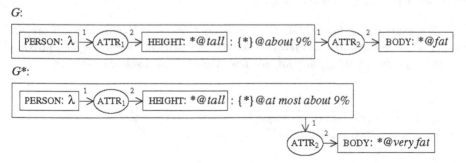

Fig. 4.2. Specialization inference rule for relative quantifiers

Proposition 4.3 A universally quantified FCG *G* entails a universally quantified FCG *G** obtained from *G* by specializing its lambda FCG.

For example, in Figure 4.3, *G* expressing *"All* people who are *tall* are *not fat"* entails *G** expressing *"All* males who are *very tall* are *not fat"*, provided that *very tall ⊆ tall*.

G:

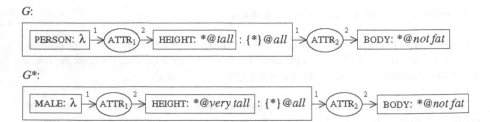

Fig. 4.3. Inference rule for the universal quantifier

In [5], CG projection, which also defines the entailment relation between two CGs, was extended for universally quantified CGs, where each CG could have more than one universally quantified concept but quantification was only on concept type labels. Here a universally quantified FCG is assumed to have only one universally quantified concept whose concept type however can be a lambda FCG. Proposition 4.3 above is in agreement with the extended CG projection in [5], with respect to the entailment relation between two involved CGs or FCGs.

4.2   Inference Rules for Absolute Quantifiers

The following propositions are obtained on the basis as noted in Section 3 that an absolutely quantified FCG is semantically equivalent to a relatively quantified one with the corresponding relative quantifier, whose defining expansion is given by Definition 3.2.

Proposition 4.4 A generally quantified FCG *G* with an absolute quantifier *Q* entails a generally quantified FCG *G** obtained from *G* by generalizing it, including its lambda FCG, and replacing *Q* with *at least Q*.

For example, in Figure 4.4, *G* expressing "*Few* people who are *tall* are *fat*" entails *G** expressing "*At least few* people who are *fairly tall* are *fairly fat*", provided that *tall* ⊆ *fairly tall*, *fat* ⊆ *fairly fat*, and *few* is used as an absolute quantifier.

G:

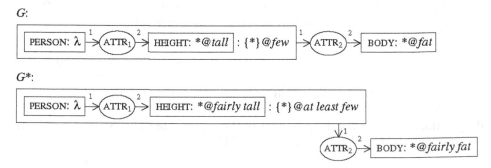

Fig. 4.4. Generalization inference rule for absolute quantifiers

Proposition 4.5 A generally quantified FCG G with an absolute quantifier Q entails a generally quantified FCG G^* obtained from G by specializing it, including its lambda FCG, and replacing Q with *at most Q*.

For example, in Figure 4.5, G expressing *"Few* people who are *tall* are *fat"* entails G^* expressing *"At most few* people who are *very tall* are *very fat"*, provided that *very tall* $\subseteq$ *tall*, *very fat* $\subseteq$ *fat*, and *few* is also used as an absolute quantifier.

G:

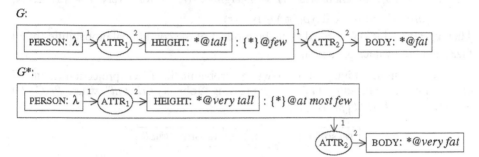

G^*:

Fig. 4.5. Specialization inference rule for absolute quantifiers

4.3 Jeffrey's Rule

In practice, it is often the case that a condition probability $v = Pr(B(x) \mid A(x))$ is obtained from statistical data meaning that, if an object x is *randomly* picked up such that $A(x)$ is **true**, then the probability for $B(x)$ being **true** is v. As such, that conditional probability value may not be applicable to a *specific* object in a universe of discourse. For example, one may have $Pr(\text{fly}(x) \mid \text{bird}(x)) = 0.9$ expressing that 90% of birds can fly, but the probability for a penguin, a specific bird, being able to fly is 0.

Therefore, for decision making, Jeffrey introduced a rule assuming such a conditional probability like $Pr(B(x) \mid A(x))$ or $Pr(B(x) \mid \neg A(x))$ to be applicable to any specific object o or, in other words, $Pr(B(o) \mid A(o)) = Pr(B(x) \mid A(x))$ and $Pr(B(o) \mid \neg A(o)) = Pr(B(x) \mid \neg A(x))$, when only the probability for $A(o)$ being **true** is known. Then, on the basis of the total probability theorem, the probability for $B(o)$ being **true** could be obtained as follows:

$$Pr(B(o)) = Pr(B(o) \mid A(o)).Pr(A(o)) + Pr(B(o) \mid \neg A(o)).Pr(\neg A(o))$$
$$= Pr(B(x) \mid A(x)).Pr(A(o)) + Pr(B(x) \mid \neg A(x)).Pr(\neg A(o)).$$

In [3], FCG projection was defined, which matches a simple FCG with another and compute the relative necessity degree of the former given the later. Here, for applying Jeffrey's rule to reasoning with generally quantified FCGs, we introduce *probabilistic FCG projection* that computes the conditional probability of a simple FCG given another one. For its definition, we apply the definition of conditional probability of fuzzy events in [1], namely, $Pr(x \text{ is } B \mid x \text{ is } A)$ whose value is a subinterval of [0, 1] treated here as a fuzzy number on [0, 1], where A and B are two fuzzy sets on the same domain.

In the following definition, $\mathbf{V}_{C_G}$ and $\mathbf{V}_{R_G}$ respectively denote the set of all concepts and the set of all relations in a simple FCG G. For a concept c, *referent*(c) and *type*(c) are respectively the referent and the type of c. For a relation r, *type*(r) is the type of *r*, *arity(r)* is the arity of *type(r)*, and *neighbour(r, i)* is the neighbour concept connected to *r* by the edge labelled *i*.

Definition 4.1 Let *G* and *H* be two simple FCGs. A *probabilistic FCG projection* from *G* to *H* is a mapping $\pi: G \to H$ such that:

1. $\forall c \in \mathbf{V}_{\mathbf{C}_G}$: *referent(c)* = * or *referent(c)* =*referent(πc)*, and *type(πc)* $\subseteq$ *type(c)*.
2. $\forall r \in \mathbf{V}_{\mathbf{R}_G}$: *neighbour($\pi r$, i)* = π*neighbour(r, i)* for every $i \in \{1, 2, ...,$ *arity(type(r))*$\}$, and *type(πr)* $\subseteq$ *type(r)*.

Then $\varepsilon_\pi = Pr(G \mid \pi G)$ is defined to be the product of the conditional probabilities of all fuzzy attribute values pairs in π.

For example, Figure 4.6 shows a probabilistic FCG projection from *G* expressing "Some person is *tall*" to *H* expressing "John is *fairly tall*", where $Pr(G \mid H)$ = $Pr(G \mid \pi G)$ = $Pr(tall \mid fairly\ tall)$.

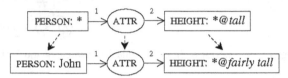

Fig. 4.6. A probabilistic FCG projection

For a probabilistic FCG projection from a lambda FCG to a simple FCG, we assume the referent λ in the lambda concept to be replaced with the generic referent *, as in the following proposition.

Proposition 4.6 Let *G* be a generally quantified FCG with a relative quantifier *Q*, and *G** be a simple FCG such that there is a probabilistic FCG projection π from the lambda FCG in *G* to *G**. Then Jeffrey's rule derives the simple FCG *H** with the probability *(at least $(Q.\varepsilon_\pi)$)* $\cap$ *(at most $(Q.\varepsilon_\pi + (1 - \varepsilon_\pi))$))* where *H** is obtained from *G* by replacing its generally quantified concept with its lambda concept *c* whose referent λ is replaced with *referent(πc)*.

For example, in Figure 4.7, *G* expresses "*Most* people who are *tall* are *not fat*" and *G** expresses "John is *fairly tall*". Then *H** expressing "John is *not fat*" can be derived with the probability *p* = *(at least $(most.\varepsilon_\pi)$)* $\cap$ *(at most $(most.\varepsilon_\pi + (1 - \varepsilon_\pi))$)*, where $\varepsilon_\pi = Pr(tall \mid fairly\ tall)$, as an answer to the query "How likely is it that John is *not fat*?".

G:

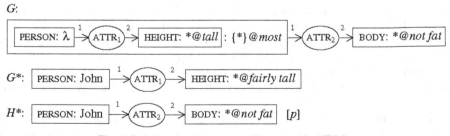

Fig. 4.7. Jeffrey's rule on generally quantified FCGs

5   Conclusion

We have formulated generally quantified FCGs where generalized quantifiers are defined by fuzzy numbers to capture their vagueness and imprecision often encountered in natural language. A generally quantified FCG with a relative quantifier, defined by a fuzzy number on [0, 1], has been interpreted as a probabilistic FCG rule with that relative quantifier as the conditional probability of the head given the body of the rule. Meanwhile, an absolutely quantified FCG is semantically equivalent to a relatively quantified one with the corresponding relative quantifier. On the basis of this semantics, we have derived generalization, specialization, and Jeffrey's inference rules performed directly on generally quantified FCGs.

This is our first step in formally integrating generalized quantifiers into conceptual graphs. The presented inference rules are not meant to be complete, but to show how some basic patterns of human reasoning on generally quantified statements in natural language could be performed as machine operations on corresponding generally quantified FCGs. In the general case, one could apply the proposed probabilistic interpretation of generally quantified FCGs to study complete inference rules on them in the framework of probabilistic logic programming.

We however believe that human reasoning is more heuristic and plausible than sound and complete. That is because one often has to make additional assumptions on given information in order to infer useful new information, and thus the inference is actually not sound with respect to the given information. Meanwhile the high complexity of a problem makes complete inference impractical. Therefore, as the theme of this work is to combine fuzzy logic and conceptual graphs, both of which emphasize the target of natural language, into a logical formalism for Artificial Intelligence approaching human expression and reasoning, our attention is focused on heuristic and plausible inference rules on FCGs.

In this paper, we considered FCGs with only one generally quantified concept whose referent is the generic set referent {*}. FCGs with more than one generally quantified concept whose set referent can include individual referents as outlined in [4] are required to represent more complex natural language expressions. Also, we did not discuss from where generally quantified statements with the proposed semantics could be obtained. For example, the statement "*Most* Swedes are *tall*" is actually a summarization of statistical data about Swedes' heights, which is a difficult induction problem for machine learning. These are among the topics that we are currently investigating.

Acknowledgements. I would like to thank the three anonymous reviewers for their comments, which helped me to improve several aspects of the paper.

References

1.   Baldwin, J.F., Martin, T.P., Pilsworth, B.W.: Fril - Fuzzy and Evidential Reasoning in Artificial Intelligence. Research Studies Press (1995).
2.   Barwise, J., Cooper, R.: Generalized Quantifiers and Natural Language. Linguistics and Philosophy 4 (1981) 159-219.
3.   Cao, T.H.: Foundations of Order-Sorted Fuzzy Set Logic Programming in Predicate Logic and Conceptual Graphs. PhD Thesis, University of Queensland (1999).

4. Cao, T.H.: Fuzzy Conceptual Graphs: A Language for Computational Intelligence Approaching Human Expression and Reasoning. In: Sincak, P. et al. (eds.): The State of the Art in Computational Intelligence. Physica-Verlag (2000) 114-120.
5. Cao, T.H., Creasy, P.N.: Universal Marker and Functional Relation: Semantics and Operations. In: Lukose, D. et al. (eds.): Conceptual Structures: Fulfilling Peirce's Dream. Lecture Notes in Artificial Intelligence, Vol. 1257. Springer-Verlag (1997) 416-430.
6. Cao, T.H., Creasy, P.N.: Fuzzy Types: A Framework for Handling Uncertainty about Types of Objects. International Journal of Approximate Reasoning 25 (2000) 217-253.
7. Dubois, D., Prade, H.: Fuzzy Sets in Approximate Reasoning, Part 1: Inference with Possibility Distributions. International Journal for Fuzzy Sets and Systems 40 (1991) 143-202.
8. Gaines, B.R.: Fuzzy and Probability Uncertainty Logics. Journal of Information and Control 38 (1978) 154-169.
9. Jeffrey, R.: The Logic of Decision. McGraw-Hill (1965).
10. Klir, G.J., Yuan, B.: Fuzzy Sets and Fuzzy Logic Theory and Applications. Prentice Hall (1995).
11. Lawry, J.: An Alternative Interpretation of Linguistic Variables and Computing with Words. In: Proceedings of the 8th International Conference on Information Processing and Management of Uncertainty in Knowledge-Based Systems (2000) 1743-1750.
12. Liu, Y., Kerre, E.: An Overview of Fuzzy Quantifiers - (I) Interpretations (II) Reasoning and Applications. International Journal for Fuzzy Sets and Systems 95 (1998) 1-21, 135-146.
13. Morton, S.K.: Conceptual Graphs and Fuzziness in Artificial Intelligence. PhD Thesis, University of Bristol (1987).
14. Peterson, P.L.: On the Logic of Few, Many and Most. Notre Dame Journal of Formal Logic XX (1979) 155-179.
15. Sowa, J.F.: Conceptual Structures - Information Processing in Mind and Machine. Addison-Wesley Publishing Company (1984).
16. Sowa, J.F.: Towards the Expressive Power of Natural Language. In: Sowa, J.F. (ed.): Principles of Semantic Networks - Explorations in the Representation of Knowledge. Morgan Kaufmann Publishers (1991) 157-189.
17. Sowa, J.F.: Matching Logical Structure to Linguistic Structure. In: Houser, N., Roberts, D.D., Van Evra, J. (eds.): Studies in the Logic of Charles Sanders Peirce. Indiana University Press (1997) 418-444.
18. Sowa, J.F.: Conceptual Graphs: Draft Proposed American National Standard. In: Tepfenhart, W., Cyre, W. (eds.): Conceptual Structures: Standards and Practices. Lecture Notes in Artificial Intelligence, Vol. 1640. Springer-Verlag (1999) 1-65.
19. Tjan, B.S., Gardiner, D.A., Slagle, J.R.: Direct Inference Rules for Conceptual Graphs with Extended Notation. In: Proceedings of the 5th Annual Workshop on Conceptual Structures (1990).
20. Tjan, B.S., Gardiner, D.A., Slagle, J.R.: Representing and Reasoning with Set Referents and Numerical Quantifiers. In: Nagle, T.E. et al. (eds.): Conceptual Structures - Current Research and Practice. Ellis Horwood (1992) 53-66.
21. Wuwongse, V., Manzano, M.: Fuzzy Conceptual Graphs. In: Mineau, G.W., Moulin, B., Sowa, J.F. (eds.): Conceptual Graphs for Knowledge Representation. Lecture Notes in Artificial Intelligence, Vol. 699. Springer-Verlag (1993) 430-449.
22. Zadeh, L.A.: Fuzzy Sets. Journal of Information and Control 8 (1965) 338-353.
23. Zadeh, L.A.: Fuzzy Logic and Approximate Reasoning. Synthese 30 (1975) 407-428.
24. Zadeh, L.A.: PRUF - A Meaning Representation Language for Natural Languages. International Journal of Man-Machine Studies 10 (1978) 395-460.
25. Zadeh, L.A.: A Computational Approach to Fuzzy Quantifiers in Natural Languages. Computers and Mathematics with Applications 9 (1983) 149-184.
26. Zadeh, L.A.: Fuzzy Logic = Computing with Words. IEEE Transactions on Fuzzy Systems 4 (1996) 103-111.

Simple Semiconcept Graphs: A Boolean Logic Approach

Julia Klinger

Technische Universität Darmstadt, Fachbereich Mathematik
Schloßgartenstr. 7, D-64289 Darmstadt, julia@mathebau.de

Abstract. The aim of this paper is to develop a logical theory of concept graphs with negation. For this purpose, we introduce semiconcept graphs as syntactical constructs and define their semantics based on power context families. Then a standard power context family is constructed which serves both as a characterization of the entailment relation and as a mechanism to translate knowledge given on the graph level to the context level. A standard graph is constructed which entails all semiconcept graphs valid in a given power context family. The possible use of semiconcept graphs in conceptual knowledge processing is illustrated by an example.

1 Introduction

We begin with a brief overview of the development of concept graphs as a contextual judgment logic. The first approach was described by R. Wille in [Wi97] when he connected the Theory of Conceptual Graphs with Formal Concept Analysis. For basic notions in Formal Concept Analysis we refer to [GW99]. A logical theory with a syntax for concept graphs was developed by S. Prediger [Pr98a]. She defined concept graphs as syntactical constructs and used a contextual semantics based on Formal Concept Analysis to interpret syntactical names as objects, formal concepts and relations of a power context family (see also [Pr98b]). This approach was extended to the nested case in [Pr00] and the case of existential quantifiers in [Pr98a]. The next challenge is to include negation. The first approach has been done in [Da00] where negation was introduced as a new syntactical element on the graph level. This is closely related to the original ideas of Peirce: Negation is described by Peirce's ovals (called *cuts*) which can be drawn around arbitrary parts of a concept graph.

However, as R. Wille pointed out in [Wi00a], it is not possible to define a negation operator on formal concepts in the sense of G. Boole. A transfer of Boole's ideas into the language of Formal Concept Analysis forces the extension of a negated concept in a given formal context to be the set-theoretic complement of the extent of the corresponding non negated concept. However, this set does not have to be the extent of a concept itself.

Therefore, it is necessary to generalize the notion of formal concepts in order to keep the correspondence between negation and set-complement. This leads

H. Delugach and G. Stumme (Eds.): ICCS 2001, LNAI 2120, pp. 101–114, 2001.
© Springer-Verlag Berlin Heidelberg 2001

$\mathbb{K}_0$: $\mathbb{K}_2$:

	relieving light pain	mucus dissolving	increasing the body's defences
Aspirin	X		
Bisolvon		X	
Ibuprofen	X		
Vitamin C			X

	Interaction	A intensified	A weakend	B intensified	B weakend
(Aspirin, Aspirin)					
(Aspirin, Bisolvon)	X	X			
(Aspirin, Ibuprofen)					
(Aspirin, Vitamin C)	X				X
(Bisolvon, Aspirin)	X	X			X
(Bisolvon, Bisolvon)					
(Bisolvon, Ibuprofen)					
(Bisolvon, Vitamin C)					
(Ibuprofen, Aspirin)	X		X		
(Ibuprofen, Bisolvon)					
(Ibuprofen, Ibuprofen)					
(Ibuprofen, Vitamin C)					
(Vitamin C, Aspirin)	X		X		
(Vitamin C, Bisolvon)					
(Vitamin C, Ibuprofen)					
(Vitamin C, Vitamin C)					

Fig. 1. The contexts $\mathbb{K}_0$ and $\mathbb{K}_2$

to the notion of semiconcepts which have already been introduced in [LW91] and further developed in [Wi00a]. Using these semiconcepts, we can introduce semiconcept graphs which include negations on the concept and relation level. In contrast to [Da00], we can assign more than only one object name to a vertex. Another advantage of using semiconcept graphs is that we can represent simultaneously that certain objects are in a relation and that others are not. Additionally, we have several operations on semiconcept graphs including negating inversion (cf. [Wi00b]).

Let us start with an example concerning important problems in the medical area. It deals with four different drugs and interactions between them. Typical questions in this field are: Are the drugs appropriate for a certain treatment? Do the substances of two or more drugs interact and, if the answer is yes, how? Do there exist different drugs with the same effects but without interaction? The first question could be dealt with the established methods of Formal Concept

Analysis. For addressing the second question we extend our language in order to be able to formalize relations. This will be done in the next paragraphs. First we give a possibility to write sentences like "the drug a does (not) interact with the drug b" as semiconcept graphs. Then it can be checked if the semiconcept graph is valid in a model consisting of a power context family together with a suitable interpretation (see Section 3). The third question can be treated with existential semiconcept graphs, which are not within the scope of this paper.

Figure 1 shows two formal contexts taken from [Am91]. The original multi-valued contexts were converted into single-valued contexts by conceptual and relational scaling (see [PW99]). Objects of the first context $\mathbb{K}_0$ are drugs and attributes are descriptions of their effects. In the second context $\mathbb{K}_2$, objects are pairs of objects of $\mathbb{K}_0$ and the attributes are different kinds of interaction. Note that this relation is not symmetric, because if and how drugs interact depends on the time each of them is consumed. In particular, (m_1, m_2) shall be read as m_1 is applied before m_2. A tuple (m_1, m_2) has the attribute "Interaction" if m_1 interacts with m_2. The other attributes specify how the two drugs interact: For example, "A intensified" means that the effect of the active agent of the first drug increases.

For a formal context $\mathbb{K} := (G, M, I)$ we define the set of ($\sqcap$-) *semiconcepts* with respect to this context as $\mathfrak{H}_\sqcap(\mathbb{K}) := \{(A, A^I) \mid A \subseteq G\}$, where I is the incidence relation between objects and attributes. Thus, a semiconcept is a pair consisting of a set A of objects and the set A^I of attributes which are shared by all elements of A. For example, the tuple ({Aspirin, Ibuprofen}, {relieving light pain}) would be a semiconcept in $\mathbb{K}_0$. We have several operations and constants on the semiconcepts of a formal context of which only one is necessary for our purpose, namely the extensional negation $\neg(A, A^I) := (G \backslash A, (G \backslash A)^I)$.

We also define an order on $\mathfrak{H}_\sqcap(\mathbb{K})$ by $(A, A^I) \sqsubseteq (B, B^I) : \Longleftrightarrow A \subseteq B$.

The set of semiconcepts of a context provided with this order is a Boolean algebra; for details please refer to [Wi00a]. The semiconcept lattice of $\mathbb{K}_0$ is shown in Figure 2.

For practical purposes it might be unwieldy to illustrate a context $\mathbb{K} := (G, M, I)$ with the entire semiconcept lattice, because the semiconcept lattice is isomorphic to the Boolean algebra of all subsets of G. Therefore, in Figure 3 we rather focus on the concept lattice of $\mathbb{K}_2$. Since every concept is also a semiconcept, this is a $\sqcap$-subsemilattice of the semiconcept lattice of $\mathbb{K}_2$. The extents of the negated concepts can be derived from the concept lattice by taking the complement of the extent.

This paper consists of five more sections. In Section 2, we introduce semiconcept graphs as syntactical constructs. Section 3 contains a semantics for semiconcept graphs based on power context families. In Section 4, we show how to translate information given on the graph level into the language of power context families. This construction is also used as a characterization of the entailment relation. In Section 5, the standard graph for a power context family is developed, which codes the same information as the power context family. By this

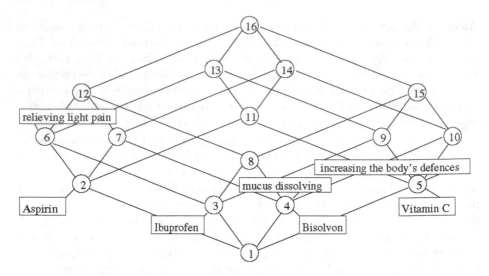

Fig. 2. The algebra of semiconcepts of $\mathbb{K}_0$

means we get a mechanism to translate the information given in a power context family to the level of semiconcept graphs as well. In Section 6, we finally sketch how to proceed with this work in the future.

2   The Syntax of Simple Semiconcept Graphs

We follow the approach in [Pr98b] where simple concept graphs were introduced as mathematically defined syntactical constructs, and extend it by assigning both a positive and a negative reference to the concept boxes (cf. [Wi00b]). We start with an alphabet of object names, semiconcept names, and relation names:

Definition 1. An alphabet of semiconcept graphs is a triple $\mathcal{A} := (\mathcal{G}, \mathcal{S}, \mathcal{R})$ where $\mathcal{G}$ is a finite set whose elements are called *object names*, $(\mathcal{S}, \leq_S)$ is a finite ordered set whose elements are called *semiconcept names* and $(\mathcal{R}, \leq_R)$ is a collection $(\mathcal{R}_k, \leq_{\mathcal{R}_k})_{k=1,\ldots,n}$ of finite ordered sets whose elements are called *relation names* .

An example for such an alphabet is given in Figure 4. The underlying structure of a semiconcept graph is a relational graph.

Definition 2. A *relational graph* is a triple (V, E, ν) with vertices V and edges E and a map $\nu : E \to \bigcup_{k=1,2,\ldots} V^k$. The expression $\nu(e) = (v_1, v_2, \ldots, v_k)$ is read as follows: $v_1, \ldots, v_k$ are *adjacent vertices* of the k-ary edge e and $|e| = k$ is called the *arity of* e. A relational graph is called *finite* if V and E are finite sets. We write $E^{(k)} := \{e \in E \mid |e| = k\}$ for $k = 1, 2, \ldots$ A relational graph is

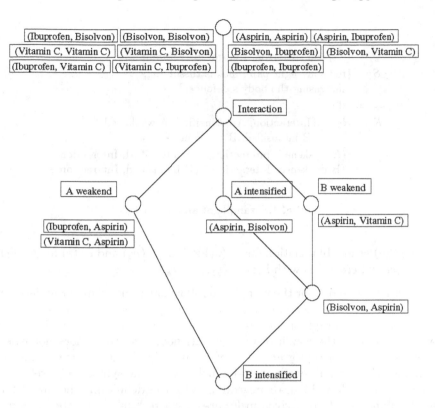

Fig. 3. The concept lattice of $\mathbb{K}_2$

said to be of *limited type* $n \in \mathbb{N}$ if $E = E^{(1)} \cup \cdots \cup E^{(n)}$, otherwise it is called unlimited.

Finally, we label the vertices and edges of a relational graph by semiconcept names and assign to each vertex a positive and a negative reference, each consisting of a set of object names:

Definition 3. A *semiconcept graph* over the alphabet $\mathcal{A} := (\mathcal{G}, \mathcal{S}, \mathcal{R})$ is a structure $\mathfrak{G} := (V, E, \nu, \kappa, \rho)$ such that

- (V, E, ν) is a finite relational graph,
- $\kappa \colon V \cup E \to \mathcal{S} \cup \mathcal{R}$ is a mapping with $\kappa(V) \subseteq \mathcal{S}$, $\kappa(E) \subseteq \mathcal{R}$ and $\kappa(e) \in \mathcal{R}_k$ for all $e \in E^{(k)}$, $k \in \{1, \ldots, n\}$,
- $\rho \colon V \to \mathcal{P}(\mathcal{G}) \times \mathcal{P}(\mathcal{G})$, $\rho \colon v \mapsto \left(\rho^+(v), \rho^-(v)\right)$ is given by two functions $\rho^+ \colon V \to \mathcal{P}(\mathcal{G})$ and $\rho^- \colon V \to \mathcal{P}(\mathcal{G})$ satisfying
  1. $\rho^+(v) \cup \rho^-(v) \neq \emptyset$ for all $v \in V$,
  2. $\rho^+(e) \cup \rho^-(e) \neq \emptyset$ for all $e \in E$,
  3. all $v, w \in V$ with $\kappa(v) \geq_\mathcal{S} \kappa(w)$ satisfy $\rho^-(v) \cap \rho^+(w) = \emptyset$,
  4. all $e, f \in E^{(k)}$ with $\kappa(e) \geq_{\mathcal{R}_k} \kappa(f)$ satisfy $\rho^-(e) \cap \rho^+(f) = \emptyset$.

$$\mathcal{A} := (\mathcal{G}, \mathcal{S}, \mathcal{R})$$
$$\mathcal{G} := \{\text{Aspirin, Bisolvon, Ibuprofen, Vitamin C}\}$$
$$\mathcal{S} := \{\text{relieving light pain, mucus dissolving,}$$
$$\text{increasing the body's defences}\}$$
$$\leq_{\mathcal{S}} := \text{id}_{\mathcal{S}}$$
$$\mathcal{R} := \mathcal{R}_2 := \{\text{Interaction, A intensified, A weakened,}$$
$$\text{B intensified, B weakened}\}$$
$$\leq_{\mathcal{R}} := \{(\text{A weakened, Interaction}), (\text{A intensified, Interaction}),$$
$$(\text{B weakened, Interaction}), (\text{B intensified, Interaction})\}$$

Fig. 4. Example of an alphabet

We use $\rho^+(e)$ as an abbreviation for $\rho^+(v_1) \times \cdots \times \rho^+(v_k)$ and $\rho^-(e)$ for $\rho^-(v_1) \times \cdots \times \rho^-(v_k)$ where $e \in E$ with $\nu(e) = (v_1, \ldots, v_k)$.

Conditions 3. and 4. for the mapping ρ follow naturally from our understanding of judgments (represented by semiconcept graphs) as valid propositions: For every semiconcept graph should exist at least one model in which it is valid. However, if one of the conditions is not satisfied, then there does not exist a model for the semiconcept graph. Therefore, in order to guarantee the existence of a model, these conditions have to be included on the syntactical level.

The three graphs in Figure 5 are examples for this definition. They are defined over the alphabet which is given in Figure 4, where "rlp" and "itbd" stand for "relieving light pain" and "increasing the body's defences", respectively.

The semiconcept graph $\mathfrak{G}_1$ consists of the relational graph $(\{u_1, u_2\}, \{e_1\}, \nu)$ where $\nu_1 \colon e_1 \mapsto (u_1, u_2)$, and the mappings κ_1 and ρ_1 with $\kappa_1(u_1) := rlp$, $\kappa_1(u_2) := itbd$, $\kappa_1(e_1) := Interaction$ and $\rho_1(u_1) := (Aspirin, Vitamin\ C)$, $\rho_1(u_2) := (Vitamin\ C, Ibuprofen)$. Then $\rho_1^+(e_1) = (Aspirin, Vitamin\ C)$ and $\rho_1^-(e_1) = (Vitamin\ C, Ibuprofen)$. The positive and negative references of a vertex are separated by a vertical stroke. We will interpret this syntactical construct in such a way that the graph $\mathfrak{G}_1$ formalizes the judgments: *Aspirin interacts with Vitamin C, Vitamin C does not interact with Ibuprofen, Aspirin falls under the semiconcept "relieves light pain" and Vitamin C does not, Vitamin C falls under the semiconcept "increases the body's defences" and Ibuprofen does not.*

The semiconcept graph $\mathfrak{G}_3$ shows a possibility to represent that Bisolvon and Ibuprofen do not interact, that Bisolvon is mucus dissolving and that Ibuprofen relieves light pain. In order to point out that we do not have to restrict ourselves to one reference, both Aspirin and Ibuprofen are assigned as positive references to w_1.

3   Semantics of Simple Semiconcept Graphs

Contextual semantics for concept graphs, as introduced in [Pr98a], has shown to be a rich and fertile approach. The contextual view is based on Formal Concept

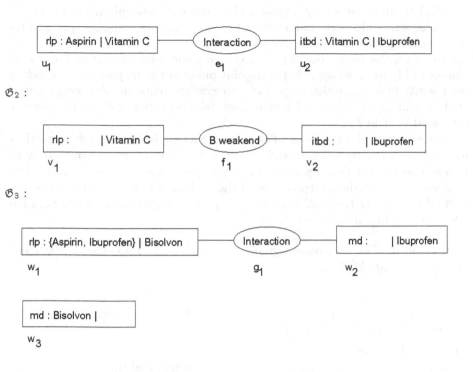

Fig. 5. Example for the Syntax

Analysis and supported by Peirce's pragmatism (see [Wi97] for details). We will now extend this approach to semiconcept graphs. First, let us recall the definition of a power context family.

Definition 4. A *power context family* $\overrightarrow{\mathbb{K}} := (\mathbb{K}_n)_{n \in \mathbb{N}}$ is a family of contexts $\mathbb{K}_k := (G_k, M_k, I_k)$ such that $G_k \subseteq (G_0)^k$ for $k \in \mathbb{N}$. The power context family $\overrightarrow{\mathbb{K}}$ is said to be *limited of type* $n \in \mathbb{N}$ if $\overrightarrow{\mathbb{K}} = (\mathbb{K}_0, \mathbb{K}_1, \dots, \mathbb{K}_n)$, otherwise it is called *unlimited*. The elements of the set $\underline{\mathfrak{H}}_\sqcap(\mathbb{K}_0)$ are called *semiconcepts*, the elements of $\mathfrak{R}_{\overrightarrow{\mathbb{K}}} := \bigcup_{k=1,2,\dots} \underline{\mathfrak{H}}_\sqcap(\mathbb{K}_k)$ are called *relation semiconcepts*.

The tuple $(\mathbb{K}_0, \mathbb{K}_1, \mathbb{K}_2)$ consisting of the two contexts from Figure 1 and the empty context $\mathbb{K}_1$ is a power context family which is limited of type 2.

With the following definition we establish a link between semiconcept graphs as syntactical constructs and power context families.

Definition 5. For an alphabet $\mathcal{A} := (\mathcal{G}, \mathcal{S}, \mathcal{R})$ and a power context family $\overrightarrow{\mathbb{K}} := (\mathbb{K}_0, \mathbb{K}_1, \dots, \mathbb{K}_n)$ the disjoint union $\lambda := \lambda_{\mathcal{G}} \dot{\cup} \lambda_{\mathcal{S}} \dot{\cup} \lambda_{\mathcal{R}}$ of the mappings $\lambda_{\mathcal{G}} : \mathcal{G} \to G_0$, $\lambda_{\mathcal{S}} : \mathcal{S} \to \underline{\mathfrak{H}}_\sqcap(\mathbb{K}_0)$ and $\lambda_{\mathcal{R}} : \mathcal{R} \to \mathfrak{R}_{\overrightarrow{\mathbb{K}}}$ is called a $\overrightarrow{\mathbb{K}}$-*interpretation* of $\mathcal{A}$ if $\lambda_{\mathcal{S}}$ and $\lambda_{\mathcal{R}}$ are order preserving and $\lambda_{\mathcal{R}}(\mathcal{R}_k) \subseteq \underline{\mathfrak{H}}_\sqcap(\mathbb{K}_k)$ holds for all $k = 1, \dots, n$. The pair $(\overrightarrow{\mathbb{K}}, \lambda)$ is called a *context-interpretation for* $\mathcal{A}$.

The interpretation λ maps syntactical elements of the semiconcept graph, i.e. object names, semiconcept and relation names, to objects, semiconcepts, and relation semiconcepts of a power context family.

For instance, the tuple $((\mathbb{K}_0, \mathbb{K}_1, \mathbb{K}_2), \lambda)$ is a context-interpretation for the alphabet in Figure 4, where λ is the disjoint union of the mappings $\lambda_{\mathcal{G}}$, which is the identity in $\mathcal{G}$, $\lambda_{\mathcal{S}}$, which maps each semiconcept name to the concept generated by it in $\mathbb{K}_0$, and $\lambda_{\mathcal{R}}$, which maps each relation name in $\mathcal{R}_k$ to the concept generated by it in $\mathbb{K}_k$.

Having assigned an element of the power context family to each syntactical name, we can specify when a semiconcept graph is valid in a model. We have to take into account that a positive and a negative reference may be assigned to each vertex. While the interpretation of the positive reference $\rho^+(v)$ of a vertex v should fall under the semiconcept $\lambda_{\mathcal{S}}\kappa(v)$, the interpretation of the negative reference should fall under $\neg\lambda_{\mathcal{S}}\kappa(v)$.

Definition 6. The semiconcept graph $\mathfrak{G} := (V, E, \nu, \kappa, \rho)$ is called a *valid semiconcept graph of* $(\vec{\mathbb{K}}, \lambda)$ if

- all $v \in V$ satisfy
  $$\lambda_{\mathcal{G}}\rho^+(v) \subseteq \operatorname{Ext}\left(\lambda_{\mathcal{S}}\kappa(v)\right), \qquad \text{(Vertex condition)}$$
  $$\lambda_{\mathcal{G}}\rho^-(v) \subseteq G_0 \backslash \operatorname{Ext}\left(\lambda_{\mathcal{S}}\kappa(v)\right)$$

- all $e \in E$ satisfy
  $$\lambda_{\mathcal{G}}\rho^+(e) \subseteq \operatorname{Ext}\left(\lambda_{\mathcal{R}}\kappa(e)\right), \qquad \text{(Edge condition)}$$
  $$\lambda_{\mathcal{G}}\rho^-(e) \subseteq (G_0)^k \backslash \operatorname{Ext}\left(\lambda_{\mathcal{R}}\kappa(e)\right)$$

If $\mathfrak{G}$ is valid in $(\vec{\mathbb{K}}, \lambda)$, then $(\vec{\mathbb{K}}, \lambda)$ is called a *model* for $\mathfrak{G}$ and $\mathfrak{G}$ is called a *semiconcept graph of* $(\vec{\mathbb{K}}, \lambda)$.

Note that the negative references along an edge have to be in the extension of the negated relation semiconcept. So we have a negation on the relation level as well.

One easily checks that the three semiconcept graphs given in Figure 5 are valid in the model $((\mathbb{K}_0, \mathbb{K}_1, \mathbb{K}_2), \lambda)$: For example, we have $\lambda_{\mathcal{G}}\rho^+(u_1) = \{\text{Aspirin}\} \subseteq \{\text{Aspirin, Ibuprofen}\} = \operatorname{Ext}\left(\lambda_{\mathcal{S}}\kappa_1(u_1)\right)$.

The following definition turns out to be a useful tool.

Definition 7. Let $\vec{\mathbb{K}}_a := \left((G_0)^k, M_k, I_{ak}\right)_{k=0,1,\dots,n}$ and $\vec{\mathbb{K}}_b := \left((G_0)^k, M_k, I_{bk}\right)_{k=0,1,\dots,n}$ be two power context families which merely differ in the incidence relations. We say $\vec{\mathbb{K}}_b$ *is an extension of* $\vec{\mathbb{K}}_a$ ($\vec{\mathbb{K}}_a$ *is contained in* $\vec{\mathbb{K}}_b$), if $I_{ai} \subseteq I_{bi}$ for all $i = 0, 1, \dots, n$.

The generalization of concepts to semiconcepts features several structural properties of the set of models for a semiconcept graph $\mathfrak{G}$: For example, if $(\vec{\mathbb{K}}, \lambda)$ is a model for $\mathfrak{G}$, then $\mathfrak{G}$ is also valid in every extension $\vec{\mathbb{K}}'$ of $\vec{\mathbb{K}}$ provided with a context-interpretation λ' that preserves the extents of λ.

4 Standard Power Context Families

When dealing with semiconcept graphs, it is desirable to have a mechanism that translates knowledge expressed on the graph level into knowledge on the context level. For this purpose, S. Prediger specified a standard model of a concept graph in which all information of the concept graph is coded. Unfortunately, it will not be possible to construct a standard model for semiconcept graphs:

Let $\mathfrak{G} := (V, E, \nu, \kappa, \rho)$ be a semiconcept graph over an alphabet $\mathcal{A} := (\mathcal{G}, \mathcal{S}, \mathcal{R})$ such that $\bigcup_{v \in V} (\rho^+(v) \cup \rho^-(v)) \neq \mathcal{G}$. Since the objects that are neither in $\bigcup_{v \in V} \rho^+(v)$ nor in $\bigcup_{v \in V} \rho^-(v)$ do not occur in the semiconcept graph, we do not have any information about them at all. This implies that in a standard model these objects should neither be in the extension of any of the semiconcepts nor in the extension of any of the negated semiconcepts. However, this is impossible, since for a semiconcept s the extension of the negated semiconcept is the complement of the extension of s. Nevertheless, we can find a power context family that codes exactly the same information as the semiconcept graph.

Let $\mathfrak{G} := (V, E, \nu, \kappa, \rho)$ be a semiconcept graph over the alphabet $\mathcal{A} := (\mathcal{G}, \mathcal{S}, \mathcal{R})$ and let n be the type of (V, E, ν). Then we define the set $\neg \mathcal{S} := \{\neg s \mid s \in \mathcal{S}\}$ of negated semiconcept names and the set $\neg \mathcal{R}_k := \{\neg r \mid r \in \mathcal{R}_k\}$ of negated relation names. We construct the *standard power context family* $\overrightarrow{\mathbb{K}}^{\mathfrak{G}} := (\mathbb{K}_0^{\mathfrak{G}}, \ldots, \mathbb{K}_n^{\mathfrak{G}})$ with $\mathbb{K}_k^{\mathfrak{G}} := (G_k^{\mathfrak{G}}, M_k^{\mathfrak{G}}, I_k^{\mathfrak{G}})$ in the following way: We take the object names in $\mathcal{G}$ as the objects in $G_0^{\mathfrak{G}}$ and the set $\mathcal{S} \,\dot{\cup}\, \neg \mathcal{S}$ as the attributes in $M_0^{\mathfrak{G}}$. Then we construct the incidence relation $I_0^{\mathfrak{G}}$ by relating the object name g and the semiconcept name $\kappa(v)$ if g is in the positive reference of the vertex v. For preserving the order, g is also put in relation to every semiconcept which is greater than $\kappa(v)$ with respect to $\leq_{\mathcal{S}}$. Additionally, for each object name h in the negative reference of v, the tuple $(h, \neg \kappa(v))$ is added to $I_0^{\mathfrak{G}}$ together with $(h, \neg t)$ for all semiconcept names t which are smaller than $\kappa(v)$. The construction of the contexts $\mathbb{K}_k^{\mathfrak{G}}$ is essentially the same as above: As objects in $\mathbb{K}_k^{\mathfrak{G}}$, we take all k-tuples of object names, the elements of $\mathcal{R}_k \,\dot{\cup}\, \neg \mathcal{R}_k$ are the attributes. Then $(g_1, \ldots, g_k)$ and a relation name R are related if there exists an edge e in the semiconcept graph such that $(g_1, \ldots, g_k)$ is an element of $\rho^+(e)$ and $\kappa(e) \leq_{\mathcal{R}_k} R$. Finally, $(h_1, \ldots, h_k)$ is set in relation to a negated relation name $\neg Q$ if $(h_1, \ldots, h_k)$ is in $\rho^-(e)$ for a k-ary edge e and $R \leq_{\mathcal{R}_k} \kappa(v)$.

We provide the standard power context family $\overrightarrow{\mathbb{K}}^{\mathfrak{G}}$ of a semiconcept graph $\mathfrak{G}$ with the natural $\overrightarrow{\mathbb{K}}^{\mathfrak{G}}$-interpretation $\lambda^{\mathfrak{G}} := \lambda_{\mathcal{G}}^{\mathfrak{G}} \,\dot{\cup}\, \lambda_{\mathcal{S}}^{\mathfrak{G}} \,\dot{\cup}\, \lambda_{\mathcal{R}}^{\mathfrak{G}}$ where $\lambda_{\mathcal{G}}^{\mathfrak{G}} := id_{\mathcal{G}}$, $\lambda_{\mathcal{S}}^{\mathfrak{G}} \colon \mathcal{S} \to \mathfrak{H}_{\sqcap}(\mathbb{K}_0)$ with $\lambda_{\mathcal{S}}^{\mathfrak{G}} \colon s \mapsto (s^{I_0^{\mathfrak{G}}}, s^{I_0^{\mathfrak{G}} I_0^{\mathfrak{G}}})$ and $\lambda_{\mathcal{R}}^{\mathfrak{G}} \colon \mathcal{R} \to \bigcup_{k=1, \ldots} \mathfrak{H}_{\sqcap}(\mathbb{K}_k)$ with $\lambda_{\mathcal{R}}^{\mathfrak{G}} \colon R \mapsto (R^{I_k^{\mathfrak{G}}}, R^{I_k^{\mathfrak{G}} I_k^{\mathfrak{G}}})$ for $R \in \mathcal{R}_k$ and $(k = 1, \ldots)$. It is easy to prove that $\mathfrak{G}$ is valid in $(\overrightarrow{\mathbb{K}}^{\mathfrak{G}}, \lambda^{\mathfrak{G}})$.

The standard power context family is also a useful tool to address questions of inference. Let us first recall the definition of semantical entailment:

Definition 8. Let $\mathfrak{G}_1$ and $\mathfrak{G}_2$ be two semiconcept graphs over the same alphabet. We say $\mathfrak{G}_1$ entails $\mathfrak{G}_2$ if $\mathfrak{G}_2$ is valid in every model for $\mathfrak{G}_1$. We denote this by $\mathfrak{G}_1 \models \mathfrak{G}_2$.

Although $\mathfrak{G}$ is valid in $(\overrightarrow{\mathbb{K}}^{\mathfrak{G}}, \lambda^{\mathfrak{G}})$, we do not have a standard model in the sense of [Pr98b] for $\mathfrak{G}$: A semiconcept graph $\mathfrak{G}_2$ which is valid in $(\overrightarrow{\mathbb{K}}^{\mathfrak{G}}, \lambda^{\mathfrak{G}})$ does not have to be valid in every model for $\mathfrak{G}$. Nevertheless, the following proposition explains how we can characterize the entailment relation by means of the standard power context family.

Proposition 9. Let $\mathfrak{G}_1 := (V_1, E_1, \nu_1, \kappa_1, \rho_1)$ and $\mathfrak{G}_2 := (V_2, E_2, \nu_2, \kappa_2, \rho_2)$ be two semiconcept graphs over the same alphabet. Then $\mathfrak{G}_1 \models \mathfrak{G}_2$ if and only if $\overrightarrow{\mathbb{K}}^{\mathfrak{G}_1}$ is an extension of $\overrightarrow{\mathbb{K}}^{\mathfrak{G}_2}$.

With Proposition 9, which is proven in the Appendix, we have a simple method to do reasoning with semiconcept graphs on the context level: For every semiconcept graph, we can decide if it is entailed in another one by checking if the standard power context family of the first one is contained in the standard power context family of the latter. For example, by using this characterization, we can check that, in Figure 5, $\mathfrak{G}_1$ entails $\mathfrak{G}_2$ and that $\mathfrak{G}_3$ does neither entail $\mathfrak{G}_1$ nor $\mathfrak{G}_2$.

We can now deduce some inference rules. For instance, most of the inference rules for simple concept graphs (cf. [Pr98a]) can be adapted to semiconcept graphs: *Double a vertex, delete an isolated vertex, double an edge, delete an edge, exchange a semiconcept name, exchange a relation name, restrict a reference, copy the concept graph.* It has to be taken into account that we have a positive and a negative reference for each vertex. For example the rule *exchange a semiconcept name* now reads: *substitute the assignment $v \mapsto \kappa(v)$ for $v \mapsto c$ for such a semiconcept name for which there are vertices $w, u \in V$ with $\kappa(w) \leq_S c \leq_S \kappa(u)$ and $\rho^+(v) \subseteq \rho^+(w)$, $\rho^-(v) \subseteq \rho^-(u)$.*

Finally, the characterization allows us to understand the entailment relation in a structural way: It is reflexive and transitive, i.e. a quasiorder on the class of all semiconcept graphs. Moreover, it implies that equivalent semiconcept graphs (i.e. two semiconcept graphs $\mathfrak{G}_1$ and $\mathfrak{G}_2$ which satisfy $\mathfrak{G}_1 \models \mathfrak{G}_2$ and vice versa) have the same standard power context family. Starting from the set of semiconcept graphs over a certain alphabet, we can consider the equivalence classes of $\models$ and characterize the induced order on them in a simple way: It is the inclusion of the incidence relations of the corresponding standard power context families.

5   Standard Graph of a Power Context Family

In the previous section we constructed a power context family for a semiconcept graph that coded all information given in the graph. In this section we will consider the opposite direction: We construct the semiconcept graph for a power context family that codes the same information as the power context family. Hence, we get a mechanism to translate the information given in a power context family into the level of semiconcept graphs as well. Again, we try an approach similar to [Pr98a]: The smallest concepts with non-empty extension were taken and the largest possible references were assigned.

The smallest non-empty semiconcepts are those semiconcepts with exactly one element in the extension. Additionally, for each object g we have to take the largest semiconcept that does not have g in its extension. This is necessary since the negative references have to be treated dually to the positive references. So, we get the following situation:

We start with a power context family $\overline{\mathbb{K}} := (G_k, M_k, I_k)_{k=0,1,\ldots,n}$ with $G_0 \neq \emptyset$. To construct a semiconcept graph, we define the alphabet $(\mathcal{G}, \mathcal{S}, \mathcal{R})$ by $\mathcal{G} := G_0$, $\mathcal{S} := \mathfrak{H}_\sqcap(\mathbb{K}_0)$ and $\mathcal{R} := \mathfrak{R}_{\overline{\mathbb{K}}}$. Now we define the set of vertices as $V_{\overline{\mathbb{K}}} := V_{\overline{\mathbb{K}}}^1 \cup V_{\overline{\mathbb{K}}}^2$ with $V_{\overline{\mathbb{K}}}^1 := \{g \mid g \in G_0\}$ and $V_{\overline{\mathbb{K}}}^2 := \{G_0 \backslash \{g\} \mid g \in G_0\}$. The set of edges with arity k is the union of G_k with $(V_{\overline{\mathbb{K}}}^2)^k$, i.e.

$$E_{\overline{\mathbb{K}}} := \bigcup_{k=1,\ldots,n} G_k \cup \{(G_0 \backslash \{g_1\}, \ldots, G_0 \backslash \{g_k\}) \mid (g_1, \ldots g_k) \in G_k\})$$

The mapping $\nu_{\overline{\mathbb{K}}} : E_{\overline{\mathbb{K}}} \to \bigcup_{k=1,2,\ldots} V_{\overline{\mathbb{K}}}^k$ is the identity function. The mapping $\kappa_{\overline{\mathbb{K}}} : V_{\overline{\mathbb{K}}} \cup E_{\overline{\mathbb{K}}} \to \mathfrak{H}_\sqcap(\mathbb{K}_0) \cup \mathfrak{R}_{\overline{\mathbb{K}}}$ maps each vertex to the semiconcept that has the vertex as extent, the edges of the form $(g_1, \ldots, g_k)$ are mapped to the relation semiconcept with $(g_1, \ldots, g_k)$ as extent and each edge of the form $(G_0 \backslash \{g_1\}, \ldots, G_0 \backslash \{g_k\})$ is mapped to $((G_0)^k \backslash (g_1, \ldots, g_k), ((G_0)^k \backslash (g_1, \ldots, g_k))^{I_k})$. Finally we choose $\rho(v)_{\overline{\mathbb{K}}}(A)$ $:= (A, \emptyset)$ if $A \in V_{\overline{\mathbb{K}}}^1$ and $\rho(v)_{\overline{\mathbb{K}}}(A) := (\emptyset, G_0 \backslash A)$ if $A \in V_{\overline{\mathbb{K}}}^2$.

The resulting semiconcept graph $\mathfrak{G}(\overline{\mathbb{K}}) := (V_{\overline{\mathbb{K}}}, E_{\overline{\mathbb{K}}}, \nu_{\overline{\mathbb{K}}}, \kappa_{\overline{\mathbb{K}}}, \rho_{\overline{\mathbb{K}}})$ is called the *standard graph* of the power context family $\overline{\mathbb{K}}$.

It is easy to see that $\mathfrak{G}(\overline{\mathbb{K}})$ is indeed a semiconcept graph, and checking the conditions of Definition 6 we see that the standard graph $\mathfrak{G}(\overline{\mathbb{K}})$ of a power context family $\overline{\mathbb{K}}$ is valid in $(\overline{\mathbb{K}}, id)$. Note that, if we omit one vertex or one edge of the standard graph, the resulting semiconcept graph does not entail the standard graph and therefore the standard graph is irredundant.

To make sure that the standard graph codes indeed all the information given in the power context family, we have to check that the standard graph entails all other valid concept graphs. The following proposition (which is proven in the Appendix) guarantees that our translation mechanism works properly:

Proposition 10. The standard graph $\mathfrak{G}(\overline{\mathbb{K}})$ of a power context family $\overline{\mathbb{K}} := (G_k, M_k, I_k)_{k=0,1,\ldots,n}$ entails every semiconcept graph $\mathfrak{G}'$ that is valid in $(\overline{\mathbb{K}}, id)$.

Thus, with the standard power context family and the standard graph we have tools for switching between the context level and the level of semiconcept graphs.

6 Further Research

Obviously, the language of the semiconcept graphs as developed in this paper can be extended to graphs with generic markers. This has already been done for semiconcept graphs with an existential quantifier, but this topic would exceed the scope of this paper. Also nestings should be included in the theory. Since our semantics is based on power context families, a situation-based contextual approach as introduced in [Pr00] can be expected to carry over naturally.

As our medical interaction example has shown, potential areas of application are those where objects not being in a relation is just as important as objects being in a relation. As all real applications need software support, the logic of semiconcept graphs should become part of a contextual logic extension of TOSCANA which is a program for building conceptual information systems. A possible direction of such an extension was discussed in [EGSW00].

7 Appendix

Proof of Proposition 9: For proving the first direction, we assume that $\overrightarrow{\mathbb{K}}^{\mathfrak{G}_2}$ is not contained in $\overrightarrow{\mathbb{K}}^{\mathfrak{G}_1}$, thus that $I_k^{\mathfrak{G}_2} \not\subseteq I_k^{\mathfrak{G}_1}$ for some $k \in \{0, 1, \ldots n\}$. Then we show that there exists a model for $\mathfrak{G}_1$ in which $\mathfrak{G}_2$ is not valid. We will only consider the case $k = 0$, for $k \geq 1$ the two arguments given below can be adapted. So we assume $I_0^{\mathfrak{G}_2} \not\subseteq I_0^{\mathfrak{G}_1}$, which means that there exists a $g \in \mathcal{G}$, $s \in \mathcal{S} \dot{\cup} \neg \mathcal{S}$ with $(g, s) \in I_0^{\mathfrak{G}_2}$, but $(g, s) \notin I_0^{\mathfrak{G}_1}$.

First, we consider the case $s \in \mathcal{S}$. Recall that the semiconcept graph $\mathfrak{G}_1$ is valid in $(\overrightarrow{\mathbb{K}}^{\mathfrak{G}_1}, \lambda^{\mathfrak{G}_1})$, and that $(g, s) \in I_0^{\mathfrak{G}_2}$ means that there exists $v \in V_2$ with $\kappa_2(v) \leq_S s$ and $g \in \rho_2^+(v)$, thus $(g, \kappa_2(v)) \in I_0^{\mathfrak{G}_2}$. Furthermore, $(g, s) \notin I_0^{\mathfrak{G}_1}$ implies (by definition of $I_0^{\mathfrak{G}_1}$) that $(g, t) \notin I_0^{\mathfrak{G}_1}$ for all $t \leq_S s$, thus, in particular, $(g, \kappa_2(v)) \notin I_0^{\mathfrak{G}_1}$. But then we have $g \in G_0^{\mathfrak{G}_1} \setminus \text{Ext}\left(\lambda_S^{\mathfrak{G}_1}(\kappa_2(v))\right)$, therefore $\mathfrak{G}_2$ is not valid in $(\overrightarrow{\mathbb{K}}^{\mathfrak{G}_1}, \lambda^{\mathfrak{G}_1})$.

Secondly, we consider the case $s \in \neg \mathcal{S}$, thus that $s = \neg t$ for some $t \in \mathcal{S}$ and that there exists a $g \in \mathcal{G}$ with $(g, \neg t) \in I_0^{\mathfrak{G}_2}$, but $(g, \neg t) \notin I_0^{\mathfrak{G}_1}$. We will now construct a model for $\mathfrak{G}_1$ which is not a model for $\mathfrak{G}_2$. Recall that $(g, \neg t) \in I_0^{\mathfrak{G}_2}$ means that there exists $v \in V_2$ with $t \leq_S \kappa_2(v)$ and $g \in \rho_2^-(v)$, thus $(g, \neg \kappa_2(v)) \in I_0^{\mathfrak{G}_2}$. The definition of $I_0^{\mathfrak{G}_1}$ implies $(g, \neg r) \notin I_0^{\mathfrak{G}_1}$ for all $t \leq_S r$. Thus, in particular, we have $(g, \neg \kappa_2(v)) \notin I_0^{\mathfrak{G}_1}$.

If $(g, \kappa_2(v)) \in I_0^{\mathfrak{G}_1}$, we would have $g \in \text{Ext}\left(\lambda_S^{\mathfrak{G}_1}(\kappa_2(v))\right)$ and $\mathfrak{G}_2$ would not be valid in $(\overrightarrow{\mathbb{K}}^{\mathfrak{G}_1}, \lambda^{\mathfrak{G}_1})$. So we assume $(g, \kappa_2(v)) \notin I_0^{\mathfrak{G}_1}$. Then the definition of $I_0^{\mathfrak{G}_1}$ implies that $(g, r) \notin I_0^{\mathfrak{G}_1}$ for all $r \leq_S \kappa_2(v)$. We set $I_0' := I_0^{\mathfrak{G}_1} \dot{\cup} \{(g, r) \mid r \geq_S \kappa_2(v)\}$ and extend λ' accordingly. Then $\lambda_{\mathcal{G}}'(g) \in \text{Ext}\left(\lambda_S'(\kappa_2(v))\right)$ and because of $g \in \rho_2^-(v) \cap \text{Ext}\left(\lambda_S'(\kappa_2(v))\right)$ we have that $\mathfrak{G}_2$ is not valid in the model $(\overrightarrow{\mathbb{K}}', \lambda')$ with $\overrightarrow{\mathbb{K}}' := (\mathbb{K}_0', \mathbb{K}_1^{\mathfrak{G}_1}, \mathbb{K}_2^{\mathfrak{G}_1}, \ldots)$ where $\mathbb{K}_0 := (\mathcal{G}, \mathcal{S} \dot{\cup} \neg \mathcal{S}, I_0')$.

The semiconcept graph $\mathfrak{G}_1$ is still valid in $(\overrightarrow{\mathbb{K}}', \lambda')$: On the one hand we added some pairs (g, u) to $I_0^{\mathfrak{G}_1}$ where $u \in \mathcal{S}$, so $\lambda_{\mathcal{G}}'(\rho_1^+(w)) \subseteq Ext(\lambda_S'(\kappa_1(w)))$ still holds for all $w \in V_1$.

On the other hand we did not remove any pairs from $I_0^{\mathfrak{G}_1}$, therefore $\lambda_{\mathcal{G}}'(\rho_1^-(w)) \subseteq G_0 \setminus Ext(\lambda_S'(\kappa_1(w)))$ for all $w \in V_1$.

Finally, the considerations above show that the constructed context is "consistent": we still have $u^{I_0'} \cap (\neg u)^{I_0'} = \emptyset$ for all $u \in \mathcal{S}$.

Now we consider the reverse direction. We suppose that $\overrightarrow{\mathbb{K}}^{\mathfrak{G}_2}$ is contained in $\overrightarrow{\mathbb{K}}^{\mathfrak{G}_1}$ and we have to show that $\mathfrak{G}_2$ is valid in every model for $\mathfrak{G}_1$. Let $(\overrightarrow{\mathbb{K}}, \lambda)$ with $\overrightarrow{\mathbb{K}} := (G_k, M_k, I_k)_{k=0,1,\ldots n}$ be a model for $\mathfrak{G}_1$. In order to show that $\mathfrak{G}_2$ is valid in $(\overrightarrow{\mathbb{K}}, \lambda)$ we have to check the two conditions of Definition 6.

We restrict ourselves to the verification of the vertex condition, the edge condition follows with similar arguments. Let $v \in V_2$ and $g \in \rho_2^+(v)$. Since $\mathfrak{G}_2$ is valid in $(\vec{\mathbb{K}}^{\mathfrak{G}_2}, \lambda^{\mathfrak{G}_2})$ and $\vec{\mathbb{K}}^{\mathfrak{G}_1}$ is an extension of $\vec{\mathbb{K}}^{\mathfrak{G}_2}$, we have that $(g, \kappa_2(v)) \in I_0^{\mathfrak{G}_1}$. This gives us that $g \in \rho_1^+(w)$ for some $w \in V_1$ with $\kappa_1(w) \leq_S \kappa_2(v)$. Since $(\vec{\mathbb{K}}, \lambda)$ is a model for $\mathfrak{G}_1$ we have $\lambda_{\mathcal{G}}(g) \in \mathrm{Ext}(\lambda_S(\kappa_1(w)))$ and, since λ_S is order preserving, we have $\lambda_S(\kappa_1(w)) \sqsubseteq \lambda_S(\kappa_2(v))$ (where "$\sqsubseteq$" is the order on $\underline{\mathfrak{H}}_{\sqcap}(\mathbb{K}_0)$). Therefore $\lambda_{\mathcal{G}}(g) \in \mathrm{Ext}(\lambda_S(\kappa_2(v)))$.

Let $v \in V_2$ and $g \in \rho_2^-(v)$. Since $\mathfrak{G}_2$ is valid in $(\vec{\mathbb{K}}^{\mathfrak{G}_2}, \lambda^{\mathfrak{G}_2})$ and $\vec{\mathbb{K}}^{\mathfrak{G}_1}$ is an extension of $\vec{\mathbb{K}}^{\mathfrak{G}_2}$, we have that $(g, \neg\kappa_2(v)) \in I_0^{\mathfrak{G}_1}$. Thus there exists a $w \in V_1$ with $g \in \rho_1^-(w)$ and $\kappa_2(v) \leq_S \kappa_1(w)$. Because $\mathfrak{G}_1$ is valid in $(\vec{\mathbb{K}}, \lambda)$ we have $\lambda_{\mathcal{G}}(g) \in G_0 \backslash \mathrm{Ext}(\lambda_S(\kappa_1(w)))$. Since λ_S is order preserving we get $\mathrm{Ext}(\lambda_S(\kappa_2(v))) \subseteq \mathrm{Ext}(\lambda_S(\kappa_1(w)))$, which implies $G_0 \backslash \mathrm{Ext}(\lambda_S(\kappa_1(w))) \subseteq G_0 \backslash \mathrm{Ext}(\lambda_S(\kappa_2(v)))$ and therefore $g \in G_0 \backslash \mathrm{Ext}(\lambda_S(\kappa_2(v)))$. $\qquad\square$

Proof of Proposition 10: We show that the standard power context family of $\mathfrak{G}'$ is contained in the standard power context family of $\mathfrak{G}(\vec{\mathbb{K}})$. Again, we restrict ourselves to the verification of $I_0^{\mathfrak{G}'} \subseteq I_0^{\mathfrak{G}(\vec{\mathbb{K}})}$, the arguments given below can be adapted for the case $k \geq 1$.

We have $H_0^{\mathfrak{G}(\vec{\mathbb{K}})} = G_0$ and $M_0^{\mathfrak{G}(\vec{\mathbb{K}})} = \underline{\mathfrak{H}}_{\sqcap}(\mathbb{K}_0) \,\dot\cup\, \{\neg m \mid m \in \underline{\mathfrak{H}}_{\sqcap}(\mathbb{K}_0)\}$. Now we look at the definition of $I_0^{\mathfrak{G}(\vec{\mathbb{K}})}$: Let $g \in G_0$. If $s \in \underline{\mathfrak{H}}_{\sqcap}(\mathbb{K}_0)$ we have $(g, s) \in I_0^{\mathfrak{G}(\vec{\mathbb{K}})} :\iff \exists v \in V_{\vec{\mathbb{K}}} : \kappa_{\vec{\mathbb{K}}}(v) \leq_S s$ and $g \in \rho_{\vec{\mathbb{K}}}^+(v)$. By the definition of $V_{\vec{\mathbb{K}}}$ and $\rho_{\vec{\mathbb{K}}}^+$, this is equivalent to $g \in \mathrm{Ext}(s)$. If $\neg t \in \{\neg m \mid m \in \underline{\mathfrak{H}}_{\sqcap}(\mathbb{K}_0)\}$ we have $(g, \neg t) \in I_0^{\mathfrak{G}(\vec{\mathbb{K}})} :\iff \exists v \in V_{\vec{\mathbb{K}}} : t \leq_S \kappa_{\vec{\mathbb{K}}}(v)$ and $g \in \rho_{\vec{\mathbb{K}}}^-(v)$. However, by the definition of $V_{\vec{\mathbb{K}}}$ and $\rho_{\vec{\mathbb{K}}}^-$ this is equivalent to $g \in G_0 \backslash \mathrm{Ext}(s)$. Since $\mathfrak{G}'$ is valid in $(\vec{\mathbb{K}}, id)$ we have that $(g, s) \in I_0^{\mathfrak{G}'}$ implies $g \in \mathrm{Ext}(s)$, thus $(g, s) \in I_0^{\mathfrak{G}(\vec{\mathbb{K}})}$. Also $(g, \neg t) \in I_0^{\mathfrak{G}'}$ implies $g \in G_0 \backslash \mathrm{Ext}(t)$, thus $(g, \neg t) \in I_0^{\mathfrak{G}(\vec{\mathbb{K}})}$.

$\qquad\square$

References

[Am91]   Hermann P. T. Ammon, Arzneimittelneben- und wechselwirkungen. Ein Handbuch und Tabellenwerk für Ärzte und Apotheker, Wissensch. VG. 1991.

[Da00]   F. Dau: Negations in Simple Concept Graphs. In: B. Ganter, G.W. Mineau (Eds.): Conceptual Structures: Logical, Linguistic, and Computational Issues, Springer Verlag, Berlin-New York 2000, 263-276.

[EGSW00]   P. Eklund, B. Groh, G. Stumme, R. Wille: A Contextual-Logic Extension of TOSCANA. In: B. Ganter, G.W. Mineau (Eds.): Conceptual Structures: Logical, Linguistic, and Computational Issues, Springer Verlag, Berlin-New York 2000, 453-467.

[GW99]   B. Ganter, R. Wille: Formal Concept Analysis: Mathematical Foundations. Springer, Berlin-Heidelberg-New York 1999.

[LW91]   P. Luksch, R. Wille: A mathematical model for conceptual knowledge systems. In: H.-H. Bock, P. Ihm (Eds.): Classification, data analysis and knowledge organisation, Springer, Heidelberg 1991, 156-162.

[Pr98a]   S. Prediger: Kontextuelle Urteilslogik mit Begriffsgraphen, Shaker Verlag, Aachen 1998.

[Pr98b]   S. Prediger: Simple Concept Graphs: A Logic Approach. In: M,-L Mugnier, M. Chein (Eds): Conceptuel Structures: Theory, Tools and Applications, Springer Verlag, Berlin-New York 1998, 225-239.

[PW99]   S. Prediger, R. Wille: The lattice of concept graphs of a relationally scaled context. In: W. Tepfenhart, W. Cyre (Eds.): Conceptual Structures: Standards and Practices, Springer, Berlin Heidelberg New York 1999, 401-414.

[Pr00]   S. Prediger: Nested Concept Graphs and Triadic Power Context Families: A Situation-Based Contextual Approach. In: B. Ganter, G.W. Mineau (Eds.): Conceptual Structures: Logical, Linguistic, and Computational Issues, Springer Verlag, Berlin-New York 2000, 249-262.

[Wi97]   R. Wille: Conceptual Graphs and Formal Concept Analysis. In: D. Lukose et. al. (Eds.): Conceptual Structures: Fulfilling Peirce's Dream, Springer Verlag, Berlin-New York 1997, 290-303.

[Wi00a]   R. Wille: Boolean Concept Logic. In: B. Ganter, G.W. Mineau (Eds.): Conceptual Structures: Logical, Linguistic, and Computational Issues, Springer Verlag, Berlin-New York 2000, 317-331.

[Wi00b]   R. Wille: Boolean Judgment Logic. This Volume.

Boolean Judgment Logic

Rudolf Wille

Technische Universität Darmstadt, Fachbereich Mathematik,
Schloßgartenstr. 7, D–64289 Darmstadt; wille@mathematik.tu-darmstadt.de

Abstract. How to introduce negations for formal (semi-)concepts of formal contexts was shown in *"Boolean Concept Logic"* [Wi00a]. For formal judgments, which are represented by concept graphs of power context families, there is the problem that the negation of a judgment is no more a judgment since judgments are understood as valid propositions. This problem is solved in *"Boolean Judgment Logic"* by introducing the *"negating inversion"*. This leads to basic algebras of (semi-)concept graphs of power context families which are investigated in this paper to obtain a mathematical foundation of Boolean Judgment Logic. The basic notions and relationships are illustrated by an example concerned with an information system for supporting the configuration of PCs.

Contents

1 Contextual Judgment Logic

Boolean Judgment Logic is seen as part of *"Contextual Logic"* which, in general, is developed as a mathematization of the traditional philosophical logic with its doctrines of concepts, judgments, and conclusions (see [Wi96], [Wi00b]); in particular, Boolean Judgment Logic is considered to be basic to the more general *"Contextual Judgment Logic"* which is aiming at the clarification of forms and relations of judgments in given contexts, where judgments are understood as valid propositions. Since judgments are philosophically conceived as combinations of concepts, Contextual Judgment Logic has to be built on *"Contextual Concept Logic"* which itself is grounded on the mathematization of context and concept elaborated in *Formal Concept Analysis* (see [Wi82], [GW99]). For this paper, we presuppose the basic notions and results of Formal Concept Analysis and refer the reader to [GW99].

The development of Boolean Judgment Logic particularily continues the work on *"Boolean Concept Logic"* [Wi00a] on the judgment level. For this development, ideas of the *Theory of Conceptual Graphs* [So84] and their mathematizations used in Contextual Judgment Logic (see [Wi97], [Pr98]) are taken over to

H. Delugach and G. Stumme (Eds.): ICCS 2001, LNAI 2120, pp. 115–128, 2001.

reach a basic mathematical theory of formal judgments which are modelled as concept graphs of power context families. New, with respect to the developed Contextual Judgment Logic, is the integration of formal *negations* founded on those used in Boolean Concept Logic.

Let us briefly recall the discussion in [Wi00a] about the problem of introducing formal negations on the concept lattice of a formal context (G, M, I). For obtaining the negation of a formal concept (A, B) of (G, M, I), it would be natural, according to the Boolean Class Logic, to form the complement $G \setminus A$ of the extent A, but $G \setminus A$ need not to be an extent again; the smallest formal concept containing $G \setminus A$ in its extent is $((G \setminus A)^{II}, (G \setminus A)^I)$. Since the intersection of A and $(G \setminus A)^{II}$ might be not empty, the assignment of (A, B) to $((G \setminus A)^{II}, (G \setminus A)^I)$ is called the *weak negation* on the concept lattice of (G, M, I). For keeping the correspondence between negation and set-complement, the only possible way is to generalize the notion of formal concepts. To obtain a powerful generalization of Boolean Algebra for mathematically treating Contextual Concept Logic, the best generalization seems to be the notion of protoconcept: (A, B) is said to be a *protoconcept* of (G, M, I) if $A \subseteq G$, $B \subseteq M$, and $A^{II} = B^I$ (which is equivalent to $A^I = B^{II}$). The *negation* of a protoconcept (A, B) of (G, M, I) is defined as the protoconcept $\neg(A, B) := (G \setminus A, (G \setminus A)^I)$.

In applications (for example see Fig. 5), protoconcepts which are not formal concepts often occur only as negated formal concepts or as meets of those. Such protoconcepts are even $\sqcap$-semiconcepts where a $\sqcap$-*semiconcept* has the form (A, A^I). In this paper, we therefore restrict our considerations to the negation in the Boolean algebras $\mathfrak{H}_\sqcap(\mathbb{K}) := (\mathfrak{H}_\sqcap(\mathbb{K}), \sqcap, \sqcup, \neg, \bot, \overline{\top})$ of the $\sqcap$-semiconcepts of formal contexts $\mathbb{K} := (G, M, I)$. As above, the negation on the set $\mathfrak{H}_\sqcap(\mathbb{K}) := \{(A, A^I) \mid A \subseteq G\}$ of all $\sqcap$-*semiconcepts* of $\mathbb{K}$ is given by $\neg(A, A^I) := (G \setminus A, (G \setminus A)^I)$; the other operations are defined by $(A, A^I) \sqcap (B, B^I) := (A \cap B, (A \cap B)^I)$, $(A, A^I) \sqcup (B, B^I) := (A \cup B, (A \cup B)^I)$, $\bot := (\emptyset, M)$, $\overline{\top} := (G, G^I)$, and the order relation $\sqsubseteq$ by

$$(A, A^I) \sqsubseteq (B, B^I) : \iff A \subseteq B \text{ (and } A^I \supseteq B^I).$$

Restricting to $\sqcap$-semiconcepts yields the advantage that, algebraically, we have only to deal with Boolean algebras of $\sqcap$-semiconcepts instead of the more complex double Boolean algebras of protoconcepts. For a fully developed Boolean Judgment Logic, the (weak) negations and oppositions, introduced in [Wi00a] for formal concepts and protoconcepts, respectively, should also be considered on the judgment level, but this needs further research.

2   Semiconcept Graphs Explained by an Example

How *(semi-)concept graphs* with negation may be introduced shall first be explained by an example. The example is taken from a cooperation with IBM Germany which was performed in the mid '80th (see [Wi89]). The task was to develop a prototype of an information system for supporting the *configuration of PCs* by IBM-components such as CPUs, screens, etc. Fig. 1 and 2 show two data

	F1	F2	F20S	F40S	F60S	S640	S2.63	S8.6	S10.5	D2	D720S	D2.4S	M8087	M80287	T4	T6	MIT	S	P	ZB	3270	370	BSC	SDL	PL
5150	X	X	X			X				X	X		X		X			X	X		X		X	X	X
5160	X	X	X			X				X	X		X			X		X	X		X	X	X	X	X
5162	X		X				X	X		X	X	X	X	X		X		X	X		X	X	X	X	X
5270	X	X	X	X	X	X	X	X	X	X	X	X	X	X		X		X	X		X	X	X	X	X
5271	X	X	X			X				X	X		X	X	X		X	X	X		X	X	X	X	X
5273	X	X	X	X		X				X	X	X	X	X			X	X	X		X	X	X	X	X
5371	X	X	X			X	X			X	X	X	X	X			X	X	X		X	X	X	X	X
5373	X	X	X	X		X	X			X	X	X	X	X			X	X	X						X

Fig. 1. Data context of CPUs

	D12	D13	D14	D19	640x200	640x340	640x480	720x512	960x1000	F2	F8	F16	F256	25x80	32x80	43x80	50x80
5151	X				X	X								X	X	X	
5272	X	X	X		X					X	X	X		X			
5279	X	X	X		X	X	X	X		X	X			X	X		
5379	X	X	X	X	X	X	X	X	X	X	X			X	X	X	X
5153	X	X			X					X	X	X		X			
4861	X	X			X	X				X	X	X		X	X	X	
5175	X	X			X	X	X			X	X	X	X				

Fig. 2. Data context of screens

contexts extracted from the full data about IBM-components using the original denotations of that time. The context in Fig. 1 has as *objects* the CPUs 5150, 5160, 5162, 5270, 5271, 5273, 5371, 5373 and as *attributes* "hard disk": F1, F2, "hard disk memory": F20S, F40S, F60S, "RAM": S640, S2.63, S8.6, S10.5, "disk drive": D2, "disk memory": D720S, D2.4S, "Co-processors": M8087, M80287, "cycle time": T4, T6, and further characteristics: MIT, S, P, ZB, 3270, 370, BSC, SDL, PL; the context in Fig. 2 has as *objects* the screens 5151, 5272, 5279, 5379, 5153, 4861, 5175 and *as attributes* "viewable size": D12, D13, D14, D19, "screen resolution": 640×200, 640×340, 640×480, 720×512, 960×1000, "colour depth": F2, F8, F16, F256, and "text mode": 25×80, 32×80, 43×80, 50×80. For each component context, the corresponding concept lattice was determined and stored within the developed information system; the concept lattices of the contexts in Fig. 1 and 2 are represented in Fig. 3 and 4, respectively.

The dependencies between the objects of different contexts concerning the fit between specific components can be expressed by *"semiconcept graphs"* as shown in Fig. 5 for the objects of the CPU- and the screen-context. Notice that, for instance, the negation of the attribute concept $\mu(ZB)$ in the first graph is not a formal concept, but a (⊓-)semiconcept of the CPU-context. Each of the five presented semiconcept graphs consists of two *"semiconcept instances"* joined by the binary relation "fit". A semiconcept instance is composed by a semiconcept and two object sets (separated by |) the first of which is contained in the extent of the semiconcept and the second in its complement; the relation "fit" combining two semiconcept instances indicates that every two objects from the first sets, respectively, are fitting in a PC-configuration and that every two objects from the second sets, respectively, do not fit in a

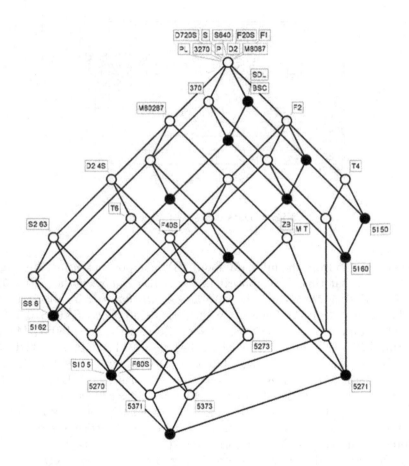

Fig. 3. The concept lattice of the CPU-context in Fig. 1

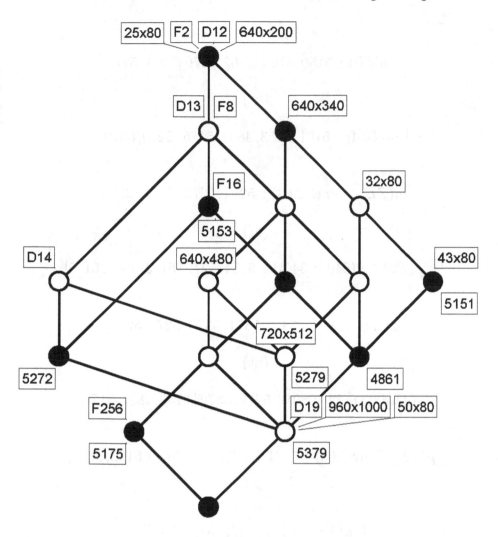

Fig. 4. The concept lattice of the screen-context in Fig. 2

PC-configuration. For the first semiconcept graph in Fig. 5, this means that
$\{5150, 5160, 5162, 5270\} \times \{5151, 5153, 4861, 5175\}$ is contained in the extent of
the relation "fit" and that $\{5271, 5273\} \times \{5279, 5379\}$ is contained in the complement of that extent. The last semiconcept graph in Fig. 5 indicates that one
of the object sets of a semiconcept instance might be empty, but at least one
must be not empty; also for the combining relation there must always be at least
one pair of objects in the extent of the relation or in its complement.

The five semiconcept graphs in Fig. 5 completely determine the extent of
the relation "fit" which can be easily checked by confirming that, according to
the five semiconcept graphs, each cell with a cross in the binary table of Fig. 6

$$[\neg\mu(ZB) : 5150, 5160, 5162, 5270 \,|\, 5271, 5273]$$

$$|$$

$$(\text{fit})$$

$$|$$

$$[\neg\mu(D14) : 5151, 5153, 4861, 5175 \,|\, 5279, 5379]$$

$$[\mu(ZB) \sqcap \neg\mu(S2.63) : 5271, 5273 \,|\, 5371, 5373]$$

$$|$$

$$(\text{fit})$$

$$|$$

$$[\neg(\mu(F8) \sqcap \mu(640 \times 340)) : 5151, 5272 \,|\, 5153, 4861, 5175]$$

$$[\mu(ZB) \sqcap \mu(S2.63) : 5371, 5373 \,|\, 5271, 5273]$$

$$|$$

$$(\text{fit})$$

$$|$$

$$[\mu(D14) \sqcap \mu(32 \times 80) : 5279, 5379 \,|\, 5153, 4861, 5175]$$

$$[\mu(ZB) \sqcap \mu(S2.63) : 5371, 5373 \,|\, 5150, 5160, 5162, 5270]$$

$$|$$

$$(\text{fit})$$

$$|$$

$$[\neg\mu(F8) : 5151 \,|\, 5272, 5279, 5379]$$

$$[\neg(\mu(ZB) \sqcap \mu(S2.63)) : \,|\, 5371, 5373]$$

$$|$$

$$(\text{fit})$$

$$|$$

$$[\neg\mu(D14) : \,|\, 5272]$$

Fig. 5. Semiconcept graphs describing the fit-dependencies between CPUs and screens

represents an object pair of the relation and each empty cell an object pair not in relation; moreover, each pair occurs exactly once in the semiconcept graphs

	5150	5160	5162	5270	5271	5273	5371	5373
5151	X	X	X	X	X	X	X	X
5272					X	X		
5279							X	X
5379							X	X
5153	X	X	X	X				
4861	X	X	X	X				
5175	X	X	X	X				

Fig. 6. Binary table representing the extent of the relation "fit"

of Fig. 5. The $\sqcap$—semiconcepts in those graphs yield further information; for instance, they indicate that the relation "fit" causes partitions on the CPUs induced by $\neg\mu(ZB)$, $\mu(ZB) \sqcap \mu(S2.63)$, $\neg\mu(ZB) \sqcap \mu(S2.63)$ and on the screens by $\neg(\mu(F8) \sqcap \mu(640 \times 340))$, $\mu(D14) \sqcap \mu(32 \times 80)$, $\neg\mu(D14) \sqcap \mu(F8)$. Thus, semiconcept graphs may represent formal judgments which yield information in a condensed manner.

The developed prototype of the information system was formed by the concept lattices for different types of PC-components which were pairwise linked by the fitting relation. A *process of establishing a PC-configuration* could then be performed stepwise by always using the knowledge about the consequences of already taken decisions. For instance, in Fig. 3, the consequences of the decision for an SDL-CPU are indicated by the blackened circles which represent all concepts having SDL in their intent; in particular, the decision restricts the choice to the CPUs 5150, 5160, 5162, 5270, and 5271. The SDL-decision effects the choice of a screen via the fitting relation so that only the concepts represented by the blackened circles in Fig. 4 may lead to a functioning configuration. This becomes clear by the first and fourth semiconcept graph showing that the screens 5279 and 5379 do not fit with any SDL-CPU (the other screens fit with at least one of the SDL-CPUs). Now, if one wants at least 8 colours (attribute F8) then only one of the screens 5272, 5153, 4861, and 5175 will fit. Any further decision may have consequences on other components which are then similarly represented by blackened circles in the line diagrams of the corresponding concept lattices.

The example stands for a type of information systems established on concept lattices for different types of components whose contextual-logic dependencies are expressible by semiconcept graphs; of course, there might be quite different relations even of higher arities and not only one binary relation. Therefore a general theory of semiconcept graphs with negation is desirable which will be developed as an extension of the existing mathematical theory of concept graphs in Contextual Judgment Logic.

3   Algebras of Semiconcept Graphs

In general, the contextual semantics of the developed Contextual Judgment Logic is based on power context families and their concept graphs representing formal judgments. A *power context family* is a sequence

$$\vec{\mathbb{K}} := (\mathbb{K}_0, \mathbb{K}_1, \ldots, \mathbb{K}_k, \ldots)$$

of formal contexts $\mathbb{K}_k := (G_k, M_k, I_k)$ with $G_k \subseteq (G_0)^k$ for $k = 0, 1, \ldots$; $\vec{\mathbb{K}}$ is said to be *limited of type* $n \in \mathbb{N}$ if $\vec{\mathbb{K}} := (\mathbb{K}_0, \mathbb{K}_1, \ldots, \mathbb{K}_n)$, otherwise it is called *unlimited*. For obtaining conformity with the *"Contextual Logic of Relations"* (see [Bu91], [Wi00c]), we extend each context $\mathbb{K}_k$ ($k = 1, 2, \ldots$) to the relational context $\dot{\mathbb{K}}_k := ((G_0)^k, M_k, I_k)$; for unifying notation, we set $\dot{\mathbb{K}}_0 := \mathbb{K}_0$. Recall that, in general, the extent A of a $\sqcap$–semiconcept $\mathfrak{b} := (A, A^I)$ is denoted by $Ext(\mathfrak{b})$.

A *relational graph* is a set structure (V, E, ν) consisting of two sets V and E and a mapping $\nu : E \to \bigcup_{k=1,2,\ldots} V^k$. The elements of V and E are called *vertices* and *edges*, respectively, and $\nu(e) = (v_1, \ldots, v_k)$ is read: $v_1, \ldots, v_k$ are the *adjacent vertices* of the *k-ary edge* e; $|e| := k$ is the *arity* of e, and $|v| := 0$ is the arity of any vertex v. Let $E^{(k)} := \{u \in V \cup E \mid |u| = k\}$ ($k = 0, 1, \ldots$); in particular, this means $E^{(0)} = V$. A relational graph (V, E, ν) is said to be *limited of type* $n \in \mathbb{N}$ if $E = E^{(1)} \cup \cdots \cup E^{(n)}$, otherwise it is called *unlimited*.

As the example in Section 2 indicates, it is desirable to generalize the definition of a concept graph of a power context family to that of a semiconcept graph (concept graphs are then special cases of semiconcept graphs). A *semiconcept graph* of a power context family $\vec{\mathbb{K}} := (\mathbb{K}_0, \ldots, \mathbb{K}_k, \ldots)$ with $\mathbb{K}_k := (G_k, M_k, I_k)$ for $k = 0, 1, \ldots$ is a set structure $\mathfrak{G} := (V, E, \nu, \kappa, \rho)$ for which (V, E, ν) is a relational graph and

- $\kappa : V \cup E \to \bigcup_{k=0,1,\ldots} \mathfrak{H}_\sqcap(\dot{\mathbb{K}}_k)$ is a mapping with $\kappa(u) \in \mathfrak{H}_\sqcap(\dot{\mathbb{K}}_k)$ for all $u \in E^{(k)}$ ($k = 0, 1, \ldots$),
- $\rho : V \to \mathfrak{P}(G_0) \setminus \{\emptyset\}$ is a mapping with $\rho^+(v) := \rho(v) \cap Ext(\kappa(v))$ and $\rho^-(v) := \rho(v) \setminus \rho^+(v)$ such that, for $\nu(e) = (v_1, \ldots, v_k)$, $\rho^+(v_j) \neq \emptyset$ for all $j = 1, \ldots, k$ or $\rho^-(v_j) \neq \emptyset$ for all $j = 1, \ldots, k$, and $\rho^+(v_1) \times \cdots \times \rho^+(v_k) \subseteq Ext(\kappa(e))$ and $\rho^-(v_1) \times \cdots \times \rho^-(v_k) \in (G_0)^k \setminus Ext(\kappa(e))$.

For example, the first semiconcept graph represented in Fig. 5 is the set structure $\mathfrak{G} := (V, E, \nu, \kappa, \rho)$ with $V := \{v_1, v_2\}$, $E := \{e\}$, $\nu(e) := (v_1, v_2)$,

- $\kappa(v_1) := \neg\mu(ZB)$, $\kappa(v_2) := \neg\mu(D14)$, and $\kappa(e) := \mathit{fit}$,
- $\rho^+(v_1) := \{5150, 5160, 5162, 5270\}$, $\rho^-(v_1) := \{5271, 5273\}$, $\rho^+(v_2) := \{5151, 5153, 4861, 5175\}$, $\rho^-(v_2) := \{5279, 5379\}$;

$\mathfrak{G}$ is founded on the power context family $\vec{\mathbb{K}} := (\mathbb{K}_0, \mathbb{K}_2)$ where $\mathbb{K}_0 := (G_0, M_0, I_0)$ is the disjoint union of the formal contexts in Fig. 1 and 2, and $\mathbb{K}_2 := (G_2, M_2, I_2)$ is the "binary" context for which $G_2 := (G_0)^2$, $M_2 := \{fit\}$, and $(g, h)I_2(fit)$ if the components g and h are linked by a cross in Fig. 6 (in Fig. 5, (fit) stands for the attribute concept $\mu(fit)$).

For a semiconcept graph $\mathfrak{G} := (V, E, \nu, \kappa, \rho)$, the triples $[\kappa(v) : \rho^+(v)|\rho^-(v)]$ for $v \in V$ are called "semiconcept instances" of $\mathfrak{G}$. In general, a *semiconcept instance* of $\mathbb{K}_0$ is defined as a triple $[\mathfrak{b} : C|D]$ with $\mathfrak{b} \in \mathfrak{H}_\sqcap(\mathbb{K}_0)$, $C \subseteq Ext(\mathfrak{b})$, and $D \subseteq G_0 \setminus Ext(\mathfrak{b})$; they are understood as elementary semiconcept graphs representing distinctive judgments (see [Wi01]). The set of all semiconcept instances of $\mathbb{K}_0$ is denoted by $\mathfrak{H}_\sqcap^{inst}(\mathbb{K}_0)$. Since it is useful to consider the mapping ρ not only on vertices but also on edges, for $\nu(e) = (v_1, \ldots, v_k)$, we define $\rho(e) := \rho^+(e) \cup \rho^-(e)$ with $\rho^+(e) := \rho^+(v_1) \times \ldots \times \rho^+(v_k)$ and $\rho(e)^- := \rho^-(v_1) \times \ldots \times \rho^-(v_k)$. Now, we can also introduce "k-ary semiconcept instances" of the semiconcept graph $\mathfrak{G}$ by $[\kappa(e) : \rho^+(e)|\rho^-(e)]$ for $e \in E^{(k)}$. In general, a *k-ary semiconcept instance* of the relational context $\mathbb{K}_k$ ($k = 1, 2, \ldots$) is defined as a triple $[\mathfrak{b} : C_1 \times \cdots \times C_k | D_1, \times \cdots \times D_k]$ with $\mathfrak{b} \in \mathfrak{H}_\sqcap(\mathbb{K}_k)$, $C_1 \times \cdots \times C_k \subseteq Ext(\mathfrak{b})$, and $D_1, \times \cdots \times D_k \subseteq (G_0)^k \setminus Ext(\mathfrak{b})$. The set of all k-ary semiconcept instances of $\mathbb{K}_k$ is denoted by $\mathfrak{H}_\sqcap^{k-inst}(\mathbb{K}_k)$ for $k = 1, 2, \ldots$; in addition, we define $\mathfrak{H}_\sqcap^{0-inst}(\mathbb{K}_0) := \mathfrak{H}_\sqcap^{inst}(\mathbb{K}_0)$. On $\mathfrak{H}_\sqcap^{k-inst}(\mathbb{K}_k)$, the *(affirmative) generalization order* $\succeq$ is defined by

$$[\mathfrak{b}_1 : C_1|D_1] \succeq [\mathfrak{b}_2 : C_2|D_2] : \Longleftrightarrow \mathfrak{b}_1 \sqsupseteq \mathfrak{b}_2, \ C_1 \subseteq C_2, \ D_1 \subseteq D_2.$$

Negations are already involved in the semiconcept instances $[\mathfrak{b} : C|D]$ by the claim $d \notin Ext(\mathfrak{b})$ for all $d \in D$. But, since the negation of a judgment is no more a judgment, a negating operation on semiconcept graphs cannot correspond to the negation of propositions. Therefore, we propose a negating inversion which can be defined for semiconcept instances by $-[\mathfrak{b} : C|D] := [\neg\mathfrak{b} : D|C]$ because $[\neg\mathfrak{b} : D|C]$ is again a semiconcept instance. For semiconcept graphs, we introduce in general the operations *conjunction* ($\bigwedge$), *disjunction* ($\bigvee$), and *negating inversion* ($-$) as follows:

- $\bigwedge_{t \in T}(V_t, E_t, \nu_t, \kappa_t, \rho_t) := (V_\wedge, E_\wedge, \nu_\wedge, \kappa_\wedge, \rho_\wedge)$ where
  $V_\wedge := \bigcup_{t \in T} V_t$ and $E_\wedge := \bigcup_{t \in T} E_t$,
  $\nu_\wedge(e_t) := \nu_t(e_t)$ for $e_t \in E_t$,
  $\kappa_\wedge(u_t) := \kappa_t(u_t)$ for $u_t \in V_t \cup E_t$,
  $\rho_\wedge(v_t) := \rho_t(v_t)$ for $v_t \in V_t$;
  instead of $\bigwedge_{t \in \{1,2\}} \mathfrak{G}_t$, we usually write $\mathfrak{G}_1 \wedge \mathfrak{G}_2$;

- $\bigvee_{t \in T}(V_t, E_t, \nu_t, \kappa_t, \rho_t) := (V_\vee, E_\vee, \nu_\vee, \kappa_\vee, \rho_\vee)$ where
  $V_\vee := \{(v_t)_{t \in T} \in \prod_{t \in T} V_t \mid \bigcap_{t \in T} \rho_t^+(v_t) \neq \emptyset \text{ or } \bigcap_{t \in T} \rho_t^-(v_t) \neq \emptyset\}$,
  $E_\vee := \bigcup_{k = 1, 2, \ldots} E_\vee^{(k)}$ with
  $E_\vee^{(k)} := \{(e_t)_{t \in T} \in \prod_{t \in T} E_t^{(k)} \mid \bigcap_{t \in T} \rho_t^+(e_t) \neq \emptyset \text{ or } \bigcap_{t \in T} \rho_t^-(e_t) \neq \emptyset\}$,
  $\nu_\vee(e) := ((v_{t1})_{t \in T}, \ldots, (v_{tk})_{t \in T})$ for $e := (e_t)_{t \in T}$ and $\nu_t(e_t) = (v_{t1}, \ldots, v_{tk})$,
  $\kappa_\vee(v) := \bigsqcup_{t \in T} \kappa_t(v_t)$ for $v := (v_t)_{t \in T} \in V_\vee$,
  $\kappa_\vee(e) := \bigsqcup_{t \in T} \kappa_t(e_t)$ for $e := (e_t)_{t \in T} \in E_\vee^{(k)}$ and $k = 1, 2, \ldots,$

$\rho_V^+(v) := \bigcap_{t \in T} \rho_t^+(v_t)$ and $\rho_V^-(v) := \bigcap_{t \in T} \rho_t^-(v_t)$ for $v := (v_t)_{t \in T} \in V_V$; instead of $\bigvee_{t \in \{1,2\}} \mathfrak{G}_t$, we usually write $\mathfrak{G}_1 \vee \mathfrak{G}_2$;

- $-(V, E, \nu, \kappa, \rho) := (V, E, \nu, \bar{\kappa}, \bar{\rho})$ where
  $\bar{\kappa}(v) := \neg\kappa(v)$ for $v \in V$ and $\bar{\kappa}(e) := \neg\kappa(e)$ for $e \in E$,
  $\bar{\rho}(v) := \rho(v)$ for $v \in V$, i.e. $\bar{\rho}^+(v) = \rho^-(v)$ and $\bar{\rho}^-(v) = \rho^+(v)$;

As extremal constants one has the *empty graph* $\mathfrak{G}_\emptyset := (\emptyset, \emptyset, \emptyset, \emptyset, \emptyset)$ and the *universal graph*

- $\mathfrak{G}_u := (V_u, E_u, \nu_u, \kappa_u, \rho_u)$ where
  $V_u := \mathfrak{H}_\cap^{inst}(\mathbb{K}_0)$ and $E_u := \bigcup_{k=1,2,\dots} \mathfrak{H}_\cap^{k-inst}(\mathbb{K}_k)$,
  $\nu_u[\mathfrak{b} : C_1 \times \cdots \times C_k | D_1, \times \cdots \times D_k]$
  $:= ([(C_1, (C_1)^{I_0}) : C_1 | D_1], \dots, [(C_k, (C_k)^{I_0}) : C_k | D_k])$,
  $\kappa_u[\mathfrak{b} : C | D] := \mathfrak{b}$ for $[\mathfrak{b} : C | D] \in V_u \cup E_u$, and
  $\rho_u^+[\mathfrak{b} : C | D] := C$ and $\rho_u^-[\mathfrak{b} : C | D] := D$ for $[\mathfrak{b} : C | D] \in V_u$.

The assertion of the following lemma can be easily confirmed:

Lemma 1 *The conjunction, disjunction, and negating inversion of semiconcept graphs are again semiconcept graphs.*

Thus, the semiconcept graphs of a power context family $\vec{\mathbb{K}}$ form an algebra with the binary conjunction ($\wedge$), the binary disjunction ($\vee$) and the negating inversion (-) as operations and with $\mathfrak{G}_\emptyset$ and $\mathfrak{G}_u$ as constants (nullary operations); this algebra is denoted by

$$\mathfrak{H}_\cap^{graph}(\vec{\mathbb{K}}) := (\mathfrak{H}_\cap^{graph}(\vec{\mathbb{K}}), \wedge, \vee, -, \mathfrak{G}_\emptyset, \mathfrak{G}_u).$$

As direct consequence of the definitions we get the following proposition:

Proposition 1 *For any power context family $\vec{\mathbb{K}}$, the algebra $\mathfrak{H}_\cap^{graph}(\vec{\mathbb{K}})$ satisfies the following equations:*

1. $\mathfrak{G}_1 \wedge \mathfrak{G}_2 = \mathfrak{G}_2 \wedge \mathfrak{G}_1$
2. $\mathfrak{G}_1 \wedge (\mathfrak{G}_2 \wedge \mathfrak{G}_3) = (\mathfrak{G}_1 \wedge \mathfrak{G}_2) \wedge \mathfrak{G}_3$
3. $\mathfrak{G}_1 \vee (\mathfrak{G}_2 \wedge \mathfrak{G}_3) = (\mathfrak{G}_1 \vee \mathfrak{G}_2) \wedge (\mathfrak{G}_1 \vee \mathfrak{G}_3)$ *if* $E_2^k \cap E_3^k = \emptyset$ $(k = 0, 1, \dots)$
4. $(\mathfrak{G}_1 \wedge \mathfrak{G}_2) \vee \mathfrak{G}_3 = (\mathfrak{G}_1 \vee \mathfrak{G}_3) \wedge (\mathfrak{G}_2 \vee \mathfrak{G}_3)$ *if* $E_1^k \cap E_2^k = \emptyset$ $(k = 0, 1, \dots)$
5. $-(-\mathfrak{G}) = \mathfrak{G}$
6. $-(\mathfrak{G}_1 \wedge \mathfrak{G}_2) = (-\mathfrak{G}_1) \wedge (-\mathfrak{G}_2)$

To obtain the commutative and associative law also for the disjunction and the distributive laws without constraints, we have to identify isomorphic semiconcept graphs: Two semiconcept graphs $\mathfrak{G}_1 := (V_1, E_1, \nu_1, \kappa_1, \rho_1)$ and $\mathfrak{G}_2 := (V_2, E_2, \nu_2, \kappa_2, \rho_2)$ of $\vec{\mathbb{K}}$ are said to be *isomorphic* (in symbols: $\mathfrak{G}_1 \cong \mathfrak{G}_2$) if there exist bijective maps $\alpha : V_1 \to V_2$ and $\beta_k : E_1^{(k)} \to E_2^{(k)}$ $(k = 1, 2, \dots)$ with $\nu_1(e) = (v_1, \dots, v_k) \Leftrightarrow \nu_2(\beta_k(e)) = (\alpha(v_1), \dots, \alpha(v_k))$, $\kappa_1(v) = \kappa_2(\alpha(v))$, $\kappa_1(e) = \kappa_2(\beta_k(e))$ if $|e| = k$, and $\rho_1(v) = \rho_2(\alpha(v))$. Obviously, $\cong$ is a congruence relation on $\mathfrak{H}_\cap^{graph}(\vec{\mathbb{K}})$ so that we obtain the quotient algebra

$$\underline{\underline{\mathfrak{H}}}_\cap^{graph}(\vec{\mathbb{K}}) := (\mathfrak{H}_\cap^{graph}(\vec{\mathbb{K}})/\cong, \wedge, \vee, -, \underline{\mathfrak{G}}_\emptyset, \underline{\mathfrak{G}}_u),$$

the elements of which are the congruence classes $\underline{\mathfrak{G}}_0 := \{\mathfrak{G} \in \mathfrak{H}_\sqcap^{graph}(\mathbb{K}) \mid \mathfrak{G}_0 \cong \mathfrak{G}\}$ represented by arbitrary semiconcept graphs $\mathfrak{G}_0$ of $\vec{\mathbb{K}}$. The next proposition can again be easily proved:

Proposition 2 *For any power context family $\vec{\mathbb{K}}$, the algebra $\underline{\underline{\mathfrak{H}}}_\sqcap^{graph}(\mathbb{K})$ is a commutative semiring, i.e., the operations $\wedge$ and $\vee$ are commutative, associative, and distributive (as in 3. and 4. of Proposition 1 without constraints); furthermore, $\underline{\underline{\mathfrak{H}}}_\sqcap^{graph}(\mathbb{K})$ also satisfies the equations 5. and 6. of Proposition 1.*

4   Semi-conceptual Content and Generalization Order

For defining the generalization order between semiconcept graphs, we first grasp the conceptual information represented in semiconcept graphs by the notion of their semi-conceptual content: For a semiconcept graph $\mathfrak{G} := (V, E, \nu, \kappa, \rho)$ of the power context family $\vec{\mathbb{K}}$ we define the *semi-conceptual content* $C(\mathfrak{G}) := (C_0(\mathfrak{G}), \ldots, C_k(\mathfrak{G}), \ldots)$ of $\mathfrak{G}$ by $C_k(\mathfrak{G}) :=$

$$\{(\bar{g}, \mathfrak{b}) \in (G_0)^k \times \mathfrak{H}_\sqcap(\dot{\mathbb{K}}_k) \mid \text{ there are } \mathfrak{c}_t \in \mathfrak{H}_\sqcap(\dot{\mathbb{K}}_k)\, (t \in T) \text{ with } \mathfrak{b} \sqsupseteq \sqcap_{t \in T} \mathfrak{c}_t$$
$$\text{and } \forall t \in T \exists u_t \in E^{(k)} : \mathfrak{c}_t = \kappa(u_t), \bar{g} \in \rho^+(u_t) \text{ or } \mathfrak{c}_t = \neg\kappa(u_t), \bar{g} \in \rho^-(u_t)\}$$

for $k = 0, 1, \ldots$; in particular, $C_k(\mathfrak{G})$ is called the *semi-conceptual k-content* of $\mathfrak{G}$. The (affirmative) generalization order between semiconcept instances can now be extended to a *generalization order* between semiconcept graphs by defining

$$\mathfrak{G}_1 \lesssim \mathfrak{G}_2 :\Leftrightarrow C_k(\mathfrak{G}_1) \supseteq C_k(\mathfrak{G}_2) \text{ for } k = 0, 1, \ldots;$$

$\mathfrak{G}_1$ is then said to be *less general* (*more specific*) than $\mathfrak{G}_2$. Semiconcept graphs $\mathfrak{G}_1$ and $\mathfrak{G}_2$ of $\vec{\mathbb{K}}$ are called *equivalent* (in symbols: $\mathfrak{G}_1 \sim \mathfrak{G}_2$) if $\mathfrak{G}_1 \lesssim \mathfrak{G}_2$ and $\mathfrak{G}_2 \lesssim \mathfrak{G}_1$, i.e. $C(\mathfrak{G}_1) = C(\mathfrak{G}_2)$. The equivalence class containing the semiconcept graph $\mathfrak{G}$ is denoted by $[\mathfrak{G}]_\sim$ or shortly by $[\mathfrak{G}]$. The set $\widetilde{\mathfrak{H}}_\sqcap^{graph}(\mathbb{K})$ of all equivalence classes of semiconcept graphs of $\vec{\mathbb{K}}$ together with the order $\leq$ induced by the quasi-order $\lesssim$ is an ordered set denoted by $\underline{\widetilde{\mathfrak{H}}}_\sqcap^{graph}(\mathbb{K})$. Notice that we have been able to extend the generalization order to semiconcept graphs even with negations.

Lemma 2 *For semiconcept graphs $\mathfrak{G}$ and $\mathfrak{G}_t := (V_t, E_t, \nu_t, \kappa_t, \rho_t)\, (t \in T)$ of a power context family $\vec{\mathbb{K}}$, the following equalities hold for $k = 0, 1, \ldots$:*

$$C_k(\bigwedge_{t \in T} \mathfrak{G}_t) \supseteq \bigcup_{t \in T} C_k(\mathfrak{G}_t), \quad C_k(\bigvee_{t \in T} \mathfrak{G}_t) = \bigcap_{t \in T} C_k(\mathfrak{G}_t), \quad and \quad C_k(-\mathfrak{G}) = C_k(\mathfrak{G}).$$

Proof: For $k = 0, 1, \ldots$ we can conclude

$$C_k(\bigwedge_{t \in T} \mathfrak{G}_t) \supseteq \{(\bar{g}, \mathfrak{b}) \mid \mathfrak{b} \sqsupseteq \mathfrak{c}_t \text{ with } (\bar{g}, \mathfrak{c}_t) \in C_k(\mathfrak{G}_t) \text{ for some } t \in T\}$$
$$= \bigcup_{t \in T} C_k(\mathfrak{G}_t).$$
$$C_k(\bigvee_{t \in T} \mathfrak{G}_t) = \{(\bar{g}, \mathfrak{b}) \mid \mathfrak{b} \sqsupseteq \bigsqcup_{t \in T} \mathfrak{c}_t \text{ with } (\bar{g}, \mathfrak{c}_t) \in C_k(\mathfrak{G}_t) \text{ for } t \in T\}$$
$$= \bigcap_{t \in T} \{(\bar{g}, \mathfrak{b}) \mid \mathfrak{b} \sqsupseteq \mathfrak{c}_t \text{ with } (\bar{g}, \mathfrak{c}_t) \in C_k(\mathfrak{G}_t)\}$$
$$= \bigcap_{t \in T} C_k(\mathfrak{G}_t).$$

The third equality follows directly from the definitions.   □

Proposition 3 *For any power context family $\vec{\mathbb{K}}$, the ordered set $\widetilde{\mathfrak{H}}_{\sqcap}^{graph}(\vec{\mathbb{K}})$ is a completely distributive complete lattice having $[\mathfrak{G}_\emptyset]$ as greatest element and $[\mathfrak{G}_u]$ as smallest element; in particular,*

$$\bigwedge_{t\in T}[\mathfrak{G}_t] = [\bigwedge_{t\in T}\mathfrak{G}_t] \quad and \quad \bigvee_{t\in T}[\mathfrak{G}_t] = [\bigvee_{t\in T}\mathfrak{G}_t]$$

for arbitrary semiconcept graphs $\mathfrak{G}_t$ of $\vec{\mathbb{K}}$ $(t \in T)$.

Proof: As defined above, $\mathfrak{G}_1 \gtrsim \mathfrak{G}_2$ is equivalent to $C_k(\mathfrak{G}_1) \subseteq C_k(\mathfrak{G}_2)$ for $k = 0, 1, \ldots$, and $\mathfrak{G}_1 \sim \mathfrak{G}_2$ is equivalent to $C(\mathfrak{G}_1) = C(\mathfrak{G}_2)$; furthermore, Lemma 2 yields that, for each k=0,1,..., the sets of semi-conceptual k-contents are closed under intersection in the power set of $\{(\bar{g}, \mathfrak{b}) \mid \bar{g} \in \mathfrak{b} \in \mathfrak{H}_{\sqcap}(\mathbb{K}_k)\}$ and so the asserted equality $\bigvee_{t\in T}[\mathfrak{G}_t] = [\bigvee_{t\in T}\mathfrak{G}_t]$ holds. Hence, arbitrary infima and suprema exist in the ordered set $\widetilde{\mathfrak{H}}_{\sqcap}^{graph}(\vec{\mathbb{K}})$, i.e., $\widetilde{\mathfrak{H}}_{\sqcap}^{graph}(\vec{\mathbb{K}})$ is a complete lattice. Because of $C_k(\mathfrak{G}_\emptyset) = \emptyset$ and $C_k(\mathfrak{G}_u) = \{(\bar{g}, \mathfrak{b}) \mid \mathfrak{b} \in \mathfrak{H}_{\sqcap}(\dot{\mathbb{K}}_k)$ and $\bar{g} \in Ext(\mathfrak{b})\}$ for $k = 0, 1, \ldots$, $[\mathfrak{G}_\emptyset]$ is the greatest element and $[\mathfrak{G}_u]$ is the smallest element of $\widetilde{\mathfrak{H}}_{\sqcap}^{graph}(\vec{\mathbb{K}})$. As $C_k(\bigwedge_{t\in T}\mathfrak{G}_t) \supseteq \bigcup_{t\in T}C_k(\mathfrak{G}_t)$ for $k = 0, 1, \ldots$ by Lemma 2, we have $[\bigwedge_{t\in T}\mathfrak{G}_t] \le \bigwedge_{t\in T}[\mathfrak{G}_t]$. For any $(\bar{g}, \mathfrak{b}) \in C_k(\bigwedge_{t\in T}\mathfrak{G}_t)$ there exist $(\bar{g}, \mathfrak{c}_t) \in C_k(\mathfrak{G}_t)$ $(t \in T)$ with $\mathfrak{b} \sqsupseteq \bigcap_{t\in T}\mathfrak{c}_t$ and therefore $(\bar{g}, \mathfrak{b}) \in C_k(\mathfrak{G})$ for any $\mathfrak{G} \in [\bigwedge_{t\in T}\mathfrak{G}_t]$. This proves $\bigwedge_{t\in T}[\mathfrak{G}_t] = [\bigwedge_{t\in T}\mathfrak{G}_t]$.

That the complete lattice $\widetilde{\mathfrak{H}}_{\sqcap}^{graph}(\vec{\mathbb{K}})$ is even completely distributive, this follows from the next proposition which internally clarifies the lattice structure of $\widetilde{\mathfrak{H}}_{\sqcap}^{graph}(\vec{\mathbb{K}})$ and has an analogous proof as that of Proposition 1 in [PW99]. □

Proposition 4 *For a power context family $\vec{\mathbb{K}}$, for $\bar{g} \in (G_0)^k$ and $k = 0, 1, \ldots$, let $L_k^{\bar{g}} := \{\mathfrak{c} \in \mathfrak{H}_{\sqcap}(\mathbb{K}_k) \mid \bar{g} \in Ext(\mathfrak{c})\} \cup \{\top_k^{\bar{g}}\}$ be the interval of all supersemiconcepts of $(\{\bar{g}\}, \{\bar{g}\}^{I_k})$ in $\mathfrak{H}_{\sqcap}(\dot{\mathbb{K}}_k)$ together with a new top-element $\top_k^{\bar{g}}$. Then $\widetilde{\mathfrak{H}}_{\sqcap}^{graph}(\vec{\mathbb{K}})$ is isomorphic to the complete subdirect product of the completely distributive complete lattices $L_k^{\bar{g}}$ with $k = 0, 1, \ldots$ and $\bar{g} \in (G_0)^k$ consisting of all elements $\vec{\mathfrak{a}} := (\mathfrak{a}_k^{\bar{g}})_{k=0,1,\ldots}^{\bar{g}\in(G_0)^k}$ of the directed product $\prod_{k=0,1,\ldots}^{\bar{g}\in(G_0)^k} L_k^{\bar{g}}$ satisfying the following conditions (k = 1, 2, ...):*

(E_k) *If $\mathfrak{a}_k^{\bar{g}} \ne \top_k^{\bar{g}}$ and $\bar{g} = (g_1, \ldots, g_k)$ then $\mathfrak{a}_0^{g_i} \ne \top_0^{g_i}$ for $i = 1, \ldots, k$.*

5   Further Research

This paper can only be considered as the beginning of developing Boolean Judgment Logic. After identifying basic algebraic structures of Boolean Judgment Logic, namely

- the algebras $\mathfrak{H}_{\sqcap}^{graph}(\vec{\mathbb{K}})$ (and $\underline{\mathfrak{H}}_{\sqcap}^{graph}(\vec{\mathbb{K}})$) of semiconcept graphs (up to isomorphism) and

– the completely distributive complete lattices $\widetilde{\mathfrak{H}}_{\sqcap}^{\,graph}\,(\overrightarrow{\mathbb{K}})$ of information equivalence classes,

the interplay of syntax and semantics of Boolean judgments have to be clarified which already found a promising start in [Kl01] by extending the syntax described in [Pr98] to Boolean Judgment Logic. Another extension of [Pr98] is elaborated in [Da00] and [Da01] to connect Contextual Judgment Logic with First Order Predicate Logic; this work contributes to Boolean Judgment Logic too. As already mentioned above, the developed semantics shall be extended by including the notions of protoconcepts together with the operations of (weak) negation and (weak) opposition. The main direction of research will concentrate on the investigation of structures which come up in applications of Boolean Judgment Logic. For instance, a successful method of knowledge acquisition is asking for distinctive judgments so that contextual-logic structures of distinctive judgments are of interest (see [Wi01]). A general aim is to extend TOSCANA so that relationally scaled databases can be investigated by methods of Boolean Judgment Logic (cf. [EGSW0]).

Acknowledgement. This paper has substantially benefited from the financial support given by the Australian Research Council (IREX-Programme) which allowed the author to work in 1999/2000 for four months at the School of Information Technology of the Griffith University (Gold Coast Campus).

References

[Bu91]   R. W. Burch: *A Peircean reduction thesis*. Texas Tech University Press, Lubbock 1991.

[Da00]   F. Dau: Negations in simple concept graphs. In: B. Ganter, G. W. Mineau (eds.): *Conceptual structures: logical, linguistic, and computational issues*. LNAI **1867**. Springer, Heidelberg 2000, 263–276.

[Da01]   F. Dau: Concept graphs and predicate logic. This volume.

[EGSW0]   P. Eklund, B. Groh, G. Stumme, R. Wille: A contextual-logic extension of TOSCANA. In: B. Ganter, G. W. Mineau (eds.): *Conceptual structures: logical, linguistic, and computational issues*. LNAI **1867**. Springer, Heidelberg 2000, 453–467.

[GW99]   B. Ganter, R. Wille: *Formal Concept Analysis: mathematical foundations*. Springer, Heidelberg 1999; German version: Springer, Heidelberg 1996.

[Kl01]   J. Klinger: Simple semiconcept graphs: a Boolean logic approach. This volume.

[Pr98]   S. Prediger: *Kontextuelle Urteilslogik mit Begriffsgraphen. Ein Beitrag zur Restrukturierung der mathematischen Logik*. Dissertation, TU Darmstadt. Shaker, Aachen 1998.

[PW99]   S. Prediger, R. Wille: The lattice of concept graphs of a relationally scaled context. In: W. Tepfenhart, W. Cyre (eds.): *Conceptual structures: standards and practices*. LNAI **1640**. Springer, Heidelberg 1999, 401–414.

[So84]   J. F. Sowa: *Conceptual structures: information processing in mind and machine*. Adison-Wesley, Reading 1984.

128 R. Wille

[Wi82]   R. Wille: Restructuring lattice theory: an approach based on hierarchies of concepts. In: I. Rival (ed.): *Ordered sets*. Reidel, Dordrecht-Boston 1982, 445–470.

[Wi89]   R. Wille: Lattices in data analysis: how to draw them with a computer. In: I. Rival (ed.): *Algorithms and order*. Kluwer, Dordrecht/Boston 1989, 33–58.

[Wi96]   R. Wille: Restructuring mathematical logic: an approach based on Peirce's pragmatism. In: A. Ursini, P. Agliano (eds.): *Logic and Algebra*. Marcel Dekker, New York 1996, 267–281.

[Wi97]   R. Wille: Conceptual Graphs and Formal Concept Analysis. In: D. Lukose, H. Delugach, M. Keeler, L. Searle, J. F. Sowa (eds.): *Conceptual structures: fulfilling Peirce's dream*. LNAI **1257**. Springer, Heidelberg 1997, 290–303.

[Wi00a]   R. Wille: Boolean Concept Logic. In: B. Ganter, G. W. Mineau (eds.): *Conceptual structures: logical, linguistic, and computational issues*. LNAI **1867**. Springer, Heidelberg 2000, 317–331.

[Wi00b]   R. Wille: Contextual Logic summary. In: G. Stumme (ed.): *Working with Conceptual Structures. Contributions to ICCS 2000*. Shaker, Aachen 2000, 265–276.

[Wi00c]   R. Wille: Lecture notes on contextual logic of relations. FB4-Preprint, TU Darmstadt 2000.

[Wi01]   R. Wille: The contextual-logic structure of distinctive judgments. FB4-Preprint, TU Darmstadt 2001.

Pattern Structures and Their Projections

Bernhard Ganter[1] and Sergei O. Kuznetsov[2]

[1] Institut für Algebra, TU Dresden
D-01062 Dresden, Germany
ganter@math.tu-dresden.de
[2] All-Russia Institute for Scientific and Technical Information (VINITI)
Usievicha 20, 125219 Moscow, Russia
serge@viniti.ru

Abstract. Pattern structures consist of objects with descriptions (called patterns) that allow a semilattice operation on them. Pattern structures arise naturally from ordered data, e.g., from labeled graphs ordered by graph morphisms. It is shown that pattern structures can be reduced to formal contexts, however sometimes processing the former is often more efficient and obvious than processing the latter. Concepts, implications, plausible hypotheses, and classifications are defined for data given by pattern structures. Since computation in pattern structures may be intractable, approximations of patterns by means of projections are introduced. It is shown how concepts, implications, hypotheses, and classifications in projected pattern structures are related to those in original ones.

Introduction

Our investigation is motivated by a basic problem in pharmaceutical research. Suppose we are interested which chemical substances cause a certain effect, and which do not. A simple assumption would be that the effect is triggered by the presence of certain molecular substructures, and that the non-occurence of the effect may also depend on such substructures.

Suppose we have a number of observed cases, some in which the effect does occur and some where it does not; we then would like to form hypotheses on which substructures are responsible for the observed results. This seems to be a simple task, but if we allow for combinations of substructures, then this requires an effective strategy.

Molecular graphs are only one example where such an approach is natural. Another, perhaps even more promising domain is that of *Conceptual Graphs* (CGs) in the sense of Sowa [21] and hence, of logical formulas. CGs can be used to represent knowledge in a form that is close to language. It is therefore of interest to study how hypotheses can be derived from Conceptual Graphs.

A strategy of hypothesis formation has been developed under the name of JSM-method by V. Finn [8] and his co-workers. Recently, the present authors have demonstrated [11] that the approach can neatly be formulated in the language of another method of data analysis: Formal Concept Analysis (FCA) [12].

H. Delugach and G. Stumme (Eds.): ICCS 2001, LNAI 2120, pp. 129–142, 2001.
© Springer-Verlag Berlin Heidelberg 2001

The theoretical framework provided by FCA does not always suggest the most efficient implementation right away, and there are situations where one would choose other data representation forms. In this paper we show that this can be done in full compliance with FCA theory.

1   Formal Contexts

From every binary relation, a complete lattice can be constructed, using a simple and useful construction. This has been observed by Birkhoff [3] in the 1930s, and is the basis of Formal Concept Analysis, with many applications to data analysis.

The construction can be described as follows: Start with an arbitrary relation between two sets G and M, i.e., let $I \subseteq G \times M$, and define

$$A' := \{m \in M \mid (g, m) \in I \text{ for all } g \in A\} \qquad \text{for } A \subseteq G,$$

$$B' := \{g \in G \mid (g, m) \in I \text{ for all } m \in B\} \qquad \text{for } B \subseteq M.$$

Then the pairs (A, B) satisfying

$$A \subseteq G, B \subseteq M, A' = B, A = B'$$

are called the **formal concepts** of the **formal context** (G, M, I). When ordered by

$$(A_1, B_1) \leq (A_2, B_2) : \iff A_1 \subseteq A_2 \quad (\iff B_2 \subseteq B_1),$$

they form a complete lattice, called the **concept lattice** of (G, M, I).

The name "Formal Concept" reflects the standard interpretation, where the elements of G are viewed as "objects", those of M as "attributes", and where $(g, m) \in I$ encodes that object g has attribute m. It has been demonstrated that the concept lattice indeed gives useful insight in the conceptual structure of such data (see [12] and references there).

That data are given in form of a formal context is a particularly simple case. If other kind of data is to be treated, the usual approach is first to bring it in this standard form by a process called "scaling". Recently, another suggestion was discussed by several authors [14], [15] [16] [17]: to generalize the abovementioned lattice construction to contexts with an additional order structure on G and/or M. This seems quite natural, since the mappings $A \mapsto A'$, $B \mapsto B'$ used in the construction above form a Galois connection between the power sets of G and M. It is well known that a complete lattice can be derived more generally from any Galois connection between two complete lattices.

On the other hand, one may argue that there is no need for such a generalization and that no proper generalization will be achieved, since the basic construction already is as general as possible: it can be shown that every complete lattice is isomorphic to some concept lattice.

Nevertheless, such a more general approach may be worthwhile for reasons of efficiency, and it seems natural as well. Several authors [2], [4], [7] have considered the case where instead of having attributes the objects satisfy certain logical formulas. In such a situation, shared attributes are replaced by common subsumers of the respective formulae.

We show how such an approach is linked to the general FCA framework. We discuss some operational and algorithmic aspects and demonstrate our results on an example.

2   Pattern Structures

Let G be some set, let $(D, \sqcap)$ be a meet-semilattice and let $\delta : G \to D$ be a mapping. Then $(G, \underline{D}, \delta)$ with $\underline{D} = (D, \sqcap)$ is called a **pattern structure**, provided that the set

$$\delta(G) := \{\delta(g) \mid g \in G\}$$

generates a complete subsemilattice $(D_\delta, \sqcap)$ of $(D, \sqcap)$.[1] Each such complete semilattice has lower and upper bounds, which we denote by **0** and **1**, respectively. The intuitive meaning of a pattern structure is the set of objects with "descriptions" (patterns) with a "similarity operation" $\sqcap$ on them, i.e., an operation that for an arbitrary set of objects gives a "description" representing similarity of objects from the subset. The similarity should be independent of the order in which the objects occur, therefore the operation should be idempotent, commutative, and associative.

The condition on the complete subsemilattice looks unpleasant. But note that there are two situations where this is automatically satisfied: when $(D, \sqcap)$ is complete, and when G is finite.

If $(G, \underline{D}, \delta)$ is a pattern structure, we define the derivation operators as

$$A^\square := \bigsqcap_{g \in A} \delta(g) \qquad \text{for } A \subseteq G$$

and

$$d^\square := \{g \in G \mid d \sqsubseteq \delta(g)\} \qquad \text{for } d \in D.$$

The elements of D are called **patterns**. The order on them is given, as usual, by

$$c \sqsubseteq d : \iff c \sqcap d = c,$$

and is called the **subsumption** order. The operators $(\cdot)^\square$ obviously make a Galois connection between the power set of G and $(D, \sqsubseteq)$. The pairs (A, d) satisfying

$$A \subseteq G, \quad d \in D, \quad A^\square = d, \quad \text{and} \quad A = d^\square$$

are called the **pattern concepts** of $(G, \underline{D}, \delta)$, with extent A and **pattern intent** d. The above notions are analogues of the corresponding notions in formal contexts. For $a, b \in D$ the **pattern implication** $a \to b$ holds if $a^\square \sqsubseteq b^\square$. A pattern implication says what patterns occur in an object "description" if a certain pattern does. Similarly, for $C, D \subseteq G$ the **object implication** $C \to D$ holds if

---

[1] By which we mean that every subset X of $\delta(G)$ has an infimum $\sqcap X$ in $(D, \sqcap)$ and that D_δ is the set of these infima.

$C^\square \sqsubseteq D^\square$. Informally, this implication says that "all patterns that occur in all objects from the set C occur also in all objects from the set D."

Since $(D_\delta, \sqcap)$ is complete, there is a (unique) operation $\sqcup$ such that $(D_\delta, \sqcap, \sqcup)$ is a complete lattice. It is given by

$$\sqcup X = \sqcap\{c \in D_\delta \mid \forall_{x \in X}\ x \sqsubseteq c\}.$$

A subset M of D is $\sqcup$-**dense** for $(D_\delta, \sqcap)$ if every element of D_δ is of the form $\sqcup X$ for some $X \subseteq M$. If this is the case, then with

$$\downarrow d := \{e \in D \mid e \sqsubseteq d\}$$

we get

$$c = \sqcup(\downarrow c \cap M) \qquad \text{for every } c \in D_\delta.$$

Of course, $M := D_\delta$ is always an example of a $\sqcap$-dense set.

If M is $\sqcup$-dense in $(D_\delta, \sqcap)$, the formal context (G, M, I) with I given as $gIm :\Leftrightarrow m \sqsubseteq \delta(g)$ is called a **representation context** for $(G, \underline{D}, \delta)$.

Theorem 1. *Let $(G, \underline{D}, \delta)$ be a pattern structure and let (G, M, I) be a representation context of $(G, \underline{D}, \delta)$. Then for any $A \subseteq G$, $B \subseteq M$ and $d \in D$ the following two conditions are equivalent*

1. *(A, d) is a pattern concept of $(G, \underline{D}, \delta)$ and $B = \downarrow d \cap M$.*
2. *(A, B) is a formal concept of (G, M, I) and $d = \bigsqcup B$.*

The **proof** is by a standard application of the basic theorem of Formal Concept Analysis [12].

Thus the pattern concepts of $(G, \underline{D}, \delta)$ are in 1-1-correspondence with the formal concepts of (G, M, I). Corresponding concepts have the same first components (called **extents**). These extents form a closure system on G and thus a complete lattice, which is isomorphic to the concept lattice of (G, M, I).

3   Computing Pattern Concepts

When a pattern structure is given, then in principle we have all the information necessary to determine its concepts. We might, for example, compute all infima of subsets of D_δ and thereby all pattern concepts. To this end we can, e.g., adapt the Next Concept algorithm [12]. In computation of even finite pattern structures, one should take into account the fact that performing a single closure may be intractable. For example, already the problem of testing the $\sqsubseteq$ relation for labeled graphs from Section 7 is NP-complete, and computing $X \sqcap Y$ is even more difficult. Thus, in designing an algorithm for computing pattern concepts, one needs first to minimize the number of operations $\sqcap$, then the number of $\sqsubseteq$ relation tests, and, in the last turn, the number of operations with Boolean vectors.

After each backtrack of the original version of the Next Concept algorithm, it performs intersection of $|G|$ object intents. To avoid this in case of "expensive"

⊓ operation, one can introduce a natural tree data structure. Each vertex of the tree corresponds to a concept (A, B) and the children of the tree correspond to concepts of the form $((A \cup \{g\})'', (A \cup \{g\})')$ (actually, only some concepts of this form). An algorithm of this kind was given in [9].

A similar algorithm of this type was described in [16] for computing with sets of graphs. Given a family $\mathcal{F}$ of graph sets and an idempotent, commutative and associative operation ⊓ on them defined as in Section 7, the algorithm constructs the semilattice w.r.t. ⊓ generated by sets from $\mathcal{F}$, and its line (Hasse) diagram. The time complexity of the algorithm is $O((\alpha + \beta|G|)|G|^2|L|)$ and its space complexity is $O((\gamma|G|\|L|))$, where α is time needed to perform ⊓ operation, β is time needed to test ⊑ relation and γ is the space needed to store the largest object from D_δ.

A similar approach to computing pattern concepts and implications between objects can be made in lines of a procedure proposed in [2]. This procedure, called the **object exploration**, is the dual of the **attribute exploration** algorithm, which is standard in Formal Concept Analysis. For a given closure operator on G it computes its **stem base**, which is an irredundant system of implications on G that generate the closure operator, and the concept lattice (isomorphic in this case to the lattice of description hierarchy, i.e., the lattice of least common subsumers). Here an implication

$$A \to B, \quad A, B \subseteq G$$

holds, like in case of implications between sets of attributes, if $B \subseteq A''$. The order studied in [2] is given by the hierarchy of descriptions in Description Logic, where the description that is an infimum of two other descriptions (their least common subsumer) can be of exponential size, e.g., for ALE logic [2].

4   Structured Attribute Sets

We have introduced pattern structures to replace sets of attributes by a sort of "descriptions". However, this does not exclude the possibility that the patterns are themselves attribute combinations. A natural situation where this occurs is when the attribute set is large, but structured, so that admissible attribute combinations can be described by generating subsets.

Consider the example mentioned in the introduction: there the observed patterns are chemical compounds described by their molecular graphs. There is no natural ⊓-operation for such graphs, except when we use a little trick: we replace each graph by the set of its subgraphs, including the graph itself. Then the meet is the set of all common substructures. When describing such sets, one will usually restrict to the *maximal* common substructures and tacitly include the sub-substructures of these.

To phrase this situation more abstractly, assume that the attribute set $(P, \leq)$ is finite and (partially) ordered, and that all attribute combinations that can occur must be order ideals (downsets) of this order. Any order ideal O can be described by the set of its maximal elements M as $O := \{x \mid \exists_{y \in M} x \leq y\}$.

The maximal elements form an antichain, and conversely, each antichain is the set of maximal elements of some order ideal. Thus, in this case, the semilattice $(D, \sqcap)$ of patterns will consist of all antichains of the ordered attribute set, and will be isomorphic to the lattice of all order ideals of the ordered set (and thus isomorphic to the concept lattice of the context $(P, P, \not\geq)$, see [12]). For given antichains C_1 and C_2, the infimum $C_1 \sqcap C_2$ then consists of all maximal elements of the order ideal

$$\{m \mid \exists_{c_1 \in C_1} \exists_{c_2 \in C_2} \quad m \leq c_1 \text{ and } m \leq c_2\}.$$

Computing $C_1 \sqcap C_2$ may however be a problem. Note, e.g., that in the introductory example of chemical compounds, even the $\leq$-relation is difficult to compute, since $x \leq y$ amounts to x *is isomorphic to a subgraph of* y, which is an NP-complete problem [13].

There is a canonical representation context for this pattern structure $(G, \underline{D}, \delta)$. It is easy to see that the set of *principal ideals* $\downarrow p$ is $\sqcap$-dense in the lattice of all order ideals. Thus

$$(G, P, I) \qquad \text{with } (g, p) \in I : \iff p \leq \delta(g)$$

is (isomorphic to) a representation context for $(G, \underline{D}, \delta)$.

Since the set of order ideals is closed under unions, the semilattice $\underline{D}$ of antichains will be a distributive one. The same approach also works in the meet-distributive case, for sets selected from a closure system with the *anti-exchange* property [12]. The anti-exchange property implies that each closed set is the closure of its *extreme points*, as it is known for the example of convex polyhedra. Again we get that each closed set has a canonical generating set, that may be used as a pattern.

5   Projections

It may happen that some of the patterns in a pattern structure are too complex and difficult to handle. In such a situation one is tempted to replace the patterns with weaker, perhaps simpler ones, even if that results in some loss of information.

We formalize this using a mapping $\psi \colon D \to D$ and replacing each pattern $d \in D$ by $\psi(d)$ such that the pattern structure $(G, \underline{D}, \delta)$ is replaced by $(G, \underline{D}, \psi \circ \delta)^2$.

It is natural to require that ψ is a kernel operator (or **projection**), i.e., that ψ is

monotone: if $x \sqsubseteq y$, then $\psi(x) \sqsubseteq \psi(y)$,
contractive: $\psi(x) \sqsubseteq x$, and
idempotent: $\psi(\psi(x)) = \psi(x)$.

In what follows we will use the following fact, well-known in order theory [6]:

---

[2] In this situation we consider two pattern structures simultaneously. When we use the symbol $\sqcap$, it always refers to $(G, \underline{D}, \delta)$, not to $(G, \underline{D}, \psi \circ \delta)$.

Proposition 1. *Any projection of a complete semilattice $(D, \sqcap)$ is $\sqcap$-preserving, i.e., for any $X, Y \in D$*

$$\psi(X \sqcap Y) = \psi(X) \sqcap \psi(Y).$$

It is easy to describe how the lattice of pattern concepts changes when we replace $(G, \underline{D}, \delta)$ by $(G, \underline{D}, \psi \circ \delta)$. First, the folowing statement establishes the invariance of subsumption relation for projected data

Proposition 2. $\psi(d) \sqsubseteq \delta(g) \Leftrightarrow \psi(d) \sqsubseteq \psi \circ \delta(g).$

Proof. If $\psi(d) \sqsubseteq \delta(g)$ then $\psi(d) = \psi(\psi(d)) \sqsubseteq \psi(\delta(g))$ by the idempotence of ψ. On the other hand, if $\psi(d) \sqsubseteq \psi \circ \delta(g)$ then $\psi(d) \sqsubseteq \delta(g)$, since ψ is contractive.

The following statement establishes the relation between projected pattern structures and their representation contexts: taking a projection is equivalent to taking a subset of attributes of the representation context of the original pattern structure.

Theorem 2. *For pattern structures $(G, \underline{D}, \delta_1)$ and $(G, \underline{D}, \delta_2)$ the following statements are equivalent:*

1. *$\delta_2 = \psi \circ \delta_1$ for some projection ψ of $\underline{D}$.*
2. *There is a representation context (G, M, I) of $(G, \underline{D}, \delta_1)$ and some $N \subseteq M$ such that $(G, N, I \cap (G \times N))$ is a representation context of $(G, \underline{D}, \delta_2)$.*

Proof. $1 \Rightarrow 2$. Let $\delta_1, \delta_2 \colon G \to D$ be mappings and let $\psi \colon D \to D$ be a projection of $\underline{D}$ such that $\delta_2 = \psi \circ \delta_1$.

Define $M := D$ and $N := \{\psi(m) \mid m \in M\} \subseteq M$. Clearly (G, M, I) with $(g, m) \in I \Leftrightarrow m \sqsubseteq \delta_1(g)$, is a representation context of $(G, \underline{D}, \delta_1)$. Moreover, (G, N, J) with $(g, n) \in J \colon\Leftrightarrow n \sqsubseteq \delta_2(g)$ is a representation context of $(G, \underline{D}, \delta_2)$. It remains to prove that $J = I \cap (G \times N)$, i.e., that the equation $n \sqsubseteq \delta_1(g) \Leftrightarrow n \sqsubseteq \delta_2(g)$ holds for arbitrary $g \in G, n \in N$. Note that for each $n \in N$ we have $\psi(n) = n$ (since $\psi(n)$ is idempotent). Thus, the equation follows from 2.
$1 \Leftarrow 2$. Having $N \subseteq M$ we define

$$\psi(d) := \{n \in N \mid n \sqsubseteq d\}.$$

This mapping is obviously contractive, monotone, and idempotent. By definition of ψ, we have $n \sqsubseteq d \Leftrightarrow n \sqsubseteq \psi(d)$ for all $n \in N$ and all $d \in D$. Therefore,

$$\psi \circ \delta_1(g) = \sqcup\{n \in N \mid n \sqsubseteq \delta_1(g)\} = \sqcup\{n \in N \mid n \sqsubseteq \psi(\delta_1(g))\} = \delta_2(g).$$

Corollary 1. *Every extent of $(G, \underline{D}, \psi \circ \delta)$ is an extent of $(G, \underline{D}, \delta)$. If d is a pattern intent of $(G, \underline{D}, \delta)$, then $\psi(d)$ is a pattern intent of $(G, \underline{D}, \psi \circ \delta)$, for which $\psi(d)^{\square\square} \sqsubseteq d$.*

Pattern structures are naturally ordered by projections: $(G, \underline{D}, \delta_1) \geq (G, \underline{D}, \delta_2)$ if there is a projection ψ such that $\delta_2 = \psi \circ \delta_1$. In this case, representation $(G, \underline{D}, \delta_2)$ can be said to be rougher than $(G, \underline{D}, \delta_1)$ and the latter to be finer than the former.

The following proposition relates implications in comparable pattern structures.

Proposition 3. *Let $a, b \in D$. If $\psi(a) \to \psi(b)$ and $\psi(b) = b$ then $a \to b$.*

Proof. By contractivity of projection we have $\psi(a) \sqsubseteq a$, hence $a^\square \subseteq (\psi(a))^\square$ and $a \to \psi(a)$. If $\psi(a) \to \psi(b) = b$, then $a \to b$ follows from the transitivity of the relation $\to$.

Thus, for a certain class of implications, it is sufficient to compute them in projected data (which can be far more efficient than to compute in original pattern structure) to establish their validity for the original pattern structure. Note that this proposition does not require a and b to be subsumed by patterns from $\delta(G)$. In the case where $a \sqsubseteq \delta(g)$ for no $g \in G$, we have $a^\square = \emptyset$ by definition and all implications in Proposition 3 hold automatically.

6   Hypothesis Generation from Projected Data

In [11] we considered a learning model from [8] in terms of Formal Concept Analysis. This model assumes that the cause of a **goal property** resides in common attributes of objects that have this property. If our objects are described by some mathematical structures, we may look for common substructures of those objects that have the goal property.

This can be transferred to pattern structures, where it is assumed that the presence and absence of the goal property can be predicted from the patterns associated with the objects. The setting can be formalized as follows. Let $(G, \underline{D}, \delta)$ be a pattern structure together with an external goal property w. The set G of all objects can be partitioned into three disjoint sets: The set G_+ of those objects that are known to have the property w (these are the *positive examples*), the set G_- of those objects of which it is known that they do not have w (the *negative examples*) and the set G_τ of *undetermined examples*, i.e., of those objects, of which it is unknown if they do have property w or not. This gives three pattern substructures of $(G, \underline{D}, \delta)$: $(G_+, \underline{D}, \delta)$, $(G_-, \underline{D}, \delta)$, $(G_\tau, \underline{D}, \delta)$.

A **positive hypothesis** h is defined as a pattern intent of $(G_+, \underline{D}, \delta)$ that is not subsumed by any pattern from $\delta(G_-)$ (for short: not subsumed by any negative example). Formally:

$$h \in D \text{ is a positive hypothesis iff } h^\square \cap G_- = \emptyset \text{ and } \exists A \subseteq G_+ : A^\square = h.$$

A **negative hypothesis** is defined accordingly. These definitions implement the general idea of machine learning in terms of formal concept analysis: "given positive and negative examples of a goal class, find generalizations (generalized descriptions) of positive examples that do not cover any negative examples."

A hypothesis in the sense of [11] is obtained as a special case of this definition when $(D, \sqcap) = (2^M, \cap)$ for some set M. Hypotheses can be used for classification of undetermined examples as introduced in [8] in the following way. If $g \in G_\tau$ is an undetermined example, then a hypothesis h with $h \sqsubseteq \delta(g)$ is **for the positive classification** of g if h is positive and **for the negative classification** of g if it is a negative hypothesis. Example $g \in G_\tau$ is **classified positively** if there is a hypothesis for its positive classification and none for its negative classification. It is **classified negatively** in the converse situation. We have no classification if there is no hypothesis for positive and negative classification or contradictory classification (if there are hypotheses for both positive and negative classification).

Hypotheses have been studied in detail elsewhere [11]. Here we focus our consideration on the following aspect. What happens when we use "weaker" data approximating the original data? What is the significance of hypotheses obtained from weak data?

On the one hand, we almost always deal with weak data that describe reality approximately. For example, in case of molecular structures, a more adequate representation can be the 3D–geometrical one. But even this representation of a molecule is already an abstraction of a quantum-mechanical one, etc. On the other hand, having some data for which computation is intractable, we would like to have their tractable reasonable approximation that would allow one to judge about hypotheses and classifications in the original representation by means of results about hypotheses and classification for weak data.

This problem becomes more precise if we describe the data weakening by means of a projection $\psi : \underline{D} \rightarrow \underline{D}$. Instead of $(G, \underline{D}, \delta)$ we than work with $(G, \underline{D}, \psi \circ \delta)$ and its three parts $(G_+, \underline{D}, \psi \circ \delta)$, $(G_-, \underline{D}, \psi \circ \delta)$, and $(G_\tau, \underline{D}, \psi \circ \delta)$, as above.

To simplify our language, let us reserve the term "hypothesis" to those obtained from $(G, \underline{D}, \delta)$ and let us refer to those obtained from $(G, \underline{D}, \psi \circ \delta)$ as ψ-**hypotheses**. Now the question to be studied is: How are hypotheses and ψ-hypotheses related? In what follows we shall try to answer this question for positive hypotheses. Results similar to that below hold also for negative hypotheses and classifications.

There is no guarantee that the ψ-image of a hypothesis must be a ψ-hypothesis. In fact, our definition allows that ψ is the "null projection" with $\psi(d) = 0$ for all $d \in D$. This corresponds to total abandoning of the data, and no interesting hypotheses are to be expected in that situation.

However, we have the following

Proposition 4. *If $\psi(d)$ is a (positive) hypothesis, then $\psi(d)$ is also a (positive) ψ-hypothesis.*

Proof. If $\psi(d)$ is a positive hypothesis, then $\psi(d)$ is not subsumed by any negative example. Moreover, $\psi(d)$ is a pattern intent of $(G, \underline{D}, \psi \circ \delta)$ according to Corollary 1. Thus $\psi(d)$ is a ψ-hypothesis.

The classification set does not shrink when we pass from d to $\psi(d)$:

Proposition 5. *If d is a hypothesis for the positive classification of g and $\psi(d)$ is a positive ψ-hypothesis, then $\psi(d)$ is for the positive classification of g.*

Proof. Obvious, since $\psi(d) \sqsubseteq d \sqsubseteq g^\square$.

Proposition 6. *If $\psi(d)$ is a (positive) ψ-hypothesis, then $\psi(d)^{\square\square}$ is a (positive) hypothesis.*

Proof. Assume that $\psi(d)$ is a positive ψ-hypothesis. Then the corresponding extent in $(G, \underline{D}, \psi \circ \delta)$,

$$E := \{g \in G \mid \psi(d) \sqsubseteq \psi(\delta(g))\},$$

is contained in G_+ and is also an extent of $(G, \underline{D}, \delta)$ (by the corollary of Theorem 2). Thus $\psi(d)^{\square\square} = E^\square$ is an intent of $(G_+, \underline{D}, \delta)$ and cannot subsume any negative example, since it is subsumed by $\psi(d)$.

The propositions show that we may hunt hypotheses starting from ψ-hypotheses. We can shoot only those that can be seen in the projected data, but those can in fact be found, as the following theorem states:

Theorem 3. *For any projection ψ and any positive hypothesis $d \in D$ the following are equivalent:*

1. $\psi(d)$ is not subsumed by any negative example.
2. There is some positive ψ-hypothesis h such that $h^{\square\square} \sqsubseteq d$.

Proof. If $\psi(d)$ is not subsumed by any negative example, then $h := \psi(d)$ is a ψ-hypothesis and $h^{\square\square} \sqsubseteq d$ by Corollary 1. If h is a ψ-hypothesis, then $h^{\square\square}$ is a hypothesis by Proposition 6 and hence d, which subsumes $h^{\square\square}$, is not subsumed by any negative example.

7   An Application to Graphs

As an application we shall consider patterns given as labelled graphs. These may be structure graphs of chemical compounds, as mentioned in the introduction, and the task may be to find the patterns responsible for a biological effect. It is natural to assume (and it is the common assumption in this applied domain) that common biological effect of chemical compounds is caused by their common substructures. The graphs could as well be conceptual graphs in the sense of Sowa [21]. Let P be the set of all graphs under consideration, and let this set be ordered by generalized subgraph isomorphism (a definition is given below). As above, we have a set G (of experiments, observations, or the like) and to each $g \in G$ there is an associated graph from P, denoted by $\delta(g)$.

To be more concrete, let $(\mathcal{L}, \preceq)$ be some ordered set of "labels". In the examples below, this set will be the one displayed in Figure 1.

Let P be the set of all finite graphs, vertex-labelled by labels from $(\mathcal{L}, \preceq)$, up to isomorphism. A typical such graph is of the form $\Gamma := ((V, l), E)$, with

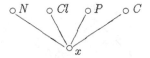

Fig. 1. The ordered set of labels for the molecular graphs in our example. x stands for "'any element"'.

vertex set V, edge set E and label assignment l. We say that $\Gamma_1 := ((V_1, l_1), E_1)$ **dominates** $\Gamma_2 := ((V_2, l_2), E_2)$, for short $\Gamma_2 \leq \Gamma_1$, if there exists a one-to-one mapping $\varphi : V_2 \to V_1$ that (for all $v, w \in V_2$)

- respects edges: $(v, w) \in E_2 \Rightarrow (\varphi(v), \varphi(w)) \in E_1$,
- fits under labels: $l_2(v) \leq l_1(\varphi(v))$.

Obviously this is an order relation. It may be called the "generalized subgraph isomorphism relation", since in the unlabelled case it reduces to the subgraph isomorphism order. For conceptual graphs it corresponds to the injective specialization relation [19] or injective morphism [18].

Example 1. Let Γ_1 and Γ_2 be molecular graphs given in Figure 2. They represent

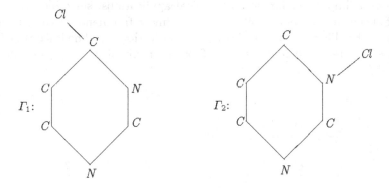

Fig. 2. Two molecular graphs representing patterns.

patterns in the described sense. Their meet, with a slight misuse of notation written as $\Gamma_1 \sqcap \Gamma_2$, is given by the set of three graphs depicted in Figure 3. Here, the disconnected graph H_1 contains more information about the cyclic structure, whereas H_2 and H_3 contain more information about the connection of the cycle with the vertex labeled by "Cl".

Example 2. Computing hypotheses for the pattern structure from the previous example may be very complex. It is therefore advisable to restrict the considerations to a subclass of the graphs, for which this is easier. This can neatly be

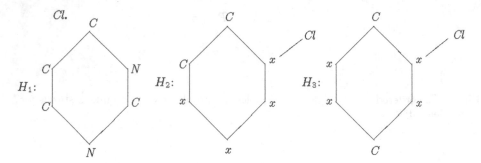

Fig. 3. $\Gamma_1 \sqcap \Gamma_2$ given by the maximal graphs in the intersection of the downsets generated by Γ_1 and Γ_2.

described using the notion of a projection: let $Q \subseteq P$ be the set of the "simple" graphs in P, then

$$(G, Q, I \cap (G \times Q))$$

represents, according to Theorem 2, some pattern structure $(G, \underline{D}, \psi \circ \delta)$ obtained from a projection $\psi : \underline{D} \to \underline{D}$.

A natural choice for Q is the class of all paths of a bounded length n. This projection is important for pharmaceutical applications, since biological activities of chemical compounds often reside in linear fragments of their molecular structures. By Theorem 2, this describes a projection ψ_n. The images of Γ_1 and Γ_2 under ψ_3, represented by sets of 3-paths maximal w.r.t. $\sqsubseteq$, are shown in Figure 4.

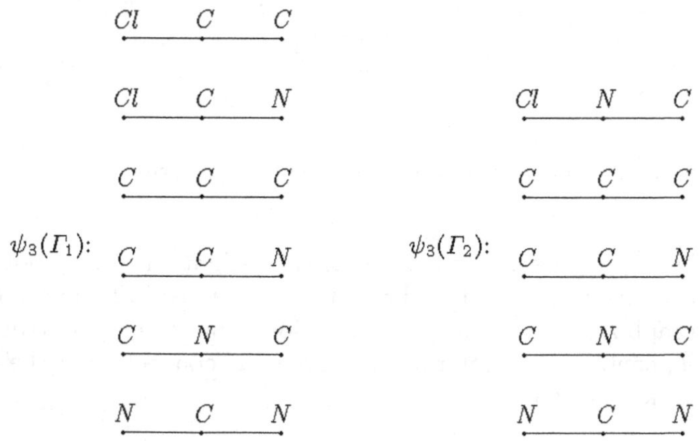

Fig. 4. The sets representing the projected graphs $\psi_3(\Gamma_1)$ and $\psi_3(\Gamma_2)$.

The number of chains of fixed length in a molecular graph is polynomial in its size. Therefore, the meet operation becomes computable in polynomial time. The maximal (w.r.t. $\sqsubseteq$) chains representing $\psi_3(\Gamma_1) \sqcap \psi_3(\Gamma_2)$ (which is, by Proposition 1, equal to $\psi_3(\Gamma_1 \sqcap \Gamma_2)$), are shown in Figure 5.

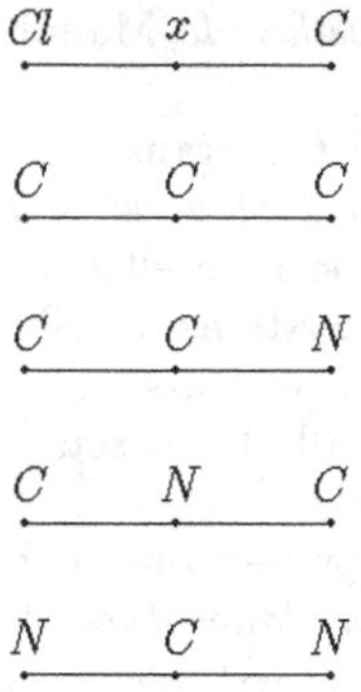

$$Cl \quad x \quad C$$

$$C \quad C \quad C$$

$$C \quad C \quad N$$

$$C \quad N \quad C$$

$$N \quad C \quad N$$

Fig. 5. $\psi_3(\Gamma_1) \sqcap \psi_3(\Gamma_2)$.

8 Conclusion

We considered analogues to formal contexts, but with a (semi)lattice of patterns instead of attributes. Such structures can easily be mapped to formal contexts, and it can be described how hypothesis generation from such pattern structures is to be organized. This becomes more intricated when we allow data weakening. We have described methods to recover those hypotheses that are reflected in the weaker data. The notion of a pattern can be applied to ordered attribute sets. Then the pattern lattice is the lattice of order ideals of the attribute order. Weakening corresponds to a restriction of the attribute set.

Acknowledgments. The authors are grateful to the members of the *Arbeitsgruppe Allgemeine Algebra und Diskrete Mathematik* at the TU Darmstadt and colleagues of *Laboratoire d'Informatique, de Robotique et de Microélectronique de Montpellier* for helpful discussions.

References

1. H.-G. Bartel, Matematische Methoden in der Chemie, Spektrum, Heidelberg, 1996.
2. F. Baader and R. Molitor, Building and Structuring Description Logic Knowledge Spaces Using Least Common Subsumers and Concept Analysis, in *Proc. 8th Int. Conf. on Conceptual Structures, ICCS'2000*, G. Mineau and B. Ganter, Eds., Lecture Notes in Artificial Intelligence, **1867**, 2000, pp. 292-305.

3. G. Birkhoff: *Lattice Theory.* AMS Colloquium Publications XXV, Providence, Rhode Island, 3rd edition 1967.
4. L. Chaudron and N. Maille, Generalized Formal Concept Analysis, in *Proc. 8th Int. Conf. on Conceptual Structures, ICCS'2000,* G. Mineau and B. Ganter, Eds., Lecture Notes in Artificial Intelligence, **1867**, 2000, pp. 357-370.
5. B. A. Davey and H. A. Priestley, *Introduction to Lattices and Order,* Cambridge University Press, 1990.
6. M. Erné, *Einführung in die Ordnungstheorie,* Mannheim, B.I.-Wissenschaftsverlag, 1982.
7. S. Férré and O. Ridoux, A Logical Generalization of Formal Concept Analysis, in *Proc. 8th Int. Conf. on Conceptual Structures, ICCS'2000,* G. Mineau and B. Ganter, Eds., Lecture Notes in Artificial Intelligence, **1867**, 2000.
8. V. K. Finn, Plausible Reasoning in Systems of JSM Type, *Itogi Nauki i Tekhniki, Seriya Informatika,* **15**, 54-101, 1991 [in Russian].
9. B. Ganter and K. Reuter, Finding all closed sets: a general approach, *Order,* **8**, 283-290, 1991.
10. B. Ganter and S. O. Kuznetsov, Stepwise Construction of the Dedekind-MacNeille Completion, *Proc. 6th Int. Conf. on Conceptual Structures, ICCS'98,* M-L. Mugnier, M. Chein, Eds., Lecture Notes in Artificial Intelligence, **1453**, 1998, pp. 295-302.
11. B. Ganter and S. O. Kuznetsov, Formalizing Hypotheses with Concepts, *Proc. 8th Int. Conf. on Conceptual Structures, ICCS'2000,* G. Mineau and B. Ganter, Eds., Lecture Notes in Artificial Intelligence, **1867**, 2000, pp. 342-356.
12. B. Ganter and R. Wille, Formal Concept Analysis. Mathematical Foundations, Berlin, Springer, 1999.
13. M. Garey and D. Johnson, *Computers and Intractability: A Guide to the Theory of NP-Completeness,* New York, Freeman, 1979.
14. R. Gugisch, Lattice Contexts: A Generalization in Formal Concept Analysis, unpublished manuscript, 2000.
15. S. O. Kuznetsov, JSM-method as a Machine Learning System, *Itogi Nauki Tekhn., ser. Informatika,* no. 15, 17-54, 1991 [in Russian].
16. S. O. Kuznetsov, Learning of Simple Conceptual Graphs from Positive and Negative Examples, in Proc. *Principles of Data Mining and Knowledge Discovery, Third European Conference, PKDD'99,* J. Zytkow, J. Rauch, Eds., Lecture Notes in Artificial Intelligence, **1704**, 1999, pp. 384-392.
17. M. Liquiere and J. Sallantin, Structural Machine Learning with Galois Lattice and Graphs, *Proc. Int. Conf. Machine Learning ICML'98,* 1998.
18. M.-L. Mugnier and M. Chein, Représenter des connaissances et raisonner avec des graphes, Revue d'Intelligence Artificielle, **10**(1), 1996, pp. 7-56.
19. M.-L. Mugnier, Knowledge Representation and Reasonings Based on Graph Homomorphisms, in *Proc. 8th Int. Conf. on Conceptual Structures, ICCS'2000,* G. Mineau and B. Ganter, Eds., Lecture Notes in Artificial Intelligence, **1867**, 2000, pp. 172-192.
20. S. Prediger, Simple Concept Graphs: A Logic Approach, in *Proc. 6th Int. Conf. on Conceptual Structures, ICCS'98,* M.-L.Mugnier, M. Chein, Eds., Lecture Notes in Artificial Intelligence, **1453**, 1998, pp. 225-239.
21. J. F. Sowa, *Conceptual Structures - Information Processing in Mind and Machine,* Reading, M.A.: Addison-Wesley, 1984.

Formal Concept Analysis Methods for Dynamic Conceptual Graphs

Bernhard Ganter and Sebastian Rudolph*

Institute of Algebra
Department of Mathematics and Natural Sciences
Dresden University of Technology
Germany

Abstract Conceptual Graphs (CG), originally developed for static data representation have been extended to cope with dynamical aspects. This paper adresses two questions connected with the topic: How can implicational knowledge about a system's states and behaviour be derived from a dynamic CG description and how can the CG specification process be supported by automatic or semiautomatic algorithms? Based on Formal Concept Analysis (FCA) we propose methods for both problems. Guided by an example we introduce two kinds of formal contexts containig the dynamic system's information: state contexts and action contexts. From these the complete implicational knowledge can be derived. Combining the techniques of attribute exploration and determination of a formal context's concepts, we demonstrate a procedure which interactively asks for the validity of implications and from this information designs a dynamic CG system with the desired properties.

1 Introduction

Reasoning about actions and planning is a central topic in AI. In the field of robotics information about environment changes as the result of executeable actions has to be aquired and processed. Often storing and handling such information in a conceptual way appears to be useful [3]. Reasoning about actions also plays an important role for specification of dynamic systems and the generation of operational models.

In recent years efforts have been made to extend the theory of conceptual graphs to dynamic aspects. The intention was not only to describe change of facts with CGs, but in a certain way to simulate such change in a CG-based system. The present paper establishes a connection between the theory of processes in CG on the one hand and methods of analyzing and exploring data from FCA on the other. It proposes a contextual representation of dynamic knowledge which supports modeling of dynamic systems and enables reasoning about situations and transitions.

In Section 2 we give a short review of CG literature relevant to this work. Section 3 comes with a rather general notion of a dynamic system on which our

* Supported by DFG/Graduiertenkolleg.

H. Delugach and G. Stumme (Eds.): ICCS 2001, LNAI 2120, pp. 143–156, 2001.

methods are based and introduces an example. How such kinds of dynamic systems can be translated into formal contexts is demonstrated in Section 4. Finally, using another example, Section 5 shows an algorithm which generates a dynamic CG description in dialogue with the user.

2 Literature

In [7], Sowa proposed *actors*, a new kind of conceptual graph nodes describing functional relations.

The CG formalism including actors covers the whole range of dealing with static knowledge. However, the description, modeling, and execution of dynamic processes is beyond its intended purpose. Therefore Delugach proposed an extension of the theory in [4] by introducing *demons*. These represent processes triggered by the existence of certain CGs (which take the role of preconditions). They act by assertion and retraction of CG concepts. Graphically they are represented by a double-lined diamond box.

Mineau further extended this approach. He allowed demons to have whole CGs as input and output. Furthermore he described processes by pre- and postcondition pairs in [6]. He showed the possibility of translating processes into sets of elementary transitions and demonstrated how an extended CG formalism could be used to represent executable processes.

3 Labeled Transition Systems with Attributes

First we give a rather general notion of a dynamic system. At any time the system is assumed to be in a certain state. Processes can be described as transitions from one of its states to another. Thereby certain more or less observable state properties may change. In order to describe such a system formally, one can record all its possible states including their properties and additionally all possible actions as well as the transitions they cause.

Definition 1. *A* LABELED TRANSITION SYSTEM WITH ATTRIBUTES *(short: LTSA)* $\mathbb{T}$ *is a 5-tuple* (S, M, I, A, T) *with*

- S *being a set of system states,*
- M *being a set of state attributes,*
- $I \subseteq S \times M$ *being a relation, where* sIm *means 'state s has attribute m',*
- A *being a set of actions, and*
- $T \subseteq (S \times A \times S)$ *being a set of transitions, where* $(s_1, a, s_2) \in T$ *means 'action a can cause the transition from state s_1 to state s_2'.*

Note that the definition of LTSAs is an extension of the classical notion of abstract automata. In both cases we have sets of states, actions and transitions specifying the system's dynamic behaviour.

Consider a simple example: a traffic light system at a crossroad. Altogether let

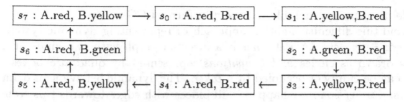

Figure 1. A state-transition graph for the traffic light scenario

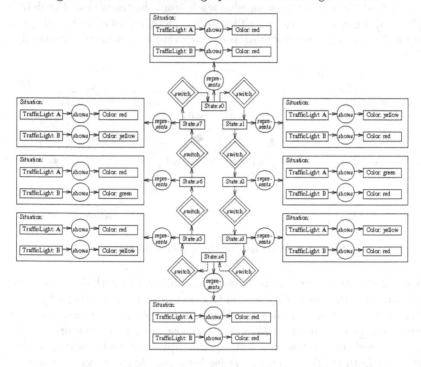

Figure 2. The CG version of the traffic light scenario's state description.

there be four traffic lights. Let A denote one pair of opposite traffic lights and B the other. In Figure 1 a graphical representation of the corresponding LTSA is shown. The possible states are drawn as boxes including their attributes. The arrows between the boxes show the transitions. Since in this case we have only one action, say *switch*, we omit the arc labelling. The system behaviour expressed by this graph can equivalently be described by a conceptual graph using the demons introduced by Delugach. The corresponding CG can be seen in Figure 2. In essence, this is a CG version of the state graph since the demons act by assertion and retraction of whole states. However, this kind of dynamic knowledge representation may be incovenient in some cases for several reasons. At first, the real structure of the system (i.e. its components) is not or only insufficiently apparent. Another (maybe even more evident) flaw appears if more complicated structures are considered. Then the state space may increase exponentially to a size which is difficult to handle computationally and impossible to represent to people in an understandable way even in form of CGs.

To avoid this difficulty, another approach of representing dynamic systems can be used: Petri nets. (A Petri net is a directed graph with two kinds of nodes: *places* drawn as circles and *transitions* represented by quadratic boxes. Each place can be empty or occupied by a *token*. The dynamic behaviour is defined as follows: a transition may happen if all places with edges towards this transition are occupied by tokens. As a result, all these tokens are removed and new tokens are put at that places, to which an edge leads from the transition. For details see e.g. [2] or [8]. Petri nets have proved useful in the theory of distributed systems.) In Figure 3 a Petri net representation of our example is shown. Eventually, also

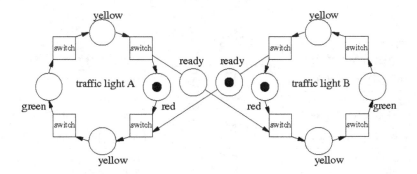

Figure 3. Petri net for the traffic light scenario.

this kind of representation can be transformed into a CG with demons. However, there is no more the possibility to identify one certain state except by the state attributes. So one has to make sure that all distinct token places in the Petri net are distinguished in the CG, otherwise the system behaviour is not completely specified. In our case, the both 'yellow'-places have to be distinguishably specified for both traffic lights, otherwise for example the succeeding state of s_7 in Figure 1 is not determined (it might be either s_8 as intended or s_6), which contradicts our purpose to specify the system behaviour exactly.

In general, a notion dealing with the possibility of behaviour specification by attributes is the following.

Definition 2. *A LTSA is called* ATTRIBUTE SPECIFIED, *iff for every four states* $s_1, s_2, s_3, s_4 \in S$ *the following holds:*

$$\forall m \in M : (s_1 Im \Leftrightarrow s_2 Im \wedge s_3 Im \Leftrightarrow s_4 Im)$$
$$\Longrightarrow \forall a \in A : ((s_1, a, s_3) \in T \Leftrightarrow (s_2, a, s_4) \in T)$$

This definition says that in an attribute specified LTSA two states 'behave' in the same way (according to the actions), if they have the same properties (according to the attributes). Conversely, this means, if two states 'behave' differently, they must be distinguishable by some attribute.

In order to avoid the unwanted indeterminism mentioned above, we have to transform the described system (with the token places - or attributes - *red, yellow, green* and *ready* for A and B respectively) into an attribute specified one. This can be done by introducing two additional token places (*beforegreen* and *aftergreen*) for each traffic light. Now we are able to 'translate' the Petri net into a CG with demons. The result is shown in Figure 4.

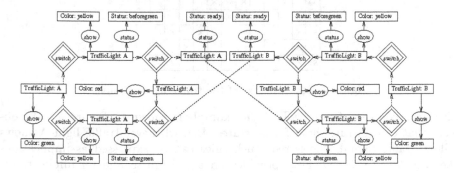

Figure 4. New CG for the traffic light scenario.

4   Introducing Formal Contexts and Deriving Information

Our purpose was to find an alternative description of the knowledge contained in the CG and Petri net representations, which enables implicational reasoning. We find it natural to apply FCA methods. We assume the reader to be familiar with the basics of FCA, for an introduction and details see e.g. [5].

FCA has mostly been applied to static data. There is some work by K.E. Wolff [9] about dynamic systems, but he considers only observation and description of state sequences without having actions triggered from outside the system. So we looked for another approach.

We intended to transform the dynamic knowledge specified in the preceding section into a contextual view, in order to enable the use of FCA methods of reasoning and data exploring. For this, we introduce two kinds of formal contexts.

4.1   The State Context

The first formal context contains only static information. It describes the system's states and their properties (attributes).

Definition 3. *The formal context* $\mathbb{K}_S := (S, M, I)$ *with the state set* S, *the attribute set* M, *and relation* $I \subseteq S \times M$ *from a LTSA* (S, M, I, A, T) *is called the* STATE CONTEXT *of this LTSA.*

	A.r	A.y	A.g	A.bg	A.ag	A.rdy	B.r	B.y	B.g	B.bg	B.ag	B.rdy
s_0	×						×					×
s_1		×	×				×					
s_2			×				×					
s_3		×			×		×					
s_4	×					×	×					
s_5	×								×	×		
s_6	×								×			
s_7	×								×		×	

Figure 5. State context of the traffic light example

The state context of our traffic light example is shown in Figure 5. We observe that certain rules hold for all states. E.g. whenever traffic light A shows yellow, traffic light B shows red. Such rules can be expressed as implications like *A.yellow → B.red*. FCA provides an algorithm to find an implicational base from the state context, containing the complete implicational knowledge of the system states according to the attributes under consideration. That is, all implications holding between these attributes (with regard to the context) can be deduced from this base. In our example it consists of the implications listed in Figure 6. Note that the symmetry between the both traffic lights is also apparent in the implicational base.

B.yellow	→ *A.red*
A.yellow	→ *B.red*
B.green	→ *A.red*
A.green	→ *B.red*
B.ready	→ *A.red, B.red*
A.ready	→ *B.red, A.red*
B.red, B.yellow	→ ⊥
A.red, A.yellow	→ ⊥
B.red, B.green	→ ⊥
A.red, A.green	→ ⊥
B.yellow, B.green	→ ⊥
A.yellow, A.green	→ ⊥
B.beforegreen, B.aftergreen	→ ⊥
A.beforegreen, A.aftergreen	→ ⊥
A.ready, B.ready	→ ⊥

Figure 6. Implicational base of the traffic light state context

4.2   The Action Context Family

Until now we only dealt with static system properties. To involve the dynamic aspect of an LTSA, another kind of context is introduced.

Definition 4. *Given a LTSA* (S, M, I, A, T), *the* ACTION CONTEXT $\mathbb{K}_a$ *of an action* $a \in A$ *is the triple* $(T_a, M_{A\Omega}, J_a)$ *with*

- $T_a := \{(s_1, s_2) \mid (s_1, a, s_2) \in T\}$
- $M_{A\Omega} := M \times \{0, 1\}$

  We ocassionally write m_A *for* $(m, 0)$ *and* m_Ω *for* $(m, 1)$ *for all* $m \in M$.
- $J_a \subseteq T_a \times M_{A\Omega}$ *being a relation with*

  $$(s_1, s_2) J_a(m, i) :\Leftrightarrow \begin{cases} s_1 I m, & \text{if } i = 0 \\ s_2 I m, & \text{if } i = 1 \end{cases}$$

The set $(\mathbb{K}_a)_{a \in A}$ *of all action contexts of a LTSA we call its* ACTION CONTEXT FAMILY.

So the action context contains all transitions which can be caused by the action as objects. The attributes of an action context are twice the state attributes: once for the initial and once for the final state of the transition.

In our example, the action context family contains only one element: the *switch*-context. It is displayed in Figure 7. Again it is possible to derive an implicational

	A.r_A	A.y_A	A.g_A	A.bg_A	A.ag_A	A.rdy_A	B.r_A	B.y_A	B.g_A	B.bg_A	B.ag_A	B.rdy_A	A.r_Ω	A.y_Ω	A.g_Ω	A.bg_Ω	A.ag_Ω	A.rdy_Ω	B.r_Ω	B.y_Ω	B.g_Ω	B.bg_Ω	B.ag_Ω	B.rdy_Ω
(s_0, s_1)	×					×						×	×		×		×							
(s_1, s_2)		×		×		×									×		×							
(s_2, s_3)			×			×									×		×	×						
(s_3, s_4)		×			×	×							×				×	×						
(s_4, s_5)	×					×	×						×							×		×		
(s_5, s_6)	×							×	×				×							×				
(s_6, s_7)	×								×				×						×			×		
(s_7, s_0)	×							×		×			×						×					

Figure 7. The *switch*-context of the traffic light example

base from this context. As one can easily see, all implications from the state context hold twice in every action context: since every action leads from one state to another, every implication holding for each state has to hold on the 'A-side' as well as on the 'Ω-side' of each transition. The implicational base is shown in Figure 8 omitting the implications already contained in the state context as explained above.

Now we have determined the implicational knowledge of the considered dynamic system. Every implication of this kind holding for the dynamic system can be inferred from the implicational base. Thus it could be used in logical programs for reasoning about the system, maybe for verifying certain system properties. Of course, the implicational base can also be coded into a non-dynamic CG-based system as production rules as described in [1]. Figure 9 shows an example.

$A.ready_A$	$\rightarrow$ $B.beforegreen_\Omega$	$B.ready_A$	$\rightarrow$ $A.beforegreen_\Omega$
$A.ready_\Omega$	$\rightarrow$ $A.aftergreen_A$	$B.ready_\Omega$	$\rightarrow$ $B.aftergreen_A$
$A.red_A, B.red_A, A.red_\Omega$	$\rightarrow$ $A.ready_A$	$B.red_A, A.red_A, B.red_\Omega$	$\rightarrow$ $B.ready_A$
$A.red_A, A.red_\Omega, B.red_\Omega$	$\rightarrow$ $B.ready_\Omega$	$B.red_A, B.red_\Omega, A.red_\Omega$	$\rightarrow$ $A.ready_\Omega$
$A.beforegreen_A$	$\rightarrow$ $A.green_\Omega$	$B.beforegreen_A$	$\rightarrow$ $B.green_\Omega$
$A.beforegreen_\Omega$	$\rightarrow$ $B.ready_A$	$B.beforegreen_\Omega$	$\rightarrow$ $A.ready_A$
$A.aftergreen_A$	$\rightarrow$ $A.ready_\Omega$	$B.aftergreen_A$	$\rightarrow$ $B.ready_\Omega$
$A.aftergreen_\Omega$	$\rightarrow$ $A.green_A$	$B.aftergreen_\Omega$	$\rightarrow$ $B.green_A$
$A.yellow_A$	$\rightarrow$ $B.red_\Omega$	$B.yellow_A$	$\rightarrow$ $A.red_\Omega$
$A.yellow_\Omega$	$\rightarrow$ $B.red_A$	$B.yellow_\Omega$	$\rightarrow$ $A.red_A$
$A.green_A$	$\rightarrow$ $A.aftergreen_\Omega$	$B.green_A$	$\rightarrow$ $B.aftergreen_\Omega$
$A.green_\Omega$	$\rightarrow$ $A.beforegreen_A$	$B.green_\Omega$	$\rightarrow$ $B.beforegreen_A$
$A.yellow_A, A.yellow_\Omega$	$\rightarrow$ $\perp$	$B.yellow_A, B.yellow_\Omega$	$\rightarrow$ $\perp$

Figure 8. Implicational base of the traffic light's *switch*-context.

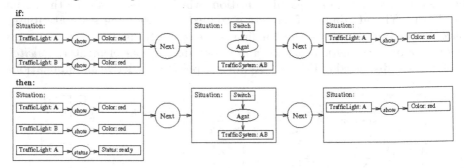

Figure 9. Production rule for one implication from the traffic light implicational base

5 From Implicit Knowledge to an Executable CG-Description

In the preceding section we showed a possibility of transferring the representation of a dynamical system (the state space and the transitions of which are already completely known) into a contextual representation in order to enable reasoning about that system.

In practice, the complete state space and the transitions usually are not explicitly known in advance. Or the system does not yet exist and an engineer wants to design it by specifying the system's static and dynamic properties.

For this case, FCA provides an appropriate tool: attribute exploration. The user just inputs the used state attributes and optionally some states and implications he already knows and then the algorithm asks 'questions' about the system, which the user has to answer.[1] In this way the user successively either makes his implicit knowledge of an existing system explicit or specifies the behaviour of

---

[1] The questions asked by the algorithm have the following form: 'Does the rule $p \rightarrow q$ hold in the system?' The user either confirms this or enters a counterexample and its attributes.

the system he wants to design. The output of the algorithm is an implicational base of the considered system.

Here the technique will be demonstrated by another example, which is already a bit more complicated. In particular, its state and action contexts are too large to be recorded and displayed completely in a convenient way. Yet the algorithm of attribute exploration is still feasible.

5.1 The Situation

Consider a dynamic system used every day: the telephone. Clearly, everyone has an implicit knowledge of its behaviour and knows how to use it correctly. Now this knowledge has to be expressed explicitly and formally correct by a CG or a Petri net, respectively. This is a task still manageable 'by hand' but already relatively intricate. Let our scope be two telephones, say A and B, which are connected to the outer world. Each of it has several states. First, it may be hung up - in this case it may be ringing or silent. If the phone is picked up, one may hear the dialling tone, the engaged tone, the call signal or one is connected to another telephone. The possible actions which can be performed are: pick up the phone, hang up, and dial a number.

5.2 Stepwise Exploration

To gain the needed information it is useful to start at an elementary level and elevate in the hierarchy of complexity. So the first step is to explore all interesting 'local' static properties of one telephone. This means an exploration involving the attributes *hungup, ringing, silent, pickedup, diallingtone, engagedtone, callsignal* and *connected*. The rule exploration yields the implicational base in Figure 10 as result. This rule base can be seen as set of constraints which ensure the consistency of one telephone's states.

hungup, pickedup	$\rightarrow \perp$
hungup, ringing, silent	$\rightarrow \perp$
pickedup, diallingtone, engagedtone	$\rightarrow \perp$
pickedup, diallingtone, callsignal	$\rightarrow \perp$
pickedup, diallingtone, connected	$\rightarrow \perp$
pickedup, engagedtone, callsignal	$\rightarrow \perp$
pickedup, engagedtone, connected	$\rightarrow \perp$
pickedup, callsignal, connected	$\rightarrow \perp$
ringing	$\rightarrow hungup$
silent	$\rightarrow hungup$
diallingtone	$\rightarrow pickedup$
engagedtone	$\rightarrow pickedup$
callsignal	$\rightarrow pickedup$
connected	$\rightarrow pickedup$

Figure 10. Implicational base for the exploration of the local static properties

attributes:			
$A.hungup$	$A.callsignal$	$B.hungup$	$B.callsignal$
$A.pickedup$	$A.callingB$	$B.pickedup$	$B.callingA$
$A.ringing$	$A.callingX$	$B.ringing$	$B.callingX$
$A.silent$	$A.connected$	$B.silent$	$B.connected$
$A.diallingtone$	$A.connectedB$	$B.diallingtone$	$B.connectedA$
$A.engagedtone$	$A.connectedX$	$B.engagedtone$	$B.connectedX$

implications: all rules from Figure 10 and:	
$A.callingB$	$\rightarrow A.calling$
$A.callingX$	$\rightarrow A.calling$
$A.callingX,\ A.callingB$	$\rightarrow \perp$
$A.connectedB$	$\rightarrow A.connected$
$A.connectedX$	$\rightarrow A.connected$
$A.connectedX,\ A.connectedB$	$\rightarrow \perp$

Figure 11. Starting information for the exploration of the system's static properties

$A.silent,\ B.callsignal$	$\rightarrow B.callingX$
$A.hungup,\ B.connected$	$\rightarrow B.connectedX$
$A.diallingtone,\ B.connected$	$\rightarrow B.connectedX$
$A.engagedtone,\ B.connected$	$\rightarrow B.connectedX$
$A.callingB$	$\rightarrow B.ringing$
$A.callingX,\ B.connected$	$\rightarrow B.connectedX$
$A.callsignal,\ B.pickedup$	$\rightarrow A.callingX$
$A.connectedB$	$\rightarrow B.connectedA$
$A.connectedX,\ B.connected$	$\rightarrow B.connectedX$

Figure 12. Implicational Base of the system's static properties

The next step is an exploration of the entire system's static properties (including two telephones in our case). For this, we introduce for each of the 'local attributes' mentioned above one attribute for telephone A and one for telephone B. Furthermore, we introduce four more attributes for each telephone, namely *callingB, callingX, connectedB, connectedX* for telephone A and *callingA, callingX, connectedA, connectedX* for telephone B, expressing whether the partner included in our scope or some other member is involved.

The background knowledge we can start with consists of two copies of the implicational base computed above: one for each telephone. We may then add some more implications to the background knowledge, linking the newly introduced attributes to the other ones. Figure 11 shows the set of attributes and the background information used as input for the second step of the exploration process. Furthermore, we can assume the behaviour of the two telephones to be equal and thus define a permutation on the set of attributes which - if applied to an implication - does not change its truth value. This permutation just 'exchanges' the two telephones. Entering this permutation in advance shortens the exploration process considerably.

The second exploration step yields implications concerning the whole system's possible states (again they can be seen as constraints) which are recorded in Figure 12. At the third exploration step the dynamical aspect is introduced. For

	$\rightarrow A.pickedup_A, A.hungup_\Omega , A.silent_\Omega$
$B.hungup_\Omega$	$\rightarrow B.hungup_A$
$B.hungup_A$	$\rightarrow B.hungup_\Omega$
$B.pickedup_\Omega$	$\rightarrow B.pickedup_A$
$B.pickedup_A$	$\rightarrow B.pickedup_\Omega$
$B.connectedX_\Omega$	$\rightarrow B.connectedX_A$
$B.connectedX_A$	$\rightarrow B.connectedX_\Omega$
$B.callingX_\Omega$	$\rightarrow B.callingX_A$
$B.callingX_A$	$\rightarrow B.callingX_\Omega$
$B.diallingtone_\Omega$	$\rightarrow B.diallingtone_A$
$B.diallingtone_A$	$\rightarrow B.diallingtone_\Omega$
$B.silent_A$	$\rightarrow B.silent_\Omega$
$A.diallingtone_A, B.silent_\Omega$	$\rightarrow B.silent_A$
$A.engagedtone_A, B.silent_\Omega$	$\rightarrow B.silent_A$
$A.callingX_A, B.silent_\Omega$	$\rightarrow B.silent_A$
$A.connectedX_A, B.silent_\Omega$	$\rightarrow B.silent_A$
$B.ringing_\Omega$	$\rightarrow B.ringing_A$
$A.diallingtone_A, B.ringing_A$	$\rightarrow B.ringing_\Omega$
$A.engagedtone_A, B.ringing_A$	$\rightarrow B.ringing_\Omega$
$A.callingX_A, B.ringing_A$	$\rightarrow B.ringing_\Omega$
$A.connectedX_A, B.ringing_A$	$\rightarrow B.ringing_\Omega$
$B.engagedtone_A$	$\rightarrow B.engagedtone_\Omega$
$A.diallingtone_A, B.engagedtone_\Omega$	$\rightarrow B.engagedtone_A$
$A.engagedtone_A, B.engagedtone_\Omega$	$\rightarrow B.engagedtone_A$
$A.callingX_A, B.engagedtone_\Omega$	$\rightarrow B.engagedtone_A$
$A.connectedX_A, B.engagedtone_\Omega$	$\rightarrow B.engagedtone_A$
$A.callingB_A$	$\rightarrow B.silent_\Omega$
$A.callsignal_A B.ringing_A, B.ringing_\Omega$	$\rightarrow A.callingX_A$
$A.connectedB_A$	$\rightarrow B.engagedtone_\Omega$
$B.ringing_A, B.silent_\Omega$	$\rightarrow A.callingB_A$
$B.connected_A, B.engagedtone_\Omega$	$\rightarrow A.connectedB_A$

Figure 13. Implicational Base of the systems dynamic properties concerning action $A.putdown$

each action we want an implicational base concerning the pre- and poststates' attributes. So, like shown before, for each state attribute m we introduce two new attributes m_A and m_Ω, indicating whether the attributes hold before respectively after the action taking place. Naturally, all static property implications can be used as background knowledge twice: for the pre- and for the post-state. The result of this exploration step concerning the action $A.putdown$ can be seen in Figure 13. The implicational bases of all action contexts together with the implicational bases concerning the static properties mentioned before represent the implicational knowledge of our dynamic system.

5.3   Transferring the Results into CGs

At the end we have to find a dynamic CG model of our described system by defining appropriate demons. The translation of a state attribute into CG description is intuitively clear - we just create a (non-dynamic) CG, which expresses the corresponding state property. To avoid confusion when considering the assertion and retraction process, we assume that the attributes are translated into distinct CGs. In order to obtain the according demons we can use another FCA technique: determining all concepts of a formal context. In our case the concepts (more exactly: the concept intents) can be derived from the implication set found in the former process. Now for each transition intent (which is a set $N \subseteq M_{A\Omega}$ with $N = \{n \mid t J n\}$ for some $t \in T_a$) we can define a demon in the following way:

Definition 5. *Let $a \in A$ be an action, $M_1, M_2 \subseteq M$ be sets of state attributes and let $\{m_A \mid m \in M_1\} \cup \{m_\Omega \mid m \in M_2\}$ a transition intent of the a-context. The demon corresponding to this transition intent retracts the CGs representing the attributes M_1 and asserts the CGs representing the attributes M_2. It is labeled with a.*

Obviously, the system designed in this way shows the intended and specified behaviour since it contains all possible state property combinations and respects all implications of the underlying implicational base.

However, this solution is still inconvenient for two reasons. First: it is not trivial to determine whether a concept intent is a transition intent because as mentioned before the whole transition set need not (and is sometimes impossible) to be known. Second: the amount of demons needed to represent one elementary action is quite large.[2] One possibility of coping with both problems is to investigate the invariants.

Definition 6. *Let $a \in A$ be an action, $M_1, M_2 \subseteq M$ be two sets of state attributes and $m \in M \setminus (M_1 \cup M_2)$ a state attribute. We call m INVARIANT UNDER ACTION $a \in A$ WITH PRECONDITIONS M_1 AND POSTCONDITIONS M_2 if the two implications $M_{1A}, M_{2\Omega}, m_A \to m_\Omega$ and $M_{1A}, M_{2\Omega}, m_\Omega \to m_A$ hold in the a-context.*

With this definition we can express in which case a certain state attribute is not changed by an action. Using this notion of invariants, the number of demons for one action can be reduced and the transition intents need not to be determined:

Proposition 1. *Let $a \in A$ be an action and $M_1, M_2 \subseteq M$ be two sets of state attributes. If all $m \in M \setminus (M_1 \cup M_2)$ are invariant under a with preconditions M_1 and postconditions M_2 then the demon retracting the information in M_1 and asserting the information in M_2 shows the specified behaviour and makes all further demons retracting (at least) M_1 and asserting (at least) M_2 obsolete.*

A demon created in that way does not 'take care' of all invariant attributes and thus may substitute several demons due to Definition 5 which do 'take care'. This proposition facilitates to develope an algorithm which creates demons

---

[2] In fact one would need as many demons for one action a as there are transitions caused by a.

Demon nr.	retracts	asserts
1	$A.pickedup, A.connected, A.connectedX$	$A.hungup, A.silent$
2	$A.pickedup, A.connected, A.connectedB$	$A.hungup, A.silent$
	$B.pickedup, B.connected, B.connectedA$	$B.pickedup, B.engagedtone$
3	$A.pickedup, A.callsignal, A.callingX$	$A.hungup, A.silent$
4	$A.pickedup, A.callsignal, A.callingB$	$A.hungup, A.silent$
	$B.hungup, B.ringing$	$B.hungup, B.silent$
5	$A.pickedup, A.engagedtone$	$A.hungup, A.silent$
6	$A.pickedup, A.diallingtone$	$A.hungup, A.silent$

Figure 14. List of demons needed to represent the action $A.putdown$

from the transition context. We pass through a lectical ordered list of concepts and check, whether the condition mentioned in the proposition above holds for the concept intent. If this is the case, then a corresponding demon is created and all subconcepts of this concept are deleted from the list (since the list is ordered lectically, all subconcepts of a concept are listed after it and thus no superfluous intent has been passed before). We proceed until the end of the list is reached. Figure 14 shows the list of demons created in this way for the $A.putdown$ action. In Figure 15 the first item of the table in Figure 14 is explicitly shown as dynamic CG.

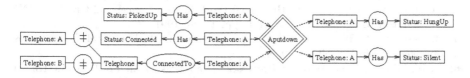

Figure 15. One dynamic CG from the list above

The complete dynamic CG for the telephone scenario would be by far too large to be displayed here. Now we have found a CG description of the dynamic system by means of FCA methods. It can be translated one to one into a petri net as well: we introduce a token place for every attribute, a transition for every demon and draw the directed edges accordingly. When used in a CG-based system, the demons have to be triggered in some way. This could be done by putting an additional retraction edge to each demon, which is linked to a CG expressing, that the corresponding action is initiated.

Note that also the case of indeterministic dynamic systems is covered by our approach.

6   Conclusion

FCA methods prove useful in both extracting processable implicational knowlededge from dynamic systems with known state space and designing CG descriptions by stepwise exploration.

There are several questions arising from this work. What inferences can be done about composite actions? How can the approach be extended to conditional actions and iterations? Given properties of an initial state and properties, which are to be achieved, how can an algorithm be found that efficiently generates a corresponding plan (i.e. a list of actions to be performed)? These questions are objects of ongoing research.

Additionally, the feasibility and efficiency of the proposed techniques have to be evaluated empirically by applying them to more complex scenarios. We expect that increasing complexity will require an elaborated theory as the basis for efficient algorithms.

References

1. J.F. Baget, D. Genest, M.L. Mugnier: A Pure Graph-Based Solution to the SCG-1 Initiative. In: W. Tepfenhart, W. Cyre (eds.): Conceptual Structures: Standards and Practices, LNAI 1640, Springer, Berlin-Heidelberg, pp. 335-376, 1999
2. B. Baumgarten: Petri-Netze. Grundlagen und Anwendungen. Spektrum, Heidelberg, 1996.
3. A. Chella, M. Frixione, S. Gaglio: Towards a Conceptual Representation of Actions. In: E. Lamma, P. Mello (eds.): AI*IA 99, LNAI 1792, Springer, Berlin-Heidelberg, pp. 333-344, 2000
4. H. S. Delugach: Dynamic Assertion and Retraction of Conceptual Graphs. In: Eileen C. Way (ed.): Proc. Sixth Annual Workshop on Conceptual Graphs. SUNY Binghamton, Binghamton, New York, pp. 15-26, 1991.
5. B. Ganter, R. Wille: Formal Concept Analysis: Mathematical Foundations. Springer, Berlin-Heidelberg 1999.
6. G. W. Mineau: From Actors to Processes: The representation of Dynamic Knowledge Using Conceptual Graphs. In: M.L. Mugnier, M. Chein (eds.): Conceptual Structures: Theory, Tools and Application, LNAI 1453, Springer, Berlin-Heidelberg, pp. 198-208, 1998.
7. J. F. Sowa: Conceptual Structures: Information Processing in Mind and Machine. Addison-Wesley, Reading, MA, 1984.
8. P. H. Starke: Analyse von Petri-Netz-Modellen. Teubner, Stuttgart, 1990.
9. K. E. Wolff: Towards a Conceptual System Theory. In: Proc. Third International Conference on Computing Anticipatory Systems. American Institute of Physics, 2000.

Many-Valued Context Analysis Using Descriptions

Ralf Gugisch

University of Bayreuth, Department of Mathematics,
D-95440 Bayreuth, Germany
ralf.gugisch@uni-bayreuth.de

Abstract. We propose an approach to many-valued contexts using formal descriptions instead of scaling. The underlying idea is the philosphical definition of a concept as a set of objects together with the *most precise description*.

We introduce a *formal description* as a mapping from the set of attributes to the power set of the values (which is extended appropriately to empty cells), assigning to each attribute the set of allowed values. Descriptions are naturally ordered by preciseness. Using this, we can introduce *extent* and *intent* according to the philosophical idea, and thus we define *concepts*. We present a way to restrict the amount of concepts for a many-valued context by *preselecting* some descriptions of interest. Furthermore, we introduce *implications* on descriptions, allowing to investigate relationships between attributes.

Within this approach, we reformulate the known theory under a different point of view. It certainly does not provide a better analysis than scaling, but it allows to avoid the generation of a huge one-valued context.

1 Many-Valued Contexts

Recall the definition of many-valued contexts:

Definition 1. *A many-valued context* $\mathbb{K} = (G, M, W, I)$ *is a set of objects* G, *a set of attributes* M, *a set of possible values* W, *and a ternary relation* $I \subseteq G \times M \times W$, *with*

$$(g, m, w) \in I, (g, m, v) \in I \Longrightarrow w = v \,.$$

$(g, m, w) \in I$ *indicates, that object* g *has the attribute* m *with value* w. *In this case, we also write* $m(g) = w$, *regarding the attribute* m *as a partial function from* G *to* W.

We can consider in particular each data base as a many-valued context, thus formal concept analysis appears as a tool of knowledge discovery and data-mining. Within this paper, we will consider the following small example (taken from [7]), representing some facts from algebra:

H. Delugach and G. Stumme (Eds.): ICCS 2001, LNAI 2120, pp. 157–168, 2001.
© Springer-Verlag Berlin Heidelberg 2001

Example 1. Let $\mathbb{K} = (G, M, W, I)$ be the many-valued context with the set of real number objects $G = \{2, \sqrt{2}, \sqrt[3]{2}, \pi\}$ together with the set of attributes $M = \{i, a, t\}$, where "i" means *irrational*, "a" *algebraic* and "t" *transcendental*. The set of values is $W = \{\times\} \cup \mathbb{N}$, and the incidence relation is specified by the table below.

	i	a	t
2		1	
$\sqrt{2}$	×	2	
$\sqrt[3]{2}$	×	3	
π	×		×

Irrational and transcendental are one-valued attributes, i.e. an "×" within the table indicates, that the given number has the corresponding property. But the attribute algebraic is many-valued: $n \in \mathbb{N}$ indicates, that this number is algebraic of degree n, i.e. it is a root of a polynomial over $\mathbb{Q}$ of degree n, n being minimal with this property. If the given number is not algebraic, then we have an empty cell in the row of this number and in the column of the attribute "a".

Formal concept analysis ([4]) was inspired by the fact that we are thinking in terms of concepts consisting of extent and intent. The extent is a set of objects and the intent is the set of attributes, which all these objects do have (according to the "logic of Port-Royal", see [1]). We want to keep this orientation also for many-valued contexts, but we want to avoid the transformation of many-valued attributes into one-valued attributes.

For this purpose we quote Arnauld and Nicole, who write at the end of chapter VI:

> "I call the intent of the idea (=concept) the (set of) attributes, which the idea comprises and which one cannot remove without destroying the idea,..."

This was meant for one-valued attributes. The known strategy to handle a many-valued context is, to scale it into a one-valued context (*plain scaling*) and to analyse the latter one considering concepts via extent and intent.

We want to present here an alternative method. It certainly yields the same results as plain scaling, but it gives a different (philosophically founded) interpretation to the concept lattices. Thus strictly speaking, we should not consider the present work as a "new strategy" or a "generalization" of formal concept analysis, rather than as a new interpretation of an already known strategy.

2   Descriptions

Considering many-valued contexts, we have, of course, to generalize the definition of concepts, and we should like to replace it by the following rough description of what we intend to make more precise later on:

> A concept is a set of objects, together with the most precise description of them.

To begin with, let us briefly describe, where the problems are. Consider, for example, the attribute "red color". It is by definition one-valued. An object either has the attribute "red color", or not.

But if you want to do knowledge discovery and data-mining within a data base, you most often need to analyse attributes having several values. Instead of saying, that an object has a red color, we can also say that the color of the object is red. In the first case, we consider the one-valued attribute "red color". In the latter form, objects can have different values in the many-valued attribute "color": red, green, blue, etc. For data analysis, such many-valued attributes are better applicable than one-valued attributes.

But what is a description? For example, we may describe a car in a way like this: "It is a blue Ford Fiesta, 60-horsepower". For computer representation, we would reformulate this information: "color: blue; model: Ford Fiesta; horsepower: 60". In this way, we could store objects of type car within a data base, using the many-valued attributes color, model and horsepower.

If we want to describe a set of objects (for example the cars, we want to take into consideration for our next buy), we could specify the set of allowed values for each attribute. If we prefer, for example, the models VW Golf, or perhaps again a Ford Fiesta, we assign to the attribute "model" the set {"VW Golf", "Ford Fiesta"}. If we state that our next car should have at least as many horsepowers as the old one, we assign an open interval $[60, \infty)$ to the attribute "horsepower". If we do not have any preferences on the color, we assign the whole set of possible colors to the attribute "color". (Normally, we would not explicitly mention the last aspect, as it is not really a restriction on the set of admissible cars.)

From the mathematical point of view, we assigned to each attribute the set of values, which are allowed for the objects we want to describe. Thus, in mathematical terms, we can define a *description* as a mapping from the set of attributes to the power set of values. (This is not the only possibility for describing objects, but the most simple one. If we need to consider relationships between attributes, we need a more powerful description, using a suitable logic. Work in this direction has been done in logical concept analysis, see [2], [3].)

With this idea of a description, we can specify, in which case an object g *fulfills* a description d, or equivalently, d *holds* for g, iff the value of g in each attribute m is contained in $d(m)$.

We also know a natural order on the set of descriptions: *preciseness*. A description is more precise (greater), iff it applies to less objects, i.e it restricts the values an object can have in some attribute.

Considering Empty Cells

The definition of descriptions so far would suffice, if each object would have in every attribute some value. Unfortunately, data bases most often have missing values. In the theory of formal concept analysis, they are called *empty cells*. In formal terms, many-valued contexts may contain pairs (g, m) with

$$(g, m, w) \notin I, \ \forall\, w \in W .$$

Such empty cells can have different interpretations: It could be, for example, that none of the values applies for the given object, or that the value is just not known.

We could try to handle empty cells just like an additional value, equivalent to the other ones. But this leads to unsatisfactory results. (In short, the effect is comparable to the consideration of the dichotomic situation of one-valued contexts, which is not wanted in every case. It would result in too many concepts. See [5] for more information.)

Thus, we need to consider empty cells in a different way: If an object g has no entry at attribute m (i.e. (g, m) is an empty cell), then a description should only hold for g if it has no restrictions on the values at attribute m. Especially, we need to distinguish between descriptions allowing all possible values at an attribute m, and descriptions additionally allowing, that there is no value given at all (i.e. allowing empty cells). The latter one is more general.

In order to handle this situation mathematically, we extend the power set of values, $\mathcal{P}(W)$, by adding a special maximal element ∞:

$$\dot{\mathcal{P}}(W) := \mathcal{P}(W) \dot{\cup} \{\infty\} .$$

We extend the order on $\mathcal{P}(W)$ given by inclusion to $\dot{\mathcal{P}}(W)$ by defining

$$A \leq B, \, \forall A, B \in \mathcal{P}(W) \text{ with } A \subseteq B, \text{ and}$$
$$A < \infty, \forall A \in \mathcal{P}(W) .$$

Furthermore, we regard each attribute $m \in M$ as a mapping from the

$$m \colon \mathcal{P}(G) \to \dot{\mathcal{P}}(W), m(A) := \begin{cases} \{m(g) | g \in A\} & \text{if } \nexists g \in A : (g, m) \text{ is empty cell,} \\ \infty & \text{otherwise.} \end{cases}$$

The notation $m(A)$ is motivated by the fact, that one often regards an attribute m as a (partial) mapping from G to W. Then, $m(A)$ denotes the image of m on the subset $A \subseteq G$. We extended this in order to handle empty cells. Furthermore, we now identify $m(g)$ with $m(\{g\})$. This is possible, as for nonempty cells (g, m), we can identify the element $w := m(g)$ of W with the singleton $\{w\} \in \mathcal{P}(W)$. For empty cells (g, m), we have therewith $m(g) = \infty$.

Finally, we can give a mathematical definition for descriptions:

Definition 2. *Let* $\mathbb{K} = (G, M, W, I)$ *be a many-valued context. A formal description* d *in* $\mathbb{K}$ *is a mapping* $d : M \to \dot{\mathcal{P}}(W)$ *from the set M of attributes to the extended power set* $\dot{\mathcal{P}}(W)$ *of values.* D *denotes the set of all descriptions in* $\mathbb{K}$*:*

$$D := \dot{\mathcal{P}}(W)^M = \{d : M \to \dot{\mathcal{P}}(W)\} .$$

This set is ordered by preciseness:

$$d_1 \leq d_2 :\Longleftrightarrow d_1(m) \geq d_2(m), \, \forall m \in M ,$$

(where $d_1(m) \geq d_2(m)$ is defined with respect to the order in $\dot{\mathcal{P}}(W)$: either $d_1(m) = \infty$, or $d_1(m) \neq \infty, d_2(m) \neq \infty$ and $d_1(m) \supseteq d_2(m)$).
An object $g \in G$ fulfills a description d, iff

$$m(g) \leq d(m), \ \forall m \in M \ .$$

(Again $\leq$ is meant with respect to the order in $\dot{\mathcal{P}}(W)$). A description $d \in D$ holds for a set of objects $A \subseteq G$, iff each object $g \in A$ fulfills d.

One can show, that the set D together with the order of preciseness forms a complete lattice. The supremum $\bigvee d_i$ of a family of descriptions $d_i, i \in I$ is given by the conjunction of the descriptions, and the infimum $\bigwedge d_i$ is given by their disjunction:

$$\bigvee d_i := m \mapsto \bigcap d_i(m) \ ,$$

$$\bigwedge d_i := m \mapsto \bigcup d_i(m) \ .$$

(The used order is dual to the order normally taken when considering conjunction and disjunction. This perhaps confusing definition is motivated by the fact, that we are now able to introduce concepts like in standard formal concept analysis, specifying a Galois connection.)

3   The Concept Lattice

Now having an idea of what a description is, we can introduce concepts as pairs of extent and intent. Therefore, we associate with each set $A \subseteq G$ of objects its *intent* $A' \in D$, the most precise description holding for each $g \in A$. Vice versa, we associate with each description $d \in D$ its *extent* $d' \subseteq G$, the set of objects fulfilling d:

$$A' := M \to \dot{\mathcal{P}}(W), m \mapsto m(A) \ ,$$
$$d' := \{g \in G \mid m(g) \leq d(m), \forall m \in M\} \ .$$

The two mappings $\varphi : A \mapsto A'$ and $\psi : d \mapsto d'$ form a Galois connection between the power set of G and the lattice D of all descriptions (see [5] for a proof of this and the following results). Thus, we have a similar situation, as with one-valued contexts, and we can introduce the concept lattice as usual.

Definition 3. *Let $\mathbb{K} = (G, M, W, I)$ be a many-valued context, and let D be the set of descriptions in $\mathbb{K}$. A formal concept of the context $\mathbb{K}$ is a pair (A, d), with $A \subseteq G, d \in D, A' = d$ and $d' = A$. We call A the extent and d the intent of the concept (A, d). $\mathcal{B}(\mathbb{K})$ denotes the set of all concepts of the context $\mathbb{K}$.*

The set of concepts is ordered by extent (which is dual to the order given by intent). In the case of $(A_1, d_1) \leq (A_2, d_2)$, we call (A_1, d_1) a *subconcept* of (A_2, d_2), and (A_2, d_2) a *superconcept* of (A_1, d_1). Furthermore, $\mathcal{B}(\mathbb{K})$ together with this order forms a complete lattice:

Theorem 1. $\mathcal{B}(\mathbb{K})$ *is a complete lattice. Infimum und supremum for a subset of concepts* $\{(A_i, d_i) \mid i \in I\} \subseteq \mathcal{B}(\mathbb{K})$ *(with an arbitrary set of indices I) is given by*

$$\bigwedge (A_i, d_i) = (\bigcap A_i, (\bigcap A_i)') = (\bigcap A_i, (\bigvee d_i)'') ,$$
$$\bigvee (A_i, d_i) = ((\bigwedge d_i)', \bigwedge d_i) = ((\bigcup A_i)'', \bigwedge d_i) .$$

Labelling the Concept Lattice

A concept lattice usually is visualized using line diagrams, labelled in a way allowing to identify the context with the concept lattice. In the following, we develop a suitable labelling of concept lattices for many-valued contexts.

The labelling with objects can be done in the same way, as with one-valued contexts: We write each object below the smallest concept, containing that object. Thus, we receive the extent of a concept as the union of all objects standing below a concept less than or equal to the given one.

In order to be able to label the concept lattice with descriptions in a clear way, we need to introduce a special subset of descriptions, let us call them *singular descriptions*. This set shall be large enough to represent each (other) description as a conjunction of singular ones. Thus, we call a description *singular*, iff

1. it has no restrictions on any attribute except for one attribute $m \in M$, and
2. for this attribute m, the only restriction is either, that it must have some value (i.e. it disallows empty cells), or that it must have a value unequal to $w \in W$ (also forbidding empty cells).

We give to these singular descriptions intuitive names: We call a description having no restrictions except that one attribute $m \in M$ must not be an empty cell, just like the attribute: m. One has to be a little bit careful with this nomenclature, as now m can stand for an attribute as well as for a description. But it should always be clear from the context, what was meant. Furthermore, if a description has no restrictions on attributes $\neq m$, but that m must neither be an empty cell, nor equal to the value w, we call that restriction $m_{\neq w}$.

$$m : M \to \dot{\mathcal{P}}(W), \quad m(n) := \begin{cases} W \text{ if } n = m , \\ \infty \text{ else.} \end{cases}$$

$$m_{\neq w} : M \to \dot{\mathcal{P}}(W), m_{\neq w}(n) := \begin{cases} W \setminus \{w\} \text{ if } n = m , \\ \infty \qquad \text{ else.} \end{cases}$$

The set of all singular descriptions $\{m \mid m \in M\} \cup \{m_{\neq w} \mid m \in M, w \in W\}$ forms a $\vee$-dense subset of D, that means, we can represent each description as conjunction of singular descriptions.

Using this, we can label the concept graph with descriptions, too: We write the singular descriptions above the greatest concept they appear in. Then we can get the intent of a concept as conjunction of all singular descriptions standing above a concept greater or equal to the given one.

The concept lattice of the many-valued context for real numbers from Example 1 can be seen in Fig. 1.

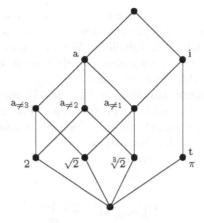

Fig. 1. The concept lattice of the context for real numbers

4 Preselection of Descriptions

An analysis of the concept lattice of the example above shows, that it contains some concepts, we are hardly interested in. For example, the context labeled by the description $a_{\neq 2}$ contains all algebraic numbers of degree 1 or of degree ≥ 3. Normally, we are only interested in concepts whose intent restricts the algebraic degree to some interval.

This effect gets worse, if we consider bigger many-valued contexts. Various unreasonable concepts would arise and make the concept lattice unnecessary complex.

The problem is, that our set D of descriptions in $\mathbb{K}$ can contain descriptions assigning attributes to abnormal sets of allowed values without any reasonable interpretation. Having such undesired descriptions, we may also get concepts using them.

But it is known from lattice theory, that we can restrict the derivation operators to a $\bigvee$-subsemilattice $\bar{D} \subseteq D$ of the set of descriptions simply by defining the intent of a set $A \subseteq G$ of objects as the most precise description in $\bar{D}$ (instead of the most precise one in D), whereas the definition of the extent remains the same. This way, we obtain another set of concepts, having only intents from $\bar{D}$. These concepts form a $\bigwedge$-subsemilattice of the original concept lattice.

Thus we just need to specify an arbitrary subset $\tilde{D} \subseteq D$, say, a *preselection* of descriptions of interest. Then we define $\bar{D}$ as the closure of $\tilde{D}$ under conjunction:

$$\bar{D} := \{\bigvee B \mid B \subseteq \tilde{D}\}\,.$$

This way, we can influence the amount of descriptions taken into consideration as intent for concepts. We can exclude a lot of unreasonable concepts, and we can do problem oriented analysis into specific aspects of a many-valued context. We call the resulting subsemilattice of concepts a *scaled concept lattice* with respect to the preselection $\tilde{D}$.

We label a scaled concept lattice with the preselected descriptions $d \in \tilde{D}$. As these are $\vee$-dense in $\bar{D}$, and as each intent in the scaled concept lattice is contained in $\bar{D}$, this labelling is sufficient. We still receive the intent of each concept as the conjunction of preselected descriptions standing above concepts greater than or equal to the considered one.

In order to have intuitive labellings in the scaled concept lattice, we define some standard descriptions usable in preselections. For most purposes, it will suffice to consider within preselections only description restricting one attribute $m \in M$, i.e. assigning ∞ to all other attributes.

The easiest kind of such descriptions are *nominal descriptions*, allowing for m only one value $w \in W$:

$$m_{=w} : M \to \dot{\mathcal{P}}(W), m_{=w}(n) := \begin{cases} \{w\} & \text{if } n = m \text{,} \\ \infty & \text{else.} \end{cases}$$

Furthermore, if an attribute has ordinal values, we can consider *ordinal descriptions*, allowing open intervals for the attribute m:

$$m_{\leq w} : M \to \dot{\mathcal{P}}(W), m_{\leq w}(n) := \begin{cases} (-\infty, w] & \text{if } n = m \text{,} \\ \infty & \text{else.} \end{cases}$$

$$m_{\geq w} : M \to \dot{\mathcal{P}}(W), m_{\geq w}(n) := \begin{cases} [w, \infty) & \text{if } n = m \text{,} \\ \infty & \text{else.} \end{cases}$$

In the same way, one could define further kinds of standard descriptions if necessary.

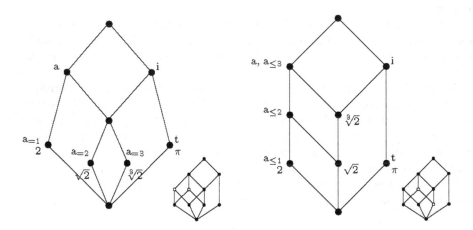

Fig. 2. Two scaled concept lattices of the context for real numbers. The embedding into the total concept lattice is indicated by the small pictures

Reconsider Example 1: Figure 2 shows two possibilities of scaled concept lattices. In the first one, we allowed only nominal descriptions for the attribute algebraic, i.e. we used the preselection $\tilde{D}_1 = \{i, a, t, a_{=1}, a_{=2}, a_{=3}\}$. The second concept lattice results, if we allow only ordinal descriptions of the type $a_{\leq w}$, i.e. we used $\tilde{D}_2 := \{i, a, t, a_{\leq 1}, a_{\leq 2}, a_{\leq 3}\}$.

Both scaled concept lattices can be embedded into the original one. This is indicated in Fig. 2 by the small pictures beside the concept lattices. As one can recognize, this embedding is infimum-preserving: the infimum of two full nodes is full, again.

Both scaled concept lattices contain less concepts as the original one, and so they are less complex. But they emphasize different aspects of the context.

5   Implications

In order to analyse relationships between attributes of a formal context, we introduce implications between descriptions. Given a many-valued context $\mathbb{K} = (G, M, W, I)$, an implication $a \to b$ between two descriptions $a, b \in D$ *holds* in $\mathbb{K}$, iff each object $g \in G$, which fulfills description a, also fulfills description b. More formally, we define:

Definition 4. *Let $\mathbb{K} = (G, M, W, I)$ be a many-valued context, and let $D = \dot{\mathcal{P}}(W)^M$ be the set of all descriptions within $\mathbb{K}$. Furthermore, let $a, b \in D$ be descriptions.*

A description $d \in D$ respects the implication $a \to b$, if $a \not\leq d$ or $b \leq d$. d respects a set $\mathcal{L}$ of implications if d respects every single implication in $\mathcal{L}$. $a \to b$ holds in a set $\{d_1, d_2, \ldots\}$ of descriptions if each of the descriptions d_i respects the implication $a \to b$. $a \to b$ holds in the context $\mathbb{K}$ if it holds in the system of object intents. In this case, we also say, that $a \to b$ is an implication of the context $\mathbb{K}$ or, equivalently, that within the context $\mathbb{K}$, a is a premise of b.

We defined an equivalent structure to implications between attributes in one-valued contexts. The known algorithms and further definitions like the stem base can be translated in a straightforward way to many-valued contexts and implications between descriptions, as shown in [5].

We can represent descriptions as conjunctions of singular or preselected ones. Thus we specify a description d just by listing all singular/preselected descriptions smaller than d.

As an implication $a \to b$ is always equivalent to $a \to a \vee b$, we don't need to repeat on the right hand side singular/preselected descriptions already standing on the left hand side of an implication. Thus, we can represent an implication $a \to b$ in a reduced way, listing on the left hand side all singular/preselected descriptions smaller than a, and on the right hand side all additional ones smaller than b.

Example 2. Reconsider Example 1 for real numbers. The stem base is:

$$t \to i$$
$$a_{\neq 1} \to i$$
$$i, a \to a_{\neq 1}$$
$$i, a_{\neq 1}, t \to M$$

As already known, we can consider specific aspects of the concept lattice using a preselection. Then, we also get a different stem base. Below we show the stem bases resulting from the two preselections using nominal or ordinal descriptions for the attribute algebraic, as already considered above (Example 4):

$$t \to i$$
$$a_{=3} \to i$$
$$a_{=2} \to i$$
$$i, a, t \to M$$
$$i, a_{=1} \to M$$

$$t \to i$$
$$a \to a_{\leq 3}$$
$$i, a_{\leq 3}, t \to M$$
$$i, a_{\leq 1} \to M$$

Using implications on descriptions, we can apply attribute exploration directly on many-valued contexts.

6   Relationship to One-Valued Contexts

It is well known, that one can generate one-valued contexts out of many-valued ones using scales. What is the relationship between these scaled contexts and the approach on many-valued contexts given in this paper?

First, we can specify for each many-valued context $\mathbb{K} = (G, M, W, I)$ a one-valued context $\mathbb{K}' = (G', M', I')$ having an isomorphic concept lattice: Just take the set of objects, $G' := G$, as set of attributes M' take the singular descriptions, and let (g, d) be in I' iff g fulfills d. If you want to use a preselection $\tilde{D}$, you get a one-valued context with isomorphic concept lattice by taking the preselected descriptions as M' instead of the singular ones.

A closer look at these one-valued contexts associated with preselections shows, that one can interpret each plain scaling as a preselection. As a consequence, we can interpret plain scaling as a subcontext of the many-valued context, having as concept lattice an $\bigwedge$-subsemilattice of the many-valued concept lattice.

On the other hand, preselections can be understood as scaling, thus it provides no better analysis results.

This approach can be seen as a natural extension of one-valued contexts from another point of view, too: If we interpret a one-valued context $\mathbb{K} = (G, M, I)$ as a many-valued one with exactly one value "$\times$": $\mathbb{K}' := (G, M, \{\times\}, I')$ with

$$(g, m, \times) \in I' \Leftrightarrow (g, m) \in I,$$

then the concept lattices of $\mathbb{K}$ and of $\mathbb{K}'$ are isomorphic.

This can be seen as follows: Considering singular descriptions in $\mathbb{K}'$, the only nontrivial ones are of the form $m, m \in M$, as the descriptions of the form $m_{\neq \times}$ do not hold for any object $g \in G$. (If an object g has a value in attribute m, i.e. (g, m) is no empty cell, it must have the only value $\times$.) Thus, as we have seen above, $\mathbb{K}'$ is isomorphic to the one-valued context $\mathbb{K}'' = (G, \{m, m \in M\}, I'')$ with

$$(g, m) \in I'' \Longleftrightarrow g \text{ fulfills } m \Longleftrightarrow (g, m, \times) \in I' \Longleftrightarrow (g, m) \in I.$$

Thus I'' is equivalent to I, and $\mathbb{K}''$ equivalent to $\mathbb{K}$.

Another point worth to consider is the comparison between implications on descriptions and implications on one-valued attributes, i.e. scaling a one valued context and considering implications within the scaled context. The second strategy may lead to trivial implications representing relationships between different states of one many-valued attribute.

Within our example for real numbers, the stem base of a scaled context using the ordinal scale for attribute algebraic needs to express implications like $a_{\leq 2} \rightarrow a_{\leq 3}$. Using descriptions, we do not need to express such implications, as we know the order on descriptions, which means that we know some implicit background information.

There exists some approach ([8]) considering background information for implications on attributes. Using this, one should obtain similar stem bases with both strategies. But it remains to do some research on this aspect.

7   Conclusion

We have introduced a theory of many-valued concept analysis using descriptions. This provides a different way to analyse concepts within many-valued contexts, besides the possibility of scaling them and analysing the resulting one-valued context. Thanks to the possibility of preselecting specific descriptions, the new strategy remains as flexible as scaling, and it turns out, that it is possible to get equivalent results with both strategies.

One advantage of using descriptions is that we avoid generating huge one-valued contexts. Furthermore, we are able to consider implications on descriptions, and therefore may apply strategies like attribute exploration directly on many-valued contexts.

Acknowledgements. Special thanks are due to A. Kerber for his continuous support, and to the referees for very helpful hints.

References

1. Arnauld, A. Nicole, P.: *La Logique ou L'Art de penser, contenant, outre les Règles communes, plusieurs observations nouvelles, propres à former le jugement.* Sixième Édition revue et de nouveau augmentée. A Amsterdam, Chez Abraham Wolfgang (1685). German translation by Axelos, C., 2. edition, Wiss. Buchges., Darmstadt (1994)
2. Chaudron, L., Maille, N.: *Generalized Formal Concept Analysis.* In: Ganter, B., Mineau, G.W., editors, Conceptual structures: Logical, Linguistic, and Computational Issues. Lecture Notes in Artificial Intelligence 1867. Springer, Berlin-Heidelberg-New York (2000) 357–370
3. Ferre, S., Ridoux, O.: *A Logical Generalization of Formal Concept Analysis.* In: Ganter, B., Mineau, G.W., editors, Conceptual structures: Logical, Linguistic, and Computational Issues. Lecture Notes in Artificial Intelligence 1867. Springer, Berlin-Heidelberg-New York (2000) 371–384

4. Ganter, B., Wille, R.: *Formal Concept Analysis* – Mathematical Foundations. Springer Verlag, Berlin (1999)
5. Gugisch, R. *Verallgemeinerung der formalen Begriffsanalyse und Anwendung auf mehrwertige Kontexte.* Diplomarbeit, Uni Bayreuth (1997)
   http://www.mathe2.uni-bayreuth.de/ralfg/papers/diplom.ps.gz
6. Gugisch, R. *Lattice Contexts – a Generatization in Formal Concept Analysis.* handout to ICCS 2000, Darmstadt (2000)
   http://www.mathe2.uni-bayreuth.de/ralfg/papers/iccs2000.ps.gz
7. Kerber, A., Lex, W.: *Kontexte und ihre Begriffe.* (1997)
   http://www.mathe2.uni-bayreuth.de/kerber/begriffe.ps
8. Stumme, G.: *Attribute exploration with background implications and exceptions.* In: Bock, H.-H., Polasek W., editors, Data analysis and information systems. Springer, Berlin-Heidelberg-New York (1996) 457-469

Mathematical Support for Empirical Theory Building

Selma Strahringer[1], Rudolf Wille[2], and Uta Wille[3]

[1] Fachhochschule Köln, Fachbereich Bibliotheks- und Informationswesen
Claudiusstr. 1, D–50678 Köln, Germany
selma.strahringer@fh-koeln.de
[2] Technische Universität Darmstadt, Fachbereich Mathematik
Schlossgartenstraße 7, D–64289 Darmstadt, Germany
wille@mathematik.tu-darmstadt.de
[3] Jelmoli AG, Information Systems
Postfach 3020, CH-8021 Zürich, Switzerland
wille_u@jelmoli.ch

Abstract. Empirical theory building in the human and social sciences may be mathematically supported by methods of Formal Concept Analysis which is the main theme of this paper. Those theories are considered which can be formalized by a contextual attribute logic. The empirical data are coded by formal contexts whose attribute logic can be used for representing scientific theories. Specific formal contexts, namely conceptual scales, are representing aspects of a theory. Their aggregation by the semiproduct and by the apposition yield a more complete representation of the theory. The gap between the theoretical and the empirical side becomes apparent by comparing the semiproduct and the apposition representation. In an iterative process of theory building this gap should be diminished. Specific support is given by algebraic representations of formal contexts which are used to represent scientific theories.

1 Empirical Theory Building in Human and Social Sciences

Empirical theory building as common in physics is successfully represented in mathematical models, for instance in euclidean vector spaces. Those formal models support the respective disciplinary theory with expressive structure which particularily stabilizes the theory at longer term. In human and social sciences such support is very seldom. A specific approach is offered by representational measurement theory (see [K+71]) which uses relational structures for mathematizing scientific theories; but the range of application is quite restricted because the relational structures are treated with the aim to reach representations by numerical structures.

What is needed for formal representations of theories in human and social sciences are non-numerical mathematical models which make the theories more structurally transparent and communicable. Those models can be derived by

H. Delugach and G. Stumme (Eds.): ICCS 2001, LNAI 2120, pp. 169–186, 2001.

methods of Formal Concept Analysis which offers a mathematical theory for formally developing and analyzing conceptual structures (see [GW99a]). Using this approach, the basic models for representing theories and their parts are formal contexts (also called conceptual scales) together with their concept lattices. A *formal context* is mathematically defined to be a set structure $\mathbb{K} := (G, M, I)$ where G is a set of (formal) objects, M is a set of (formal) attributes, and I is a binary relation between G and M indicating when an object has an attribute. By $X^I := \{m \in M \mid \forall g \in X : gIm\}$ for $X \subseteq G$ and $Y^I := \{g \in G \mid \forall m \in Y : gIm\}$ for $Y \subseteq M$, a Galois connection between the power sets $\mathfrak{P}(G)$ and $\mathfrak{P}(M)$ is given. Consequently, the pairs (A, B) with $A \subseteq G$, $B \subseteq M$, $A^I = B$, and $B^I = A$ (called *formal concepts* of $\mathbb{K}$) form a complete lattice $\mathfrak{B}(\mathbb{K})$ (called the *concept lattice* of $\mathbb{K}$) with respect to the ordering $(A_1, B_1) \leq (A_2, B_2) :\Leftrightarrow A_1 \subseteq A_2$ ($\Leftrightarrow B_2 \subseteq B_1$); the sets in $\mathfrak{U}(\mathbb{K}) := \{A \mid (A, B) \in \mathfrak{B}(\mathbb{K})\}$ are called *extents* and in $\mathfrak{I}(\mathbb{K}) := \{B \mid (A, B) \in \mathfrak{B}(\mathbb{K})\}$ are called *intents*. Within the concept lattice $\mathfrak{B}(\mathbb{K})$, the formal context $\mathbb{K}$ may be recognized by $gIm \Leftrightarrow \gamma g \leq \mu m$ where $\gamma g := (\{g\}^{II}, \{g\}^I)$ is the so-called *object concept* of g and $\mu m := (\{m\}^I, \{m\}^{II})$ is the so-called *attribute concept* of m. $\mathbb{K}$ is said to be *object clarified* (*attribute clarified*) if $\gamma g_1 = \gamma g_2$ ($\mu m_1 = \mu m_2$) always implies $g_1 = g_2$ ($m_1 = m_2$), and *object reduced* (*attribute reduced*) if, in addition, all object concepts (attribute concepts) of $\mathbb{K}$ are $\bigvee$-irreducible ($\bigwedge$-irreducible) in $\mathfrak{B}(\mathbb{K})$. For further notions and results of Formal Concept Analysis, we refer to the monograph [GW99a].

Before presenting a general approach to the formal representation of scientific theories, we first discuss the formalization by an example of a philosophical theory of categories. In his book "Knowledge Representation: Logical, Philosophical, and Computational Foundations" [So00], John Sowa reports about the most influential theories of categories, namely those of Heraclitus, Aristotle, Kant, Peirce, Husserl, Whitehead, and Heidegger. A careful study of those theories together with the aim to support knowledge representation led Sowa to his own theory of top-level categories which he based on the three distinctions "Thing - Relation - Mediation" (Peirce), "Physical - Abstract" (Heraclitus), and "Continuant - Occurrent" (Whitehead). Sowa represents the hierarchical structure of his categories by an ordered set $(S, \leq)$ shown in Figure 1 (see Figure 2.6 in [So00]).

Since $(S, \leq)$ is not a lattice, it is desirable, for representing all logical relationships of Sowa's theory, to form the Dedekind-MacNeille-Completion of the ordered set $(S, \leq)$. This completion is given by the concept lattice $\mathfrak{B}(S, S, \leq)$ whose line diagram is presented in Figure 2. The completion construction seems to suggest to consider the formal context $(S, S, \leq)$ as a basic model for representing Sowa's theory. Since the theory has the aim to classify ontological entities, it is not appropriate to take the whole set S as set of formal objects; instead of S, the set A of upper covers of the smallest element $\bot$ is more adequate as object set because the γg with $g \in A$ are exactly the $\bigvee$-irreducible object concepts, which yields, in particular, the isomorphy $\mathfrak{B}(S, S, \leq) \cong \mathfrak{B}(A, S, \leq)$.

The object and attribute reduced context $(A, \{Thing, Relation, Mediation, Physical, Abstract, Continuant, Occurrent\}, \leq)$, whose concept lattice is still

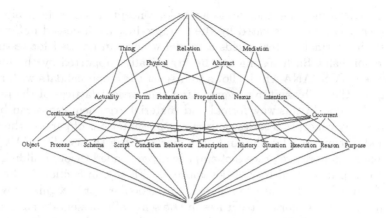

Fig. 1. Hierarchy of Sowa's top-level categories

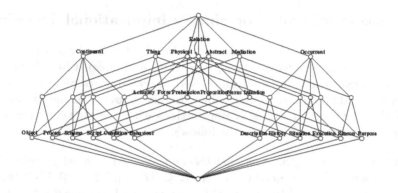

Fig. 2. Dedekind-MacNeille-Completion of ordered set shown in Figure 1

isomorphic to $\mathfrak{B}(S, S, \leq)$, is (up to isomorphism) an example of a *simply implicational standard scale*. These scales form a useful type of models for formalizing those parts of scientific theories which are expressible by implications between attributes (properties) with one-element premise and by inconsistency conditions for sets of attributes. The concept lattices of simply implicational standard scales are (up to isomorphism) the finite distributive lattices truncated by identifying all elements of an order ideal with 0. Simply implicational theories and their realization by simply implicational standard scales are discussed in detail in Section 2. For representing logically more complex theories, means of Contextual Attribute Logic [GW99b] are used; especially, the clause logic of conceptual scales is presented as entailment completion of clausal theories in Section 3. How clause logics of conceptual scales can be described by better recognizable subparts of those logics is discussed by means of standardized conceptual scales of ordinal type.

For representing richer theories as a whole, conceptual scales describing parts of the theory can be aggregated by the semiproduct as discussed in Section 4. The resulting formal context yields a model which can be used for testing the theory empirically. Such testing can be graphically supported by the management system TOSCANA which allows to inspect the empirical data with respect to the given theory by suitably aggregating the concept lattices of the involved conceptual scales. In this way, formal and material argumentations can be activated to justify conceptual scales as proper representations of parts of the theory or to modify conceptual scales and even parts of the theory. TOSCANA may even support an iterative process of empirically grounded theory building which is able to keep its own history of development. Finally, in Section 5, the clause logics of ordinal contexts are considered, in particular, for recognizing whether representations by algebraic structures in the sense of representational measurement theory are available for improving the mathematical support of empirical theory building.

2   Conceptual Scales for Simply Implicational Theories

Parts of scientific theories are often described by implications between properties (attributes) of a fixed set of properties (attributes), where implications with a one-element premise are mainly used. Inconsistencies between properties are seldom stated explicitly, since they usually follow from the literal meaning of the properties. Theories of implications with one-element premises and inconsistencies may be formalized as follows:

Definition: A *simply implicational theory* is defined as a set structure $T :=$ $(M, R, \mathfrak{A})$ where M is a non-empty set, $R \subseteq M^2$, and $\mathfrak{A} \subseteq \mathfrak{P}(M) \setminus \{\emptyset\}$ with $\mathfrak{P}(\{p, q\}) \cap \mathfrak{A} = \emptyset$ for all $(p, q) \in R$; the elements of M are called *attributes*, the pairs (p, q) in R *simple implications* (alternative denotation: $p \to q$), and the sets in $\mathfrak{A}$ *inconsistencies*.

For each simply implicational theory $T := (M, R, \mathfrak{A})$ there is an associated closure system on the attribute set M, which is determined by the implications $p \to q$ with $(p, q) \in R$ and $A \to M$ with $A \in \mathfrak{A}$ (see [GW99a], p.79ff):

$$\mathfrak{H}(T) := \{X \subseteq M \mid \forall (p, q) \in R : p \notin X \vee q \in X) \text{ and }$$
$$(\forall A \in \mathfrak{A} : A \not\subseteq X \vee X = M)\}$$

A formal context $\mathbb{K} := (G, M, I)$ is said to be a *realization* of the simply implicational theory $T := (M, R, \mathfrak{A})$ if $\mathfrak{I}(\mathbb{K}) = \mathfrak{H}(T)$ and $A^I = \emptyset$ for all $A \in \mathfrak{A}$.

A *basic question* is how to describe the realizations of simply implicational theories. In this section we give a structural answer to this question, while in the next section a logical framework is established from which a logical answer can be derived.

Definition: An *ordered set with inconsistent subsets* is defined as a set structure $\underline{P} := (P, \leq, \mathfrak{A})$ where $(P, \leq)$ is an ordered set and $\mathfrak{A}$ is a set of (non-empty) subsets of P having no lower bound in $(P, \leq)$. In the finite case we consider, for any $p \in P$, the set $\mathfrak{I}_p$ of all minimal order ideals D of $(P, \leq)$ with $p \in D$ and $A \cap D \neq \emptyset$ for all $A \in \mathfrak{A}$; let $\mathfrak{I}_{\underline{P}} := \bigcup_{p \in P} \mathfrak{I}_p$. Then the finite formal context $\mathbb{O}_{\underline{P}} := (\mathfrak{I}_{\underline{P}}, P, \not\ni)$ is called a *simply implicational standard scale*.

Proposition 1 *The simply implicational standard scale $\mathbb{O}_{\underline{P}}$ is a realization of the simply implicational theory $T_{\underline{P}} := (P, \leq, \mathfrak{A})$. The intents of $\mathbb{O}_{\underline{P}}$ are exactly the order filters of $(P, \leq)$ equal to P or not containing any $A \in \mathfrak{A}$. $\mathfrak{I}(\mathbb{O}_{\underline{P}}) \setminus \{P\}$ is therefore an order ideal $\mathcal{I}$ of the distributive lattice of all order filters of $(P, \leq)$ which satisfies $F^{\mathcal{I}} = \emptyset$ for all order filters F not in $\mathcal{I}$. Conversely, each such order ideal may be represented by the proper intents of a simply implicational scale.*

Proof: For $D \in \mathfrak{I}_{\underline{P}}$ we have the object intent $D^{\not\ni} = P \setminus D$. Thus, the intents of $\mathbb{O}_{\underline{P}}$ are just the sets $P \setminus \bigcup_{t \in T} D_t$ with $D_t \in \mathfrak{I}_{\underline{P}}$ ($t \in T$), which are obviously order filters of $(P, \leq)$ equal to P or not containing any $A \in \mathfrak{A}$. Now, let F be any such order filter unequal P. Then $P \setminus F$ is an order ideal of $(P, \leq)$ with $A \cap (P \setminus F) \neq \emptyset$ for all $A \in \mathfrak{A}$. For each $p \notin F$ we choose a minimal order ideal D_p in $P \setminus F$ with $p \in D_p$ and $A \cap D_p \neq \emptyset$ for all $A \in \mathfrak{A}$; then $F = P \setminus \bigcup_{p \in P \setminus F} D_p$. Thus, the intents of $\mathbb{O}_{\underline{P}}$ are exactly the order filters of $(P, \leq)$ equal to P or not containing any $A \in \mathfrak{A}$. It immediately follows that $\mathfrak{I}(\mathbb{O}_{\underline{P}}) = \mathfrak{H}(T_{\underline{P}})$. Clearly, $\mathfrak{I}(\mathbb{O}_{\underline{P}}) \setminus \{P\}$ is an order ideal of the lattice $\mathcal{F}(P, \leq)$ of all order filters of the finite ordered set $(P, \leq)$. Conversely, let $\mathcal{I}$ be an order ideal of $\mathcal{F}(P, \leq)$ satisfying $F^{\mathcal{I}} = \emptyset$ for all order filters F not in $\mathcal{I}$, and let $\mathfrak{A}_{\mathcal{I}} := \{F \in \mathcal{F}(P, \leq) \setminus \{\emptyset\} \mid F \notin \mathcal{I}\}$. Then $\underline{P} := (P, \leq, \mathfrak{A}_{\mathcal{I}})$ is an ordered set with inconsistent subsets and $\mathcal{I} = \mathfrak{I}(\mathbb{O}_{\underline{P}}) \setminus \{P\}$.
□

Proposition 1 yields a realization by an object reduced context for all finite simply implicational theories $T := (M, R, \mathfrak{A})$ which can be seen as follows: Let $\hat{R}$ be the reflexive transitive closure of R on M, let $E_R := \hat{R} \cap \hat{R}^{-1}$, and let $\mathfrak{A}/E_R := \{A/E_R \mid A \in \mathfrak{A}\}$; then $\underline{P}_T := (M/E_R, \hat{R}/E_R, \mathfrak{A}/E_R)$ is an ordered set with inconsistent subsets for which $\mathfrak{I}(\mathbb{O}_{\underline{P}_T}) = \mathfrak{H}(T)/E_R := \{X/E_R \mid X \in \mathfrak{H}(T)\}$. If we define, more generally, also for T the *simply implicational standard scale* by $\mathbb{Q}_T := (\mathfrak{I}_{\underline{P}_T}, M, J)$ with $\overline{D} J m :\Leftrightarrow m \notin B$ for all $B \in \overline{D}$, then we obtain the following result:

Proposition 2 *The simply implicational standard scale $\mathbb{Q}_T$ is a realization of the simply implicational theory T and has the same extents as the simply implicational standard scale $\mathbb{O}_{\underline{P}_T}$.*

Proposition 1 also indicates how an appropriate line diagram of the concept lattice of a simply implicational standard scale may be established: As a frame, the dual of the distributive lattice consisting of all order filters of the appertaining ordered set is considered. Since it is isomorphic to a cover-preserving sublattice of the direct product of chains obtained by any chain decomposition of the ordered set, it is recommendable to decompose the ordered set into a minimal number of chains and locate the grid representing of the direct product of those chains in such a way that the nodes of the grid representing the non-zero elements of the concept lattice are best readable; finally, the node representing zero has to be added properly (cf. [L+97]).

Let us close this section by discussing again the example of Section 1. Although Sowa postulates that each category is a synonym for the conjunction formed by categories of the three distinctions, it would make sense to consider them as subcategories and not as synonyms (e.g. "Script" could be understood as a proper subcategory of "Continuant Form"). Then we would obtain a simply implicational theory $(S, \leq, \mathfrak{A}_S)$ with $\mathfrak{A}_S := \{\{Thing, Relation\}, \{Thing, Mediation\}, \{Relation, Mediation\}, \{Physical, Abstract\}, \{Continuant, Ocurrent\}\}$, the simply implicational standard scale of which has the concept lattice shown in Figure 3.

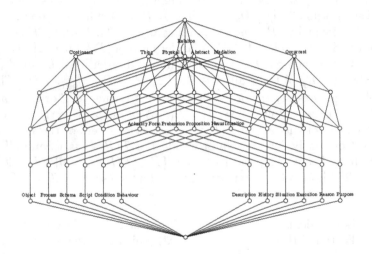

Fig. 3. Concept lattice representing the simply implicational theory of Sowa's top-level categories

3   Conceptual Scales Described by Their Clause Logic

The implications (and inconsistencies) between the attributes of a formal context determine the concept lattice of the context up to isomorphism, but they give no information about what $\bigvee$-reducible concepts are object concepts; in other words, they determine the formal context up to object reduction, but not up to object clarification. Since conceptual scales are usually understood as object clarified contexts, we consider also clauses between attributes by which a formal context can be determined up to object clarification. For explaining this, we give first a brief introduction to the *clause logic of formal contexts* (cf. [BS97],[GW99b]).

Let $\mathbb{K} := (G, M, I)$ be a formal context. A *sequent* of $\mathbb{K}$ is a pair (A, S) with $A, S \subseteq M$ which is understood as a compound attribute having the extent

$$(A, S)^I := \{g \in G \mid g \notin A^I \text{ or } gIm \text{ for some } m \in S\}.$$

For sequents we consider the order $\leq$ defined by $(A_1, S_1) \leq (A_2, S_2) :\Leftrightarrow A_1 \subseteq A_2$ and $S_1 \subseteq S_2$. A sequent (A, S) of $\mathbb{K}$ is said to be *all-extensional* if $(A, S)^I = G$, which is equivalent to say that any object of $\mathbb{K}$ having all attributes of A must also have some attribute of S (such logical statement is usually called a *clause*). The *clause logic* of a formal context $\mathbb{K}$ is the set $\mathcal{C}(\mathbb{K})$ of all all-extensional sequents of $\mathbb{K}$.

Definition: A *clausal theory* is defined as a set structure $T := (M, \mathcal{C})$ where M is a non-empty set and $\mathcal{C} \subseteq \mathfrak{P}(M) \times \mathfrak{P}(M)$; the elements of M are called *attributes* and of $\mathcal{C}$ *sequents*. T is said to be *regular* if $(A, S) \in \mathcal{C}$ is equivalent to $(X, M \setminus X) \in \mathcal{C}$ for all $X \subseteq M$ with $(A, S) \leq (X, M \setminus X)$. The *regular closure* of a clausal theory $T := (M, \mathcal{C})$ is the regular clausal theory $\overline{T} := (M, \overline{\mathcal{C}})$ with $\overline{\mathcal{C}} := \{(B, T) \in \mathfrak{P}(M) \times \mathfrak{P}(M) \mid \forall X \subseteq M : (B, T) \nleq (X, M \setminus X))$ or $(\exists (A, S) \in \mathcal{C} : (A, S) \leq (X, M \setminus X))\}$; $\overline{\mathcal{C}}$ is called the *regular closure* of the sequent set $\mathcal{C}$.

For each clausal theory $T := (M, \mathcal{C})$ there is an appertaining closure system on the attribute set M defined by

$$\mathfrak{H}(T) := \{X \subseteq M \mid \forall (A, S) \in \mathcal{C}(\forall m \in S : (A, S \setminus \{m\}) \nleq (X, M \setminus X) \vee m \in X)\}.$$

Note that a simply implicational theory $(M, R, \mathfrak{A})$ may be understood as the clausal theory $(M, \mathcal{C}(R, \mathfrak{A}))$ with $\mathcal{C}(R, \mathfrak{A}) := \{(\{p\}, \{q\}) \mid (p, q) \in R\} \cup \{(A, \{m\}) \mid A \in \mathfrak{A} \text{ and } m \in M\}$; clearly, $\mathfrak{H}(M, R, \mathfrak{A}) = \mathfrak{H}(M, \mathcal{C}(R, \mathfrak{A}))$. A formal context $\mathbb{K} := (G, M, I)$ is said to be a *realization* of a clausal theory $T := (M, \mathcal{C})$ if $\mathcal{C} \subseteq \mathcal{C}(\mathbb{K})$ and $\{X \subseteq M \mid (X, M \setminus X) \ngeq (A, S) \text{ for all } (A, S) \in \mathcal{C}\} = \{g^I \mid g \in G\}$ (cf. [BS97]).

Theorem 1 *Let $T := (M, \mathcal{C})$ be a clausal theory and let*

$$G_T := \{X \subseteq M \mid (X, M \setminus X) \ngeq (A, S) \text{ for all } (A, S) \in \mathcal{C}\}.$$

Then the conceptual scale $\mathbb{S}_T := (G_T, M, \ni)$ is a realization of T with $\mathcal{C}(\mathbb{S}_T) = \overline{\mathcal{C}}$.

Proof: Let $(A, S) \in \mathcal{C}$ and $X \in G_T$. Since $X \in (A, S)^{\ni} \Leftrightarrow (X \nsupseteq A \text{ or } X \cap S \neq \emptyset) \Leftrightarrow (X, M \setminus X) \ngeq (A, S)$, the definition of G_T yields $(A, S)^{\ni} = G_T$, i.e., (A, S) is all-extensional in $\mathbb{S}_T$. Hence $\mathcal{C} \subseteq \mathcal{C}(\mathbb{S})$. Since $X^{\ni} = X$ for each $X \in G_T$, it follows with the above equivalences that

$$\{X^{\ni} \mid X \in G_T\} = \{X \subseteq M \mid (X, M \setminus X) \ngeq (A, S) \text{ for all } (A, S) \in \mathcal{C}\}.$$

This shows that $\mathbb{S}_T$ is a realization of T. It is left to prove that the clause logic of $\mathbb{S}_T$ is the regular closure of $\mathcal{C}$. For any sequent (B, T) we have the following equivalences:

$$(B, T) \in \mathcal{C}(\mathbb{S}_T) \Leftrightarrow (B, T)^{\ni} = G_T \Leftrightarrow \{X \in G_T \mid X \ngeq (B, T)\} = G_T \Leftrightarrow$$

$$\Leftrightarrow \forall X \subseteq M : (\forall (A,S) \in C : (X, M \setminus X \not\geq (A,S)) \to (X, M \setminus X) \not\geq (B,T) \Leftrightarrow$$
$$\Leftrightarrow (\forall X \subseteq M : (B,T) \not\leq (X, M \setminus X)) \text{ or } (\exists (A,S) \in C : (A,S) \leq (X, M \setminus X)) \Leftrightarrow$$
$$\Leftrightarrow (B,T) \in \overline{C}.$$

Therefore $C(\mathbb{S}_T) = \overline{C}$. □

There is a number of general conceptual scales which have been proven useful in fixing theoretical assumptions about empirical relationships (see [GW99a], p.57). For recognizing possible applications of those scales, it is desirable to determine their respective clause logics. Most of them are realizations of simply implicational theories: these are the nominal scales, the multiordinal scales, the Boolean scales, the grid scales, the contraordinal scales, the dichotomic scales, and the one-dimensional interordinal scales. Using the results of Section 2, only the specific sequents for distinguishing the objects of those scales have to be determined. Here we demonstrate this only for the case of simply implicational standard scales:

Proposition 3 *The clause logic of a simply implicational standard scale* $\mathbb{O}_P :=$ $(\mathfrak{I}_{\underline{P}}, P, \not\geq)$ *is the regular closure of the sequent set* $C := C_1 \cup C_2 \cup C_3$ *with*

$$C_1 := \{(\{p\}, \{q\}) \mid p, q \in P \text{ with } p \prec q\},$$
$$C_2 := \{(A, \{p\}) \mid A \in \mathfrak{A} \text{ and } p \in P\},$$
$$C_3 := \{(min \bigcap_{p \in B} \lfloor p), B) \mid B \text{ is an antichain with } \downarrow B \notin \mathfrak{I}_{\underline{P}} \setminus \{\emptyset\}\}.$$

Proof: By Theorem 1, we have to prove that $\mathbb{O}_P$ is a realization of the clausal theory $T := (P, C)$. Using Proposition 1 it can be easily verified that $C \subseteq C(\mathbb{O}_P)$. So it is left to prove that $\{X \subseteq P \mid (X, P \setminus X) \not\geq (A,S) \text{ for all } (A,S) \in C\} = \{D^{\not\geq} \mid D \in \mathfrak{I}_{\underline{P}}\}$. If $(X, P \setminus X) \not\geq (A,S)$ for all $(A,S) \in C_1 \cup C_2$ then X must be an order filter of $(P, \leq)$ equal to P or not containing any $A \in \mathfrak{A}$. In addition, $(X, P \setminus X) \not\geq (A,S)$ for all $(A,S) \in C_3$ forces that $P \setminus X \in \mathfrak{I}_{\underline{P}}$ and therefore $\{X \subseteq P \mid (X, P \setminus X) \not\geq (A,S) \text{ for all } (A,S) \in C\} \subseteq \{D^{\not\geq} \mid D \in \mathfrak{I}_{\underline{P}}\}$. The converse inclusion can be easily checked. □

By Proposition 3, the simply implicational theory $(S, \leq, \mathfrak{A}_S)$ of Sowa's top-level categories (see Section 2) can be described by the sequent set $C_{\underline{S}} := C_1 \cup C_2 \cup C_3$ with

$$C_1 := \{(\{p\}, \{q\}) \mid p, q \in S \text{ with } p \prec q\},$$
$$C_2 := \{(A, \{p\}) \mid A \in \mathfrak{A}_S \text{ and } p \in S\},$$
$$C_3 := \{(min \bigcap_{p \in B} \lfloor p), B) \mid B \text{ is an antichain with } \downarrow B \notin \mathfrak{I}_{\underline{S}} \setminus \{\emptyset\}\}.$$

To respect Sowa's synonym postulate, one has to add for each synonym the appertaining implication having a two-element premise (e.g. ($\{Continuant, Form\}, \{Script\}$)).

From the standardized conceptual scales which are not realizations of simply implicational theories, the convex-ordinal scales are the most important. Therefore we close this section by discussing how those scales may be recognized as realizations of clausal theories. Let us recall that a *convex-ordinal scale* for an ordered set $(P, \leq)$ is defined as the context

$$\mathbb{C}_{\underline{P}} := (P, \{\not\geq, \not\leq\} \times P, \square)$$

with $p\square(\not\geq, q) :\Leftrightarrow p \not\geq q$ and $p\square(\not\leq, q) :\Leftrightarrow p \not\leq q$ for $p, q \in P$. We will often write $\not\geq p$ and $\not\leq p$ instead of $(\not\geq, p)$ and $(\not\leq, p)$, and $p \not\geq q$ and $p \not\leq q$ instead of $p\square \not\geq q$ and $p\square \not\leq q$. The extents of $\mathbb{C}_{\underline{P}}$ are exactly the convex subsets of $(P, \leq)$.

Proposition 4 *The clause logic of a convex-ordinal scale $\mathbb{C}_{\underline{P}}$ is the regular closure of the sequent set $\mathcal{C} := \mathcal{C}_1 \cup \mathcal{C}_2 \cup \mathcal{C}_3$ with*

$$\begin{aligned}
\mathcal{C}_1 &:= \{(\{\not\geq p\}, \{\not\geq q\}) \mid p, q \in P \text{ with } p \prec q\} \cup \\
&\quad \{(\{\not\leq p\}, \{\not\leq q\}) \mid p, q \in P \text{ with } p \succ q\}, \\
\mathcal{C}_2 &:= \{(\{\varphi(p)p \mid p \in P\}, \emptyset) \mid \varphi : P \to \{\not\geq, \not\leq\}\}, \\
\mathcal{C}_3 &:= \{(\emptyset, \{\not\geq p, \not\geq q \not\leq p, \not\leq q\} \mid p, q \in P \text{ with } p \neq q\}.
\end{aligned}$$

Proposition 4 immediately follows from the following more general Proposition 5 which characterizes the clausal theories having a convex-ordinal scale as realization.

Proposition 5 *Let $(M \dot\cup M^*, \mathcal{C}_R)$ be a clausal theory with $|M| \geq 2$, a bijection $a \mapsto a^*$ $(a \in M)$, $R \subseteq M^2$, and $\mathcal{C}_R := \mathcal{C}_1 \cup \mathcal{C}_2 \cup \mathcal{C}_3$ where*

$$\begin{aligned}
\mathcal{C}_1 &:= \{(\{a\}, \{b\}) \mid (a, b) \in R\} \cup \{(\{b^*\}, \{a^*\}) \mid (a, b) \in R\}, \\
\mathcal{C}_2 &:= \{(\{a^\epsilon \mid a \in M\}, \emptyset) \mid \epsilon : M \to \{0, *\} \text{ and } a^0 = a\}, \\
\mathcal{C}_3 &:= \{(\emptyset, \{a, b, a^*, b^*\}) \mid a, b \in M \text{ and } a \neq b\}.
\end{aligned}$$

Then $(M \dot\cup M^*, \mathcal{C}_R)$ has a convex-ordinal scale as realization.

Proof: By Theorem 1, the conceptual scale $\mathbb{S}_T := (G_T, M \dot\cup M^*, \ni)$ is a realization of T with $\mathcal{C}(\mathbb{S}_T) = \overline{\mathcal{C}}$. It remains to show that $\mathbb{S}_T$ is isomorphic to a convex-ordinal scale. Let $\leq_R$ be the reflexive transitive closure of R on M, let $E_R := \leq_R \cap \geq_R$. The relation $\leq_R$ is a quasi-order on M. Without loss of generality we assume that $\leq_R$ is an order relation, because we can always obtain an order relation by factorizing M with E_R. With $\underline{M} := (M, \leq_R)$ we can define the convex-ordinal scale

$$\mathbb{C}_{\underline{M}} := (M, \{\not\geq_R, \not\leq_R\} \times M, \square).$$

We will show that the convex-ordinal scale $\mathbb{C}_{\underline{M}}$ is isomorphic to the conceptual scale $\mathbb{S}_T$. Let

$$\alpha : M \to G_T \text{ and } \beta : \{\not\geq_R, \not\leq_R\} \times M \to M \dot\cup M^* \text{ with}$$

$$\alpha(m) := \{r \in M \mid m \not\geq_R r\} \cup \{s^* \in M^* \mid s \not\leq_R m\} \text{ and}$$
$$\beta(\not\geq_R m) := m \text{ and } \beta(\not\leq_R m) := m^* \text{ for all } m \in M.$$

Since the object intents of $\mathbb{C}_{\underline{M}}$ are $m^{\square} = \{\not\geq_R r \mid r \in M, m \not\geq_R r\} \cup \{\not\leq_R s \mid r \in M, m \not\leq_R s\}$ for $m \in M$, we have $\alpha(m) \in G_{\mathcal{T}}$. The mapping α is injective because $\leq_R$ is an order relation. We have to show that α is surjective. We denote the inverse of the bijection $*$ by $\bullet$, i.e. $a^{*\bullet} = a$. Let $E \in G_{\mathcal{T}}$. Then $E = (E \cap M) \dot\cup (E \cap M^*)$. By the clauses in $\mathcal{C}_1$ the set $E \cap M$ is an order filter of $\underline{M}$ and the set $(E \cap M^*)^{\bullet}$ is an order ideal of $\underline{M}$. The equality $(E \cap M) \cup (E \cap M^*)^{\bullet} = M$ cannot hold by the inconsistencies in $\mathcal{C}_2$. Hence $(E \cap M) \cup (E \cap M^*)^{\bullet}$ is a proper subset of M. Now the sequents in $\mathcal{C}_3$ imply that there can only be one element in M that is not in $(E \cap M) \cup (E \cap M^*)^{\bullet}$. We denote this element by p. We have shown that $E \cap M = M \setminus (p]$ and $(E \cap M^*)^{\bullet} = M \setminus [p)$. This proves $\alpha(p) = E$. Hence, α is surjective and therefore also bijective. The bijectivity of β is obvious. The equivalences $m \not\geq_R n \Leftrightarrow \beta(\not\geq_R n) \in \alpha(m)$ and $m \not\leq_R n \Leftrightarrow \beta(\not\leq_R n) \in \alpha(m)$ follow directly from the definition of α and β. This means we have shown that the pair (α, β) is an isomorphism between $\mathbb{C}_{\underline{M}}$ and $\mathbb{S}_{\mathcal{T}}$ which completes the proof.

$\square$

A general method to find generating subsets for clause logics of finite formal contexts is described in [Kr98] (see also [Ga00]).

4   Aggregation of Conceptual Scales

According to Anselm Strauss and Juliet Corbin [SC90]), empirically grounded theory building uses a systematic set of procedures to develop an inductively derived theory of a phenomenon; it starts from data which are broken down, conceptualized, and put back together in new ways to generate a rich, tightly woven, explanatory theory that closely approximates the reality it represents. Such empirical theory building may be mathematically supported by suitable data models and formal methods for decomposing those models into conceptually structurable parts that can be suitably aggregated to allow formal representations of significant scientific theories. Conceptual scales, which are discussed in the previous sections, have been proven useful for conceptually structuring parts of data models. Therefore it is desirable to describe methods of aggregation for conceptual scales which, in particular, keep the connections between conceptual scales and clausal theories represented by those scales. For the analysis of aggregations with respect to such connections, we first introduce an appropriate notion of aggregation for clausal theories:

Definition: For clausal theories $\mathcal{T}_k := (M_k, \mathcal{C}_k)$ $(k \in K)$ the *disjoint union* is defined by

$$\dot\bigcup_{k \in K} \mathcal{T}_k := (\bigcup_{k \in K} \{k\} \times M_k, \dot\bigcup_{k \in K} \mathcal{C}_k)$$

where $\dot\bigcup_{k \in K} \mathcal{C}_k$ is the disjoint union $\bigcup_{k \in K} \{(\{k\} \times A, \{k\} \times S) \mid (A, S) \in \mathcal{C}_k\}$.

In Formal Concept Analysis, the mostly used construction of aggregating conceptual scales is the semiproduct of formal contexts; it particularly yields the structural frame for representing larger concept lattices by nested line diagrams (see [GW99a]).

Definition: For conceptual scales $\mathbb{S}_k := (G_k, M_k, I_k)$ $(k \in K)$ the *semiproduct* is defined by

$$\boxtimes_{k \in K} \mathbb{S}_k := (\bigtimes_{k \in K} G_k, \bigcup_{k \in K} \{k\} \times M_k, \nabla)$$

with $(g_k)_{k \in K} \nabla (k_0, m) \Leftrightarrow g_{k_0} I_{k_0} m_{k_0}$.

Theorem 2 *The clausal logic of the semiproduct of the conceptual scales $\mathbb{S}_k := (G_k, M_k, I_k)$ $(k \in K)$ is the regular closure of the disjoint union of the clause logics $\mathcal{C}(\mathbb{S}_k)$ $(k \in K)$.*

Proof: For $(A, S) \in S_{k_0}$, we have $(\{k\} \times A, \{k\} \times S)^\nabla = \{(g_k)_{k \in K} \mid (\exists m \in A : (g_{k_0}, m) \notin I_{k_0})$ or $\exists m \in S : (g_{k_0}, m) \in I_{k_0})\} = \{(g_k)_{k \in K} \mid g_{k_0} \in (A, S)^{I_{k_0}}\} = \bigtimes_{k \in K} G_k$ because $(A, S)^{I_{k_0}} = G_{k_0}$. Thus, $(\{k\} \times A, \{k\} \times S)$ is all-extensional in $\boxtimes_{k \in K} \mathbb{S}_k$ and hence $\bigcup_{k \in K} \mathcal{C}(\mathbb{S}_k) \subseteq \mathcal{C}(\boxtimes_{k \in K} \mathbb{S}_k)$. Using Theorem 1, we obtain for $X_k \subseteq M_k$ $(k \in K)$ the following equivalences: $(\bigcup_{k \in K} \{k\} \times X_k, \bigcup_{k \in K} \{k\} \times M_k \setminus \bigcup_{k \in K} \{k\} \times X_k) \not\geq (A_{k_0}, S_{k_0})$ for all $(A_{k_0}, S_{k_0}) \in \bigcup_{k \in K} \mathcal{C}_k \Leftrightarrow (\{k\} \times X_k, \{k\} \times M_k \setminus \{k\} \times X_k) \not\geq (A_k, S_k)$ for all $(A_k, S_k) \in \mathcal{C}_k$ with $k \in K \Leftrightarrow X_k = g_k^{I_k}$ for some $g_k \in G_k$ and for all $k \in K \Leftrightarrow \bigcup_{k \in K} \{k\} \times X_k = (g_k)_{k \in K}^\nabla$ for some $(g_k)_{k \in K} \in \bigtimes_{k \in K} G_k$. Using again Theorem 1, we conclude that the clausal logic of the semiproduct of the conceptual scales $\mathbb{S}_k := (G_k, M_k, I_k)$ $(k \in K)$ is the regular closure of $\dot{\bigcup}_{k \in K} \mathcal{C}(\mathbb{S}_k)$. □

Corollary 3 *Let $T_k := (M_k, C_k)$ $(k \in K)$ be clausal theories. The regular closure of the disjoint union $\dot{\bigcup}_{k \in K} T_k$ is equal to $(\bigcup_{k \in K} M_k, \mathcal{C}(\boxtimes_{k \in K} \mathbb{S}_{T_k}))$.*

How the presented method of aggregation of conceptual scales may be used for empirically grounded theory building shall be explained through an example from political sciences. At the end of the '80th, a multitude of studies of German researchers about international regimes have been compiled (under the guidance of Beate Kohler-Koch) and comparatively analysed. The authors classified the matters of their case studies by a uniform scheme of interpretation which led to the data table shown in Figure 4 (see [Ko89],[VW91],[KV00]). The attributes and attribute values of the data table have been chosen on the base of the actual theory of regimes for grasping the relationships concerning different aspects and so gaining better insights into the conditions of development and success of international regimes. The data table is already a result of breaking the wealth of data of the studies into parts corresponding to the choosen attributes. Therefore the further theory building should be continued with the conceptualization of those parts.

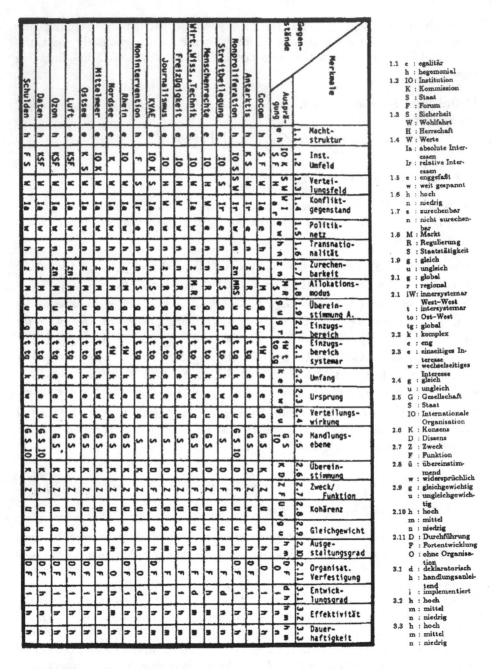

Fig. 4. Data table reporting a classification of international regimes

Formal Concept Analysis is prepared to support this by its method of *conceptual scaling* (see [GW89]). This method starts with the mathematization of a data table (as in Figure 4) by a so-called *many-valued context* $\mathbb{K} := (G, M, W, I)$ where G, M, and W are sets of (formal) objects, attributes, and attribute values, respectively, and $I \subseteq G \times M \times W$ satisfying $(g, m, w_1), (g, m, w_2) \in I \Rightarrow w_1 = w_2$; an attribute $m \in M$ can therefore be understood as a (partial) map from G into W with $m(g) = w \Leftrightarrow (g, m, w) \in I$. To obtain formal concepts, for each $m \in M$, a conceptual scale $\mathbb{S}_m := (G_m, M_m, I_m)$ is chosen with $m(G) \subseteq G_m$. In case of the many-valued context described by the data table in Figure 4, an extensive discussion with Kohler-Koch and one of her colleagues, which was guided by the described aim of their research, led to the conceptual scales whose concept lattices are presented in Figure 5; the blackened circles in the diagrams represent the object concepts given by the data.

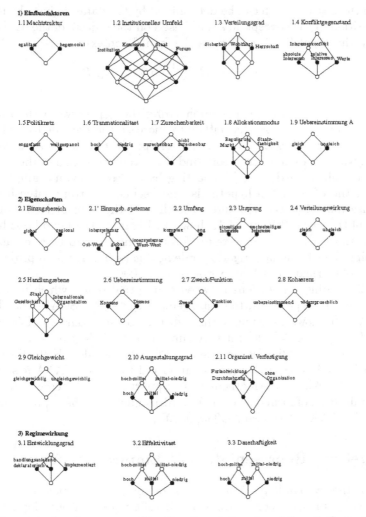

Fig. 5. Conceptual scales representing the clausal theories of the attributes in Fig. 4

The conceptual scales of Figure 5 code the clausal theories which the political scientists derived from the theory of regimes for structuring the attribute values of the respective attributes. The object concepts given by the data indicate that some of the clausal theories should eventually be extended. For instance, the diagram of scale 2.5 suggests to discuss, for the field of actions, the implications *Internationale Organisation* $\longrightarrow$ *Gesellschaft* and *Gesellschaft* $\longrightarrow$ *Staat*; obviously, the clausal theory coded by scale 1.2 might be enriched too.

The gap between theory and data becomes even more challenging after aggregating the clausal theories and the appertaining conceptual scales, respectively. Theoretically, as the Corollary 3 states, the aggregation of the clausal theories by the disjoint union and the aggregation of the conceptual scales by the semiproduct yield, up to the formation of the regular closure, the same clausal theories. But, empirically, not all concepts of the semiproduct are determined by the data. The concept intents which can be determined by the data are just the intents of the apposition of the conceptual scales based on the objects of the data context; generally, the *apposition* is defined as the formal context

$$\mid_{m \in M} \mathbb{S}_m := (G, \bigcup_{m \in M} \{m\} \times M_m, J)$$

with $gJ(m,n) :\Leftrightarrow m(g)I_m n$. For diminishing the gap between the theoretical and the empirical side, one has either to increase the data or to find new sequents which are all-extensional in the semiproduct of the conceptual scales. But even with a gap between theory and data, the apposition of the conceptual scales is useful, in particular, for rejecting hypotheses. As an example for this, we choose the often stated hypothesis that regimes of strong intensity usually have a hegemonial power structure; according to the concept lattice of the apposition of the scales 2.10 and 1.1, shown in Figure 6, the hypothesis has to be rejected. Another interesting example for studying the gap between apposition and semiproduct is the investigation of lexical gaps in a research project on the semantics of German speech-act verbs (see [GH00]).

For discovering connections between conceptual scales which may suggest, justify, or falsify hypotheses, the management system TOSCANA has been developed which allows to navigate through nested line diagrams of (even large) concept lattices (see [VW95]); it may support the whole process of empirically grounded theory building as proposed in [SC90] (see also [Ke94],[Wü98]). An example for such a research process supported by a TOSCANA system is described in [MW99]. For the mathematical support of more heterogenous theories, the generalization of formal contexts to multicontexts has been proposed and investigated in [Wi96] (see also [Ga96],[Dö99]).

5   Algebraic Representation of Empirical Theories

Empirical theory building often focusses on properties which are adequately formalized by many-valued attributes. Therefore the attribute logic of many-valued contexts has also to be considered. How such attribute logic may be developed, this shall be discussed in this section for *ordinal contexts* $\mathbb{K} :=$ $(G, M, (W_m, \leq_m)_{m \in M}, I)$ which modify a many-valued context (G, M, W, I) by replacing the attribute value set W by a family of ordered sets $(W_m, \leq_m)$ $(m \in M)$ such that $m(G) \subseteq W_m$ for $m \in M$ (see [SW92]). It is appropriate to define the *clause logic* of the ordinal context by $\mathcal{C}(\mathbb{K}) := \mathcal{C}(\mathbb{K}_\mathbb{O})$ where $\mathbb{K}_\mathbb{O} := (G^2, M, I_\mathbb{O})$ with $(g_1, g_2) I_\mathbb{O} m :\Leftrightarrow m(g_1) \leq_m m(g_2)$. The consequence of this definition is that the ordinal dependencies between attributes of $\mathbb{K}$ are just the implications between attributes of $\mathbb{K}_\mathbb{O}$ (see [GW99a], Proposition 28); let us recall that $(X, \{n\}) \in \mathfrak{P}(M)^2$ is (by definition) an *ordinal dependency* of $\mathbb{K}$ if $m(g_1) \leq_m m(g_2)$ for all $m \in X$ always imply $n(g_1) \leq_n n(g_2)$. The *functional dependencies* between attributes in M are just the ordinal dependencies of $(G, M, (W_m, =)_{m \in M}, I)$.

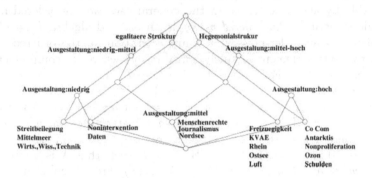

Fig. 6. Concept lattice of the apposition of the conceptual scales "regime intensity" and "power structure"

To each ordinal context $\mathbb{K} := (G, M, (W_m, \leq_m)_{m \in M}, I)$ corresponds a relational structure $\underline{S}(\mathbb{K}) := (G, (\lesssim_m)_{m \in M})$ with $g_1 \lesssim_m g_2 :\Leftrightarrow m(g_1) \leq_m m(g_2)$ as considered in representational measurement theory (see [K+71]). This opens the possibility to devote this final section to the aim of representational measurement theory, namely to study representations of such relational structures by algebraic (numerical) structures with respect to its support of empirical theory building. Again we explain our approach through an example from political sciences.

In his book "Political Order in Changing Societies" [Hu68] Samuel P. Huntington expresses basic relationships of certain political and social concepts by three equations such as

$$\frac{\text{Social mobilization}}{\text{Economic development}} = \text{Social frustration.}$$

The equations gave rise to a strong controversy having in its center the question: Are the equations meaningful or not? (cf. [WW93]) Although the disputants kept to their controversial positions, they aggreed on the same judgment that Huntington's equations at least express ordinal relationships between the named quantities, namely, when one quantity remains the same and a second quantity changes, then the third quantity always changes either in the same or the opposite direction. Let us imagine that an ordinal context $\mathbb{K} := (G, M, (W_m, \leq_m)_{m \in M}, I)$ could be derived from data concerning the mentioned quantities so that the quantities would be represented by the attributes of $\mathbb{K}$. Then the ordinal relationships, on which the disputants agreed, could be mathematically described by the following axioms (A_i) $(i = 0, 1, \ldots, n)$ concerning attributes $m_0, m_1, \ldots, m_n$ of $\mathbb{K}$ ($n = 2$ for the above equation):

$$(A_i) \quad \forall g_1, g_2 \in G : (\forall j \in \{0, 1, \ldots, n\} \setminus \{i\} : m_j(g_1) \leq_{m_j} m_j(g_2))$$
$$\longrightarrow m_i(g_2) \leq_{m_i} m_i(g_1).$$

Surprisingly, if $\mathbb{K}$ satisfies $(A_0), (A_1), \ldots, (A_n)$, the object set G can be furnished with an algebraic structure such that the ordinal relationships become describable by algebraic terms. In the general case we use ordered n-loops as algebraic structures. An *ordered n-loop* is an ordered algebra $\underline{L} := (L, f, 0, \leq)$ for which f is an order-preserving n-ary operation on the ordered set $(L, \leq)$ with 0 as neutral element uniquely solvable in each of its components always respecting the order. For each ordered n-loop $\underline{L}$, there is an appertaining ordinal context $\mathbb{K}_{\underline{L}} := (L^n, \{m_0, m_1, \ldots, m_n\}, (L, \leq_i)_{i=1,\ldots,n}, I_{\underline{L}})$ with $m_0 := f$, $\leq_0 := \geq$, and $m_i(x_1, \ldots, x_n) := x_i$ and $\leq_i := \leq$ for $i = 1, \ldots, n$. An ordinal context $\mathbb{K} := (G, M, (W_m, \leq_m)_{m \in M}, I)$ with $n + 1$ attributes has a *representation* in $\mathbb{K}_{\underline{L}}$ if there is a *quasi-embedding* of $\mathbb{K}$ into $\mathbb{K}_{\underline{L}}$, i.e., mappings $\kappa : G \to L^n$, $\lambda : M \to \{0, 1, \ldots, n\}$, and $\nu_m : W_m \to L$ such that λ is a bijection and $m(g) = w \Leftrightarrow m_{\lambda m}(\kappa g) = \nu_m w$ ($i = 0, 1, \ldots, n$). From results in [WW96] we obtain the following theorem:

Theorem 4 *An ordinal context $\mathbb{K}$ has a representation into an ordinal context $\mathbb{K}_{\underline{L}}$ for some ordered n-loop $\underline{L}$ if and only if $\mathbb{K}$ has $n + 1$ attributes and satisfies $(A_0), (A_1), \ldots, (A_n)$.*

In [WW95], the uniqueness of the representation in Theorem 4 is characterized by partial isotopies so that the algebraic descriptions of ordinal relationships in the represented context which are invariant under partial isotopies, are meaningful in the sense of representational measurement theory. If the fraction on the left side of Huntington's equations is understood as a fraction in an ordered 2-loop and if the axioms (A_0), (A_1), and (A_2) can be assumed (as the disputants did), then we can conclude that Huntington's equations are meaningful in the sense of representational measurement theory.

Theorem 4 can be specialized to characterize those ordinal contexts which have a representation by ordered Abelian groups and ordered vector spaces, respectively (see [Wl96],[WW96]). A general discussion of the representation problem emphazising the geometrical view can be found in [Wl97]. Of course, as richer the algebraic structures are as better will be the mathematical support for the represented theory. A specific problem is how to recognize which algebraic structures can be used for the representation. Theorem 4 yields a method to find out whether a representation by ordered n-loops are possible (cf. [WW96]): For an ordinal context $\mathbb{K} := (G, M, (W_m, \leq_m)_{m \in M}, I)$ with $n+1$ attributes the first step is to form an extended ordinal context

$$\mathbb{K}|\mathbb{K}^d := (G, M \dot\cup M^d, (W_n, \leq_n)_{n \in M \dot\cup M^d}, I \dot\cup I^d)$$

where $M^d := \{m^d \mid m \in M\}$, $W_{m^d} := W_m$, $w_1 \leq_{m^d} w_2 :\Leftrightarrow w_2 \leq_m w_1$, and $I^d := \{(g, m^d, w) \mid (g, m, w) \in I\}$. Now, the clause logic of the ordinal context $\mathbb{K}|\mathbb{K}^d$ allows to recognize whether $\mathbb{K}$ has a representation into an ordinal context $\mathbb{K}_{\underline{L}}$ for some ordered n-loop $\underline{L}$ because $\mathbb{K}$ satisfies $(A_0), (A_1), \ldots, (A_n)$ for some bijection $\lambda : M \to \{0, 1, \ldots, n\}$ if and only if the sequents $(M \setminus \{m\}, m^d)$ are all-extensional in $(\mathbb{K}|\mathbb{K}^d)_0$ for all $m \in M$; this property may be checked by existing computer programs of Formal Concept Analysis. An open question is which other representations could be recognized by the clause logic of formal contexts derivable from the given ordinal context.

References

[BS97]   J. Barwise, J. Seligman: Informationflow: the logic of distributed systems. Cambridge University Press, Cambridge 1997.

[Dö99]   S. Dörflein: Coherence networks of concept lattices. Dissertation, TU Darmstadt 1999. Shaker Verlag, Aachen 1999.

[Ga00]   B. Ganter: Begriffe und Implikationen. In: G. Stumme, R. Wille (eds.): Begriffliche Wissensverarbeitung: Methoden und Anwendungen. Springer, Berlin-Heidelberg 2000, 1–24.

[GW89]   B. Ganter, R. Wille: Conceptual scaling. In: F. Roberts (ed.): Applications of combinatorics and graph theory to the biological and social sciences. Springer, Berlin-Heidelberg-New York 1989, 139–167.

[GW99a]   B. Ganter, R. Wille: Formal Concept Analysis: mathematical foundations. Springer, Berlin-Heidelberg-New York 1999.

[GW99b]   B. Ganter, R. Wille: Contextual Attribute Logic. In: W. Tepfenhart, W. Cyre (eds.): Conceptual structures: standards and practices. LNAI **1640**. Springer, Berlin-Heidelberg 1999, 377–388.

[Ga96]   P. Gast: Begriffliche Strukturen mehrwertiger Multikontexte. Diplomarbeit. FB4, TU Darmstadt 1996.

[GH00]   A. Grosskopf, G. Harras: Begriffliche Erkundung semantischer Strukturen von Sprechaktverben. In: G. Stumme, R. Wille (eds.): Begriffliche Wissensverarbeitung: Methoden und Anwendungen. Springer, Berlin-Heidelberg 2000, 273–295.

[Hu68]   S. P. Huntington: Political order in changing societies. Yale Univ. Press, New Haven 1968.

[Ke94]   U. Kelle: Empirisch begründete Theoriebildung: zur Logik und Methodologie interpretativer Sozialforschung. Deutscher Studien Verlag, Weinheim 1994.

[Ko89]   B. Kohler-Koch: Zur Empirie und Theorie internationaler Regime. In: B. Kohler-Koch (ed.). Regime in den internationalen Beziehungen. Nomos Verlagsgesellschaft, Baden-Baden 1989, 17–85.

[KV00]   B. Kohler-Koch, F. Vogt: Normen und regelgeleitete internationale Kooperationen. In: G. Stumme, R. Wille (eds.): Begriffliche Wissensverarbeitung: Methoden und Anwendungen. Springer, Berlin-Heidelberg 2000, 325–340.

[K+71]   D. Krantz, R. D. Luce, P. Suppes, A. Tversky: Foundations of measurement. Vol. 1-3. Academic Press, San Diego 1971, 1989, 1990.

[Kr98]   R. Krause: Kumulierte Klauseln als aussagenlogische Sprachmittel für die Formale Begriffsanalyse. Diplomarbeit. Institut für Algebra, TU Dresden 1998.

[L+97]   R. Langsdorf, M. Skorsky, R. Wille, A. Wolf: An approach to automatic drawing of concept lattices. In: K. Denecke, O. Lüders (eds.): General algebra and applications in discrete mathematics. Shaker Verlag, Aachen 1997, 125–136.

[MW99]   K. Mackensen and U. Wille: Qualitative text analysis supported by conceptual data systems. Quality & Quantity 33 (1999), 135–156.

[So00]   J. F. Sowa: Knowledge representation: logical, philosophical, and computational foundations. Brooks Cole Publ. Comp., Pacific Grove, CA, 2000.

[SW92]   S. Strahringer, R. Wille: Towards a structure theory for ordinal data. In: M. Schader (ed.): Analyzing and modeling data and knowledge. Springer, Berlin-Heidelberg 1992, 129–139.

[SC90]   A. Strauss, J. Corbin: Basics of qualitative research: grounded theory procedures and techniques. Sage Publ., Newbury Park 1990.

[VW91]   F. Vogt, C. Wachter, R. Wille: Data analysis based on a conceptual file. In: H.-H. Bock, P. Ihm: Classification, data analysis, and knowledge organization. Springer, Berlin-Heidelberg 1991, 131–140.

[VW95]   F. Vogt, R. Wille: TOSCANA - a graphical tool for analyzing and exploring data. In: R. Tamassia, I.G. Tollis (eds.): Graph Drawing '94. Lecture Notes in Computer Science 894. Springer, Berlin-Heidelberg-New York 1995, 226–233.

[Wi96]   R. Wille: Conceptual structures of multicontexts. In: P. W. Eklund, G. Ellis, G. Mann (eds.): Conceptual structures: representation as interlingua. Springer, Berlin-Heidelberg-New York, 1996, 23–39.

[WW93]   R. Wille, U. Wille: On the controversy over Huntington's equations: when are such equations meaningful? Mathematical Social Sciences 25 (1993), 173–180.

[WW95]   R. Wille, U. Wille: Uniqueness of coordinatizations of ordinal structures. In: Contribution to General Algebra, Bd. 9. Hölder-Pichler-Tempsky, Wien 1995, 321–324.

[WW96]   R. Wille, U. Wille: Coordinatizations of ordinal structures. Order 13 (1996), 281–294.

[Wl96]   U. Wille: Geometric representations of ordinal contexts. Dissertation, Universität Gießen 1995. Shaker Verlag, Aachen 1996.

[Wl97]   U. Wille: The role of synthetic geometry in representational measurement theory. J. Math. Psychology 41 (1997), 71–78.

[Wü98]   G. Wünsche: Begriffliche Datenanalyse zur Unterstützung qualitativer Theoriebildung in den Sozialwissenschaften. Diplomarbeit. FB4, TU Darmstadt 1998.

Searching for Objects and Properties with Logical Concept Analysis

Sébastien Ferré* and Olivier Ridoux

IRISA, Campus Universitaire de Beaulieu, 35042 RENNES cedex,
{ferre,ridoux}@irisa.fr

Abstract. Logical Concept Analysis is Formal Concept Analysis where logical formulas replace sets of attributes. We define a Logical Information System that combines navigation and querying for searching for objects. Places and queries are unified as formal concepts represented by logical formulas. Answers can be both extensional (objects belonging to a concept) and intensional (formulas refining a concept). Thus, all facets of navigation are formalized in terms of Logical Concept Analysis. We show that the definition of being a refinement of some concept is a specific case of Knowledge Discovery in a formal context. It can be generalized to recover more classical KD operations like machine-learning through the computation of necessary or sufficient properties (modulo some confidence), or data-mining through association rules.

1 Introduction

Information systems offer means for organizing data, and for navigating and querying. Though navigation and querying are not always distinguished because both involve queries and answers, we believe they correspond to very different paradigms of human-machine communication. In fact, the difference can be clarified using the intension/extension duality.

Navigation implies a notion of place, and of a relation between places (e.g., file system directories, and links or subdirectory relations). Through navigation, a user may ask for the contents of a place, or ask for related places. The ability to ask for related places implies that answers in the navigation-based paradigm belong to the same language as queries. In terms of the intension/extension duality, a query is an intension, and answers are extensions for the contents part, and intensions for the related places.

In very casual terms. we consider navigation with possibly no map, i.e., no *a priori* overview of the country. Related places form simply the landscape from a given place as shown by a "viewpoint indicator". However, our proposal is compatible with any kind of *a priori* knowledge from the user.

With querying, answers are extensions only. A simulation of navigation is still possible, but forces the user to infer what could be a better query from the unsatisfactory answer to a previous query; i.e., infer an intension from an

---

* This author is supported by a scholarship from CNRS and Région Bretagne

H. Delugach and G. Stumme (Eds.): ICCS 2001, LNAI 2120, pp. 187–201, 2001.
© Springer-Verlag Berlin Heidelberg 2001

extension. This is difficult because there is no simple relation between a variation in the query, and the corresponding variation in the answer. The experience shows that facing a query whose extension is too vast, a user may try to refine it, but the resulting extension will often be either almost as vast as the former or much too small. In the first case, the query lacks of precision (i.e., number of relevant items in the answer divided by total number of items in the answer), whereas in the second case, the query recall (i.e., number of relevant items in the answer divided by number of relevant items in the system) is too low.

Godin *et al.* [GMA93] have shown that Formal Concept Analysis is a good candidate for reconciliating navigation and querying. We follow this opinion, but we believe that care must be taken to make formal contexts as close to the description languages of the end-users, and we have proposed Logical Concept Analysis (LCA) where formal descriptions are logical formulas instead of being sets of attributes [FR00b]. So doing, one may consider that the contents of an information system is a formal context in which items are associated to formulas that describe them in a user-oriented way. We call this a Logical Information System (LIS). Then, Concept Analysis automatically organizes the contents of the information system as a lattice of concepts.

Our goal in this article is to show how a form of navigation and querying can be defined, so that a user who knows neither the contents of a Logical Information System, nor the logic of its descriptions, can navigate in it and discover the parts of the contents and the parts of the logic that are relevant to his quest. Note that a more expert user may know better and may navigate more directly to his goal, but since almost everybody has his *Terra Incognita*, the no-knowledge assumption is the safest one to do.

We will present in the sequel formal means for navigating in a Logical Information System in order to find relevant objects (answers in the extensional language), and relevant properties of the formal context (answers in the intensional language). We will show how this latter point is related to data-mining, knowledge-engineering, and machine-learning. The core of the formal means used in this work is (Logical) Concept Analysis.

The article is organized as follows. Section 2 presents a guided tour of Logical Concept Analysis. Section 3 presents the notion of Logical Information System. Sections 4 and 5 present the details of the navigation in a LIS, and of the extraction of properties. Sections 6 and 7 present conclusions and perspectives.

2   Logical Concept Analysis

We recall the main definitions and results about Logical Concept Analysis (LCA). More explanations and results can be found in [FR00b].

Definition 1 (context) *A logical formal context is a triple $(\mathcal{O}, \mathcal{L}, i)$ where:*
  - $\mathcal{O}$ *is a finite set of objects,*
  - $\langle \mathcal{L}; \models \rangle$ *is a lattice of formulas, whose supremum is $\dot{\vee}$, and whose infimum is $\dot{\wedge}$; $\mathcal{L}$ denotes a logic whose deduction relation is $\models$, and whose disjunctive and conjunctive operations are respectively $\dot{\vee}$ and $\dot{\wedge}$,*

    – i *is a mapping from* $\mathcal{O}$ *to* $\mathcal{L}$ *that associates to each object a formula that describes it.*

Given a formal context K, one can form a Galois connection between sets of objects (extents) and formulas (intents) with two applications σ and τ.

Definition 2 *Let* $K = (\mathcal{O}, \mathcal{L}, i)$ *be a logical context,*
$$\sigma^K : \mathcal{P}(\mathcal{O}) \to \mathcal{L}, \ \sigma^K(O) := \bigvee_{o \in O} i(o)$$
$$\tau^K : \mathcal{L} \to \mathcal{P}(\mathcal{O}), \ \tau^K(f) := \{o \in \mathcal{O} | i(o) \models f\}$$

Formal concepts can be derived from logical contexts.

Definition 3 (concept) *In a context* $K = (\mathcal{O}, \mathcal{L}, i)$, *a concept is a pair* $c = (O, f)$ *where* $O \subseteq \mathcal{O}$, *and* $f \in \mathcal{L}$, *such that* $\sigma^K(O) \dot{\equiv} f$ *and* $\tau^K(f) = O$. *The set of objects* O *is the concept* extent *(ext(c)), whereas formula* f *is its* intent *(int(c)).*

The set of all concepts that can be built in a context K is denoted by $\mathcal{C}(K)$, and is partially ordered by $\leq^c$ defined as follows.

Definition 4 $(O_1, f_1) \leq^c (O_2, f_2) :\Longleftrightarrow O_1 \subseteq O_2 \ \ (\Longleftrightarrow f_1 \dot{\models} f_2)$

Definitions 3 and 4 lead to the following *fundamental theorem.*

Theorem 1 *Let* $K = (\mathcal{O}, \mathcal{L}, i)$ *be a context. The ordered set* $\langle \mathcal{C}(K); \leq^c \rangle$ *is a finite lattice with supremum* $\vee^c$ *and infimum* $\wedge^c$.

It is possible to label concept lattices with formulas (resp. objects) through a labelling map μ (resp. γ).

Definition 5 *Let* K *be a logical context,*
$$\mu^K : \mathcal{L} \to \mathcal{C}(K), \ \ \mu^K(f) := (\tau^K(f), \sigma^K(\tau^K(f)))$$
$$\gamma^K : \mathcal{O} \to \mathcal{C}(K), \ \ \gamma^K(o) := (\tau^K(\sigma^K(\{o\})), \sigma^K(\{o\})) = (\tau^K(i(o)), i(o))$$

We introduce a *contextualized deduction relation* as a generalization of the implications between attributes that are used in FCA for knowledge acquisition processes [GW99,Sne98].

Definition 6 (contextualized deduction) *Let* $K = (\mathcal{O}, \mathcal{L}, i)$ *be a context, and* $f, g \in \mathcal{L}$. *One says that* f contextually entails g *in context* K, *which is noted* $f \dot{\models}^K g$, *iff* $\tau^K(f) \subseteq \tau^K(g)$, *i.e., iff every object that satisfies* f *also satisfies* g.

Relation $\dot{\models}^K$ is a preorder, whose associated equivalence relation is noted $\dot{\equiv}^K$. Formulas ordered by this deduction relation form a new logic adapted to the context: the *contextualized logic*. It is connected to the concept lattice by the following theorem.

Theorem 2 $\langle \mathcal{L}/_{\dot{\equiv}^K}; \dot{\models}^K \rangle$ *and* $\langle \mathcal{C}(K); \leq^c \rangle$ *are isomorphic, with* μ^K *as an isomorphism from formulas to concepts.*

In summary, there are 3 ways of considering a concept lattice: (1) extents ordered by set inclusion $(\tau(\sigma(\mathcal{P}(\mathcal{O}))), \subseteq)$, (2) intents ordered by logical deduction $(\sigma(\tau(\mathcal{L})), \dot{\models})$, (3) formulas ordered by contextualized deduction $(\mathcal{L}, \dot{\models}^K)$.

3   A Logical Information System

A Logical Information System (LIS) is essentially a logical formal context (see Definition 1) equipped with navigation and management tools. In this article, we are mostly interested in the end-user perspective. So, we will insist on navigation tools, and will only briefly allude to the management of a logical formal context.

For illustration purpose, we will present a bibliographical reference system whose principle is the following. Objects are bibliographical references, whose contents are BibTEX entries, and descriptions are composed of titles, lists of authors, etc, extracted from the contents, and appreciations (e.g., "theoretical" or "practical") and status (e.g., "read") given by a user. Logical formulas used for descriptions and queries are sets of valued attributes. For instance, numerical fields can be described in an interval logic, and string fields can be described in a boolean logic based on the absence/presence of substrings in a string: e.g., `title: Logic & -(Concept|Context)/year: 1990..1995`. In an advanced version of this system, we also introduced regular expressions and modalities for representing incomplete knowledge.

Our LIS needs an end-user interface. We will use a shell-based interface, though this is not the most modern thing to do. This is because we believe that shell interfaces (like in UNIX or MS-DOS) are familiar to many of us, and because this abstraction level exposes properly the dialogue of queries and answers. A higher-level interface like a graphical one would hide it, whereas lower-level interfaces, like a file system, would expose irrelevant details.

A prototype of the bibliographical system has been built for experimentation purpose. It has been implemented in Prolog as a generic system in which a theorem-prover and a syntax analyzer can be plugged-in for every logic used in descriptions. It is not meant to be efficient, though it can handle several hundred entries. Contrary to other tools based on concept analysis, it does not create the concept lattice. It only manages a Hasse diagram of the formulas used so far [FR00a]. In our experiments with the logic presented above, this diagram has an average of 5 nodes per object, 3 arcs per node, and a height of about 5.

3.1   Building a Logical Context

The first task one needs to do in a LIS is to build a Logical Context $K = (\mathcal{O}, \mathcal{L}, i)$; this amounts to create and logically describe objects. To each object o is associated a *content* $c(o)$ (here, the BibTEX reference) and a logical *description* $i(o)$. We distinguish content and description in order to hide some BibTEX fields (e.g., `publisher`) and add some non-BibTEX fields (e.g., `status`). We give an example of an object by displaying its content and its description in the following table. A "minus" sign before string formulas represents negation, and the "less" and "greater" signs around strings denote an "all I know" modality that means that the strings are closed (they do not contain anything else): e.g., `title: <"logic">` $\models$ `title: - "context"` whereas `title: "logic"` $\not\models$ `title: - "context"`.

```
c(o) = @InProceedings{FerRid2000b,
    author = {Sébastien Ferré and Olivier Ridoux},
    title = {A Logical Generalization of Formal Concept Analysis},
    booktitle = {International Conference on Conceptual Structures},
    editor = Guy Mineau and Bernhard Ganter,
    series = LNCS 1867,
    publisher = Springer,
    year = {2000},
    keywords = {concept analysis, logic, context, information system} }
i(o) =
    type: <"InProceedings">/
    author: <"Sébastien Ferré and Olivier Ridoux">/
    title: <"A Logical Generalization of Formal Concept Analysis">/
    year: 2000/
    keywords: <"concept analysis, logic, context, information system">/
    status: "read"
```

For all examples given in the following sections, we consider as context K all ICCS publications until the year 1999, which consists in 209 objects. For this context, the Hasse diagram has 954 nodes and 2150 arcs. In the following experiments of this paper, all response times are shorter than 5 seconds.

3.2 Navigating in a Logical Context

Once objects have been logically described and recorded in a logical context K, one wants to retrieve them. One way to do this is *navigating* in the context. As already said, this way of searching is particularly useful in a context where the logic or the contents are unknown. The aim of navigation is thus to guide the user from a *root place* to a *target place*, which contains the object(s) of interest. For this, a LIS offers to the user 3 basic operations (the corresponding UNIX-like command names are placed between parenthesis): (1) to ask to LIS what is the current place (command pwd), (2) to go in a certain "place" (command cd), (3) to ask to LIS ways towards other "places" (command ls).

In a hierarchical file system, a "place" is a directory. But in our case, a "place" is a concept, which can be seen as a coherent set of objects (extent) and properties (intent) (cf. Definition 3). In large contexts, concepts cannot be referred to by enunciating either their extent or their intent, because both are generally too large. Formulas of the logic $\mathcal{L}$ can play this role because every formula refers to a concept through the labelling map μ (cf. Definition 5), and every concept is referred to by one or several formulas, which are often simpler than the intent.

We now describe the 3 navigation operations listed above. First of all, going from place to place implies to remember the current place, which corresponds to the working directory. In our LIS, we introduce the *working query*, wq, and the *working concept*, $wc := \mu^K(wq)$; we say that wq refers to wc. This working query is taken into account in the interpretation of most of LIS commands, and is initialized to the formula $\top$, which refers to the concept whose extent is the set of all objects. Command pwd displays the working query to the user.

Command cd takes as argument a query formula q saying in which place to go, and it changes the working query accordingly. We call l_{wq} (elaboration of wq) the mapping that associates to the query q a new working query according to the current working query wq. The query q can be seen as a *link* between the current and the new working query. Usually, cd is used to refine the working concept, i.e., to select a subset of its extent. In this case, the mapping l_{wq} is defined by $l_{wq}(q) := wq \dot\wedge q$, which is equivalently characterized by

$$\mu^K(l_{wq}(q)) =^c wc \wedge^c \mu^K(q) \text{ and } \tau^K(l_{wq}(q)) = \tau^K(wq) \cap \tau^K(q).$$

However, it is useful to allow other interpretations of the query argument. For instance, we can allow the distinction between *relative* and *absolute* query, similarly to relative and absolute paths in file systems. The previous definition of the mapping l_{wq} concerns relative queries, but can be extended to handle absolute queries by $l_{wq}(/q) := q$, where '/' denotes the absolute interpretation of queries. This allows to forget the working query. We can also imagine less usual interpretations of queries like $l_{wq}(| q) := wq \dot\vee q$. Finally, the special argument .. for the command cd enables to go back in the history of visited queries/concepts. This works much like the "Back" button in *Web* browsers.

Command ls is intended to guide the user towards his goal. More precisely, it must suggest some relevant links that could act as queries for the command cd to refine the working query. These links are formulas of $\mathcal{L}$. A set of links given by ls should be finite, of course (whereas $\mathcal{L}$ is usually infinite), even small if possible, and complete for navigation (i.e., each object of the context must be accessible by navigating). We postpone the development of this issue to Section 4.

As navigation aims at finding objects, command ls must not only suggest some links to other places, but also present the object belonging to the current place, called the *object of wq* or the *local object*. It is defined as the object labelling the working concept through the labelling map γ (cf. Definition 5). Formally, the object of a query q is defined by

$$t^K(q) := o \in \mathcal{O} \text{ such that } \gamma^K(o) =^c \mu^K(q).$$

There can be no local object, and there cannot be several ones because this would be equivalent to have two different objects described by exactly the same formula, which would make them undistinguishable. Another interesting thing to notice is that the working query can be, and is often, much shorter than the whole description of the local object (which is also the intent of the working concept), as in the following example where the formula on the first line is contextually equivalent to the description on the four other lines for accessing the object.

```
i(t(author: Mineau & Missaoui)) =
   type: <InProceedings>/
   title: <"The Representation of Semantic Constraints in Conceptual Graph Systems">/
   author: <"Guy W. Mineau and Rokia Missaoui">/
   year: 1997
```

3.3   Updating and Querying a Logical Context

Updating is done *via* shell commands like mv or cp. With option -r, every object of the working concept is concerned, while in the opposite case, only the local object is (considering it exists, cf. Section 3.2).

```
(1) cd /author: Mineau & Missaoui
(2) mv . status: "to be read"
(3) cd /keywords: FCA | GC
(4) mv -r status: "to be read"      status: "read"
```

Contents can also be changed with a LIS-command `chfile`. This changes indirectly descriptions, but the ensuing reorganization of the formal concept lattice is automatic and transparent. In fact, it costs not so much since the concept lattice is not actually represented.

Extensional queries can be submitted to a logical information system using the `-r` option with command `ls`. The answer to query `ls -r q` is simply $\tau^K(l_{wq}(q))$, i.e., the extent of the concept refered to by $l_{wq}(q)$ (cf. Section 3.2).

```
(1) ls -r /title: Logic & -(Concept | Context)/year: 1990..1995
3 B. R. Gaines. "Representation, discourse, logic and truth: situating knowledge
technology". INPROC, 1993.
2 J. F. Sowa. "Relating diagrams to logic". INPROC, 1993.
72 H. van den Berg. "Existential Graphs and Dynamic Predicate Logic". INPROC, 1995.
3 object(s)
```

4   Searching for Objects

We now define more precisely than in Section 3.2 the dialogue between a user and our LIS through commands `cd` and `ls`. Let us recall that `cd` enables the user to traverse a link from the working concept to another one, and that `ls` enables him to get from LIS a set of relevant links to refine the working concept.

Navigation Links. The following notion of refinement corresponds to the case where the elaboration mapping satisfies $l_{wq}(q) = wq \dot\wedge q$. To avoid to go in the concept $\perp^c$ whose extent is empty, we must impose the following condition on a link x: $\tau^K(wq \dot\wedge x) \neq \emptyset$. As $\mathcal{L}$ is a too wide search space, we consider a finite subset X of $\mathcal{L}$ in which links are selected. The content of X is not strictly determined but it should contain simple formulas, some frequently used formulas, and more generally, all formulas that users expect to see in `ls` answers. X can be finite because terminology and used formulas are (because the context is finite). Furthermore, we keep only greatest links (in the deduction order) as they correspond to smallest refinement steps. We can now define the set of links of a working query wq by $Link^K(wq) := \lceil \{x \in X | \tau^K(wq \dot\wedge x) \neq \emptyset\} \rceil$, where $\lceil E \rceil$ denotes the set of greatest elements of E according to the order $\models$.

Increments vs. Views. We can distinguish two kinds of links: *increments* that strictly restrict the working concept (i.e., $\tau^K(wq \dot\wedge x) \neq \tau^K(wq)$), and *views* that are properties shared by all the objects of the working concept (i.e., $\tau^K(wq \dot\wedge x) = \tau^K(wq)$). Only increments are useful as arguments of `cd`, because the application of `cd` to a view would not change the working concept.

Why not use only increments if views are not useful for refinement? Because sets of increments appear to be too large and heterogeneous: in the BibTEX example, they mix author names, title words, years, etc. Sets of links are smaller because many increments are subsumed by views, and then are not returned as links. For instance, the view (`author:` *) is returned as a summary of a large set of author name increments. So, if views are not useful for selecting objects, they are useful for selecting increments. We introduce a working view wv, similar to the working query, under which links must be searched for. For instance, if the working view is (`author:` `*`), links will be author name increments. We modify the definition of *Link* to take this into account.

Definition 7 $Link^K(wq, wv) := \lceil \{ x \in X | x \not\models wv, wv \not\models x, \tau^K(wq \wedge x) \neq \emptyset \} \rceil$.

Definition 8 *The completeness of Link is formally expressed by*
$\forall wq \in \mathcal{L} : \forall o \in \tau^K(wq) : o \neq t^K(wq) \Rightarrow \exists x \in Link^K(wq, wv) : o \in \tau^K(wq \wedge x)$.

In English, this means that for all working query wq and for all object o of its extent, if o is not yet the local object then it exists a link that enables to restrict the working extent while keeping o in it.

Theorem 3 $Link^K(wq, wv)$ *is complete for all wq and for all wv being a view for wq iff every object description belongs to X.*

In fact, X acts as the vocabulary that LIS uses in its answers. In our prototype, X is in a large part automatically generated from the context (i.e., from object descriptions) every time an object is created (command `mkfile`) or updated (commands `chfile` and `mv`); and from queries in order to incorporate the user vocabulary in the LIS one. This automatic generation guarantees the completeness of the navigation according to the above condition, while favouring small links such as author names, title words, keywords (in fact, whole object descriptions very rarely appear as links, cf. Table 1). Furthermore, the user can adjust the LIS vocabulary by adding and removing formulas in X with commands `mkdir` and `rmdir`.

To summarize, the view-based variant of command `ls` takes as argument a view v, sets the working view to $l_{wv}(v)$ (where l_{wv} works similarly to l_{wq}), shows the local object if it exists, displays each link x of the set $Link^K(wq, wv)$ along with the size of its selected extent $\tau^K(wq \wedge x)$, and finally displays the size of the working extent $\tau^K(wq)$. Increments and views are distinguished according to their cardinality compared to the working size; views simply have the same size as the working concept, whereas increments have strictly smaller sizes.

User/LIS Dialogue. We now show how commands `cd` and `ls` compose a rather natural dialogue between the user and LIS. The user can refine the working concept with command `cd`, and asks for suggested links with the command `ls`. LIS displays to the user relevant increments for forthcoming `cd`'s, and relevant views for forthcoming `ls`'s. Commands `cd` (resp. increments) are *assertions* from the user (resp. from LIS): "I want this kind of object!" (resp. "I have this kind of object!"). Commands `ls` (resp. views) are *questions* from the user (resp. from LIS): "What kind of object do you have?" (resp. "What kind of object do you want?"). It should also be noticed that both the user and LIS can answer to questions both by assertions and by questions.

A complete example of a dialogue is given in Table 1. The left part of this table shows what is really displayed by our prototype, and the right part is an English translation of the dialogue. Notice that this translation is rather systematic and could thus be made automatic. (n) is the prompt for the n-th query from the user. On the 2nd query, the question of the user is so open, that LIS only answers by questions. On the 3rd query, the user replies to one of these questions (title: *) by an assertion; but on the 4th query, he sends back to LIS another of these questions (author: *) to get some relevant suggestions. On the 5th query, he just selects a suggested author, "Wille", and then gets his co-authors on Concept Analysis with the 6th query. On the 7th query, he selects a co-author and finally finds an object at the 8th query.

Table 1. Example of User/LIS Dialogue in the BibTEX context.

`(1) pwd`	`(1) What is currently selected?`
`/`	`    All objects.`
`(2) ls`	`(2) What do you have?`
`  209 type: *`	`    What kind of type do you want?`
`  209 author: *`	`    What kind of author do you want?`
`  209 year: ..`	`    What kind of year do you want?`
`  209 title: *`	`    What kind of title do you want?`
`  209 object(s)`	`    209 objects are currently selected.`
`(3) cd title: "Concept Analysis"`	`(3) I want objects whose title contains "Concept Analysis"!`
`(4) ls author: *`	`(4) What kind of author do you have (for this)?`
`  1 author: "Mineau"`	`    I have 1 object with author "Mineau"!`
`  1 author: "Lehmann"`	`    I have 1 object with author "Lehmann"!`
`  1 author: "Stumme"`	`    I have 1 object with author "Stumme"!`
`  1 author: "Prediger"`	`    I have 1 object with author "Prediger"!`
`  3 author: "Wille"`	`    I have 3 objects with author "Wille"!`
`  4 object(s)`	`    4 objects are currently selected.`
`(5) cd author: Wille`	`(5) I want objects with author "Wille"!`
`(6) ls`	`(6) What kind of author do you have (yet)?`
`  1 author: "Mineau"`	`    I have 1 object with author "Mineau"!`
`  1 author: "Lehmann"`	`    I have 1 object with author "Lehmann"!`
`  1 author: "Stumme"`	`    I have 1 object with author "Stumme"!`
`  3 author: "Wille"`	`    What kind of author "Wille" do you want?`
`  3 object(s)`	`    3 objects are currently selected.`
`(7) cd author: Mineau`	`(7) I want objects with author "Mineau"!`
`(8) ls`	`(8) What do you have?`
`  200 Guy W. Mineau and Gerd Stumme and Rudolf Wille.`	
`  "Conceptual Structures Represented by Conceptual Graphs`	
`    and Formal Concept Analysis". INPROCEEDINGS, 1999.`	
`  1 object(s)`	`    1 object is currently selected.`
`(9) pwd`	`(9) What is currently selected?`
`  author: "Wille" & "Mineau"/`	`    Objects with authors "Wille" and "Mineau",`
`  title: "Concept Analysis"`	`    and whose title contains "Concept Analysis".`

Related Work. We finish this section by comparing our LIS navigation with other kinds of navigation based on Concept Analysis. Lindig [Lin95] designed a concept-based component retrieval based on sets of *significant keywords* which are equivalent to our increments for the logic of attributes underlying FCA. Godin et al. [GMA93] propose a direct navigation in the lattice of concepts, which is in fact very similar to Lindig's approach except that only greatest significant keywords, according to the contextualized deduction on attributes, are displayed to the user. They have also notions common to our LIS such as working query, direct query specification, and history of selected queries. Cole and Stumme [CS00] developed a Conceptual Email Manager (CEM) where the navigation is based on Conceptual Scales [Pre97,PS99]. These scales are similar to our views in the sense that they select some attributes acting as increments and displayed, as for us, with the size of the concept they select. A difference with our LIS is that these increments are ordered according to concept lattices of scales, but it could also be done in LIS by a post-treatment on answers of command ls if we had a GUI. But the main difference with all of these approaches is that we use an (almost) arbitrary logic to express properties. This enables us to have automatic subsumption relations (e.g., (author: Wille & -Mineau) $\models$ (author: Wille) $\models$ (author: *)), and thus some implicit views (e.g., author: *, year: ..).

5   Searching for Properties

In previous sections, Concept Analysis (CA) is used to specify navigation and querying in a LIS. However, in the past CA has been often applied in domains such as *data-analysis*, *data mining*, and *learning*.

Data-analysis consists in structuring data in order to help their understanding. These data are often received as tables or relations and structured by partitions, hierarchies, or lattices. With CA, formal contexts (binary relations between objects and attributes) are structured in concept lattices [GW99]. This is applied for instance in software engineering for configuration analysis [KS94]. *Data-mining* is used to extract properties from large amount of data. These properties are association rules verified (exactly or approximately) by the data. This is analogous to implications between attributes in FCA (cf. p. 79 in [GW99]), and to contextualized logic in LCA [FR00b]. *Unsupervised learning* is similar to data-analysis in the sense that one tries to discover some properties, and to understand some data, whereas *supervised learning* is similar to data-mining as some rules are searched for between known properties and the property to be learned. For instance, Kuznetsov applied CA to the learning of a positive/negative property from positive and negative instances [Kuz99].

The issue of this section is to show whether these features of Knowledge Discovery (KD) can be incorporated in our LIS, and how. Our aim is not to fully implement them in the LIS itself, but to offer primitives that could be combined for building more sophisticated KD features. First, we show how each of the three above kinds of KD can be formally expressed with the only notion of contextualized logic (cf. Section 2).

KD through Contextualized Logic. A context K plays the role of a theory by extending the deduction relation and enabling new entailments (e.g., $bird \models^{\overset{.}{K}} fly$ when every bird flies in the context). All these contextual entailments are gathered with logical entailments to form the contextualized logic, which is thus a means for extracting some knowledge from the context. Two kinds of knowledge can be extracted: knowledge about the context by deduction ("Every bird of this context *do* fly"), and knowledge about the domain (which the context belongs to) by induction ("Every bird of the domain *may* fly").

Concept lattices produced by data-analysis are isomorphic to contextualized logics (cf. Theorem 2). Associations rules produced by data-mining or supervised learning match the contextualized deduction relation, possibly qualified by a confidence defined by $conf(f \models^K g) = \frac{|\tau^K(f) \cap \tau^K(g)|}{|\tau^K(f)|}$.

Considering two properties $f, g \in \mathcal{L}$, their contextual relation is determined by the sizes of 3 sets of objects $\pi_l^K(f, g) := |\tau^K(f) \setminus \tau^K(g)|$, $\pi_c^K(f, g) := |\tau^K(f) \cap \tau^K(g)|$ and $\pi_r^K(f, g) := |\tau^K(g) \setminus \tau^K(f)|$. For instance, f contextually entails g iff $\pi_l^K(f, g) = 0$, f and g are contextually separated iff $\pi_c^K(f, g) = 0$, or x is an increment of wq (cf. Section 4) iff $\pi_c^K(x, wq) \neq 0$ and $\pi_r^K(x, wq) \neq 0$.

Generalizing the LIS navigation to KD. The links of navigation defined in Section 4 can be defined on such contextual relations w.r.t. the working query, as it can be seen on the following reformulation of *Link*:
$Link^K(wq, wv) =$
$\quad \lceil \{(x, \pi_c^K(x, wq)) \in X \times \mathbb{N} \mid x \models wv, wv \not\models x, \pi_c^K(x, wq) > 0\} \rceil$.
We propose to generalize the search for links of navigation into the search for some contextual properties *Propr*.

Definition 9 $Propr^K(wq, wv) :=$
$\quad \lceil \{(x, label^K(x, wq)) \in X \times Label \mid x \models wv, wv \not\models x, propr^K(x, wq)\} \rceil$.

where *propr* and *label* are respectively a predicate and an application defined with π_l, π_c, and π_r. The predicate *propr* specifies which contextual properties are searched for. The application *label* associates to each property a value belonging to *Label* and indicating to the user the relevance of this property. Properties can then be ordered according to their label. Now, for each kind of contextual property of interest, we can define a new LIS command similar to the command `ls` for searching for this kind of properties in the context. As an illustration, the following paragraph defines two such commands. It is important to notice that sets of properties displayed by these new commands depends strongly on the working query. So, they cannot produce an exhaustive knowledge like data-analysis, but are only useful for discovering some knowledge underlying a context through a navigation-like process. This *human-centered knowledge discovery process* is rather different from the most common approach of data analysis and data-mining, but has already been advocated [HSWW00,KGLB00].

Searching for necessary or sufficient properties. First, we define some common notions such as *support* and *confidence* [HSWW00] of a rule $f \to g$ in a context K:
$$supp^K(f \to g) := \pi_c^K(f, g) \text{ and } conf^K(f \to g) := \frac{\pi_c^K(f, g)}{\pi_l^K(f, g) + \pi_c^K(f, g)}.$$
We define a *necessary property* as a property x entailed by the working query wq with a confidence greater than $conf_{min}$. This leads to the following instantiations of *propr* and *label* in the definition of *Nec*:
$\quad propr^K(x, wq) :\Longleftrightarrow conf^K(wq \to x) \geq conf_{min}$,
$\quad label^K(x, wq) := conf^K(wq \to x) \in [0, 1]$.

This results in the new command **nec** that takes as argument $conf_{min}$ (set to > 0.0 by default) in addition to a view. This command enables the user to navigate among properties common to all objects of the working concept, and so in its intent. In the left part of Table 2, we search for the necessary properties of articles written by Sowa or Mineau. The 2nd query shows that all of these articles have fields **type**, **author**, **year**, and **title** defined. The 3rd query lists title words with confidence greater than 0.15: "Conceptual Graph" or "CG" appears in more than half of the considered articles (note that the formula **title: "Conceptual Graph" | "CG"** has been added to X manually with command **mkdir**, unlike other formulas), which is not very surprising, but "Formal" and "Context" also appear in more than a quarter of them.

Table 2. Examples of Knowledge Extraction in the BibTEX context.

```
(1) cd /author: Sowa | Mineau          (1) cd /title: Knowledge
(2) nec /                              (2) suf /
  1.000 type: *                          27 0.129 type: *
  1.000 author: *                        27 0.129 author: *
  1.000 year: ..                         27 0.129 year: ..
  1.000 title: *                         27 0.129 title: *
  18 object(s)                           27 object(s)
(3) nec 0.15 title: *                  (3) suf 2 0.2 author: *
  0.167 title: "Definition"             2 0.286 author: "Ellis"
  0.167 title: "Constraints"            2 0.400 author: "Angelova"
  0.222 title: "Represent"              2 0.667 author: "Bontcheva"
  0.222 title: "Processes"              2 1.000 author: "Gaines"
  0.278 title: "Context"                3 0.375 author: "Dick"
  0.278 title: "Formal"                 3 0.429 author: "Lukose"
  0.500 title: "Graph"                  3 0.500 author: "Cyre"
  0.556 title: "Concept"                4 0.667 author: "Martin"
  0.556 title: "Conceptual Graph" | "CG"  27 object(s)
  18 object(s)
```

We define a *sufficient property* as a property x that entails the working query wq with a support greater than $supp_{min}$, and with a confidence greater than $conf_{min}$. This leads to the following instantiations of $propr$ and $label$ in the definition of Suf:

$$propr^K(x, wq) :\Leftrightarrow supp^K(x \to wq) \geq supp_{min} \wedge conf^K(x \to wq) \geq conf_{min},$$
$$label^K(x, wq) := (supp^K(x \to wq), conf^K(x \to wq)) \in \mathbb{N} \times [0, 1].$$

This results in the new command **suf** that takes as argument $supp_{min}$ (set to 1 by default) and $conf_{min}$ (set to > 0.0 by default) in addition to a view. This command extracts some properties with which the working query property is always, or at least often, associated. Therefore, it can help to build some decision procedure for the working property, which is the issue of supervised learning. In the right part of Table 2, we search for the sufficient properties of articles whose title contains the word "Knowledge". The 2nd query just shows these articles represent 12.9% of the whole context, and serves mainly as a menu of the available fields. Then, the 3rd query lists author names with support greater than 2, and confidence greater than 0.2: e.g., Martin has written 4 articles about

"Knowledge", which consists in 2/3 of his articles, and Gaines has written 2 articles, both talking about knowledge.

Commands nec and suf both enable the search for association rules whose left or right part is fixed on the working query. In an application built on top of LIS, they could be combined to find all association rules between two sets of properties (ex. between authors and title words, in the BibTEX context).

6   Future Work

Our most practical perspective is to design a *logical file system*, which would implement the ideas we have presented in this article. The expected advantage is to offer the services described here at a standard system level that is accessible for every application. So doing, even applications that do not know about logical information systems (like e.g., compilers) would benefit from it.

A graphical user-interface to logical file systems would allow to display in an integrated fashion the working query, the working view, and the corresponding extent and set of links. For instance, a graphical interface for keeping trace of navigation, like what is becoming standard for file browsers, has been already experimented for a simple logic (attributes with values) but should be developed further. This amounts to keep a trace of the path from the start of the navigation to the current place. Moreover, the set of links could be presented graphically as a diagram of ordered formulas. A further refinement is to take into account the contextualized deduction, to get something similar to concept lattices derived from scales [CS00]. This amounts to represent an overview of possible future navigations.

The *Web* can also be explored using our techniques if one considers answers to web-queries as a formal context into which to navigate.

There is also a connection with natural language that we wish to explore further (see Section 4). We believe that a logical information system can provide the rational for a human-machine interface in natural language, in which both the human being and the machine could submit assertions and queries. In this case, logics that have been widely used for representing natural language semantics seem to be the right choice. One such logic is the deduction relation of Conceptual Graphs [Sow84,Sow99].

7   Conclusion

We have presented the specifications of a Logical Information System based on (Logical) Concept Analysis. As opposed to previous attempt of using Concept Analysis for organizing data, we do not propose to navigate directly in the concept lattice. Instead, we use the contextualized logic (i.e., the logical view of the concept lattice) to evaluate the relevance of navigation links. Those that do not narrow the focus of the search are called *views*. They only restrict the language of available navigation links. Other links, that do narrow the focus of the search, are called *increments*. They can be used to come closer to some place of interest.

In this way, standard commands of a file system shell can be mimicked in a logical context. However, a simple generalization of the definition of links forms a framework in which operations of data-analysis or data-mining can be expressed. Using this framework, purely symbolic navigation as well as statistical exploration can be integrated smoothly as variants of the same generic operation.

References

[CS00]   Richard Cole and Gerd Stumme. CEM - a conceptual email manager. In Guy Mineau and Bernhard Ganter, editors, *International Conference on Conceptual Structures*, LNCS 1867, pages 438–452. Springer, 2000.

[FR00a]   Sébastien Ferré and Olivier Ridoux. A file system based on concept analysis. In Yehoshua Sagiv, editor, *International Conference on Rules and Objects in Databases*, LNCS 1861, pages 1033–1047. Springer, 2000.

[FR00b]   Sébastien Ferré and Olivier Ridoux. A logical generalization of formal concept analysis. In Guy Mineau and Bernhard Ganter, editors, *International Conference on Conceptual Structures*, LNCS 1867, pages 371–384. Springer, 2000.

[GMA93]   R. Godin, R. Missaoui, and A. April. Experimental comparison of navigation in a Galois lattice with conventional information retrieval methods. *International Journal of Man-Machine Studies*, 38(5):747–767, 1993.

[GW99]   B. Ganter and R. Wille. *Formal Concept Analysis – Mathematical Foundations*. Springer, 1999.

[HSWW00]   Joachim Hereth, Gerd Stumme, Rudolf Wille, and Uta Wille. Conceptual knowledge discovery and data analysis. In Guy Mineau and Bernhard Ganter, editors, *International Conference on Conceptual Structures*, LNCS 1867, pages 421–437. Springer, 2000.

[KGLB00]   Pascale Kuntz, Fabrice Guillet, Rémi Lehn, and Henri Briand. A user-driven process for mining association rules. In D.A. Zighed, J. Komorowski, and J. Zytkow, editors, *Principles of Knowledge Discovery and Data mining*, LNAI 1910, pages 483–489. Springer-Verlag, 2000.

[KS94]   M. Krone and G. Snelting. On the inference of configuration structures from source code. In *International Conference on Software Engineering*, pages 49–58. IEEE Computer Society Press, May 1994.

[Kuz99]   S. O. Kuznetsov. Learning of simple conceptual graphs from positive and negative examples. In Jan M. Żytkow and Jan Rauch, editors, *Principles of Data Mining and Knowledge Discovery*, LNAI 1704, pages 384–391, Berlin, 1999. Springer.

[Lin95]   C. Lindig. Concept-based component retrieval. In *IJCAI95 Workshop on Formal Approaches to the Reuse of Plans, Proofs, and Programs*, 1995.

[Pre97]   Susanne Prediger. Logical scaling in formal concept analysis. *LNCS 1257*, pages 332–341, 1997.

[PS99]   Susanne Prediger and Gerd Stumme. Theory-driven logical scaling. In *International Workshop on Description Logics*, volume 22, Sweden, 1999.

[Sne98]   G. Snelting. Concept analysis – A new framework for program understanding. *ACM SIGPLAN Notices*, 33(7):1–10, July 1998.

[Sow84]   John Sowa. *Conceptual structures. Information processing in man and machine*. Addison-Wesley, Reading, US, 1984.

[Sow99]   John Sowa. Conceptual graphs: Draft proposed American National Standard. In William Tepfenhart and Walling Cyre, editors, *International Conference on Conceptual Structures*, LNAI 1640, pages 1–65, Berlin, 1999. Springer.

Reverse Pivoting in Conceptual Information Systems

Jo Hereth[1] and Gerd Stumme[2]

[1] Technische Universität Darmstadt, Fachbereich Mathematik
Schlossgartenstraße 7, D-64289 Darmstadt, hereth@mathematik.tu-darmstadt.de
[2] Universität Karlsruhe (TH), Institut für Angewandte Informatik und Formale
Beschreibungsverfahren (AIFB), D-76128 Karlsruhe, stumme@aifb.uni-karlsruhe.de

Abstract. In database marketing, the behavior of customers is analyzed by studying the transactions they have performed. In order to get
a global picture of the behavior of a customer, his single transactions
have to be composed together. In On-Line Analytical Processing, this
operation is known as *reverse pivoting*. With the ongoing data analysis
process, reverse pivoting has to be repeated several times, usually requiring an implementation in SQL. In this paper, we present a construction
for conceptual scales for reverse pivoting in Conceptual Information Systems, which directly interacts with the visualization. The construction
allows the reuse of previously created queries without reprogramming
and offers a visualization of the results by line diagrams.

1 Introduction

With increasing ease of gathering data about customer behavior, Database Marketing has become an essential issue for many companies. Its aim is to facilitate
decision support, customer-tailored production, and target-group specific advertising. These goals often require a clustering on the set of customers in order
to retrieve groups of 'similar' customers which can then be treated in a uniform
manner. The clustering is based on attributes of the customers (like age, gender, address, etc.) on the one hand, and their behavior (in terms of credit card
transactions, web access patterns, etc.) on the other hand. The data are usually
stored in data warehouses [Ki96]. The classical conceptual database model is the
star schema. It consists of a *fact table* and several *dimension tables*. The fact
table of a star schema may for instance contain credit card transactions; and the
dimension tables provide information about the structure of attributes like the
time of purchase, the point of sales, the amount of the transaction, etc. Another
star schema may consist of a fact table containing the customers, surrounded
by dimension tables for the age, the gender, the address, etc. If the second star
schema replaces a vertex of the first star, we obtain a *snowflake schema*. A variety of On-Line Analytical Processing (OLAP) [Th97] tools are on the market,
which allow an intuitive access to data stored in such star and snowflake schemes.

The situation becomes more difficult when data has to be analyzed which is
not in the center of a star/snowflake. This is for instance the case when clusters of customers are to be determined who are not solely 'similar' according to

H. Delugach and G. Stumme (Eds.): ICCS 2001, LNAI 2120, pp. 202–215, 2001.
© Springer-Verlag Berlin Heidelberg 2001

their demographic attributes, but also according to their buying behavior. As each transaction contains a customer ID, basically all information is available. The difficulty lies in the point that there are many transactions related to one customer, and it is *a priori* not clear in which way the information has to be extracted and aggregated. For instance, there are obvious queries like the total amount of money spent, but also more sophisticated queries like the distribution of the purchase dates over time. Usually, such queries require an explicit implementation in SQL. As the requests of the data analysts change over time, queries have to be implemented several times. Therefore, constructions are needed which allow to derive, from existing queries, queries against a table which is not in the center of a snowflake.

Conceptual Information Systems [SSVWW93] have been extensively applied to data where the conceptual database model is similar to a star schema. The implementation of the management system TOSCANA for Conceptual Information Systems [VW95] is based on a relational database management system, and offers different views on the data by means of *conceptual scales*. Conceptual scales can be considered as conceptual hierarchies analogue to the dimensions of the star schema [St00]. At the moment, Conceptual Information Systems do not support reverse pivoting. In this paper, we discuss how reverse pivoting can be done within Conceptual Information Systems. The idea is to construct new scales for the analysis of the customers based on the already existing scales for the transactions. We present a new scale construction and discuss its use. The advantage of this mathematically founded construction is two-fold: (*i*) the construction goes along with the visualization technique offered by line diagrams of concept lattices, and (*ii*) it allows to bypass SQL programming by offering a predefined operation.

This work has been done in cooperation with the database marketing department of a Swiss department store. In the project, we observed that reverse pivoting is not the only feature required for Conceptual Information System in a database marketing application. For instance, an important functionality in database marketing is the comparison of the same attributes for different groups, e. g. the comparison of sales in consecutive quarters or in different regions. In order to enable this functionality in Conceptual Information Systems, we have introduced *scale prisms* which allow to display a conceptual scale (or different ones) several time, each time applied to different data.

The remainder of this paper is organized as follows: The database marketing scenario is described in the next section (see also [HSWW00]). In Section 3 we recall the basics of the star and the snowflake schema as conceptual database model and explain the idea of reverse pivoting. In Section 4 we discuss reverse pivoting in Conceptual Information Systems.

In Section 5, we present the scale construction and discuss its use and its visualization. Here we also give the definition of scale prisms and provide an example. Section 6 concludes the paper.

2   The Database Marketing Scenario

The results presented in this paper have been inspired by a joint research project of Darmstadt University of Technology and the database marketing department of a Swiss department store (cf. [HSWW00,He00]). The database marketing department bases its market analysis on data collected from bonus card holders. The bonus card is a credit card which can be used in the shops of about a dozen companies. The operational systems of the companies collect information from more than 300,000 customers and has historical data reaching back five years, resulting in a database of 9,000,000 transactions. The database marketing department has to incorporate the data from different sources and create a coherent database. The department analyzes these data in order to improve the marketing by so-called *direct mailings*, i. e. personalized advertisement by mail. For this purpose the database marketing department has to select the corresponding target group. The information is also used to strengthen the cooperation between the companies. Here are some typical questions in this context:

1. How high were the scales in the shops of partner X?
2. Which customers are between 45 and 65 years old?
3. How many male card holders live near Zurich?
4. Which customers of Partner X are not yet customers of Partner Y?
5. How many customers have been inactive for more than twelve months?
6. Which card holders were active in a shop from Partner X last year?

Query 1 can be answered within a standard Conceptual Information System which is focused on the transactions. Queries 2 and 3 can be answered within a standard Conceptual Information System which is focused on the customers. Queries 4, 5, and 6 — which are the most interesting for the database marketing department —, however, need reverse pivoting.

If the set of questions to be asked is predefined, reverse pivoting can be performed once, and the necessary data can be stored in the data table underlying the Conceptual Information System. However, in the process of analysis, the users sometimes wants to see yet another aspect of the data. This implies a rework of the underlying data table and a rewrite of the SQL-statements. In the project with the database marketing department, reverse pivoting was performed manually, in order to gain experiences about the types of queries needed. The resulting system is still in use today in the database marketing department. Based on the experiences made, we developed the constructions which are presented in this paper.

3   Snowflake Schemes and Reverse Pivoting

In a date warehouse like the one of our database marketing department, the standard approach to model the data is by *star schemes* and *snowflake schemes.* (cf. [Ki96]). Figure 1 shows the snowflake schema for the transaction data. The transaction data themselves are stored in a large, central *fact table* (the table

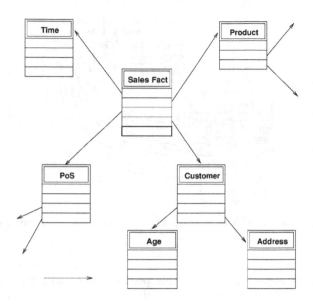

Fig. 1. Example of a snowflake schema

Sales Fact); and information about the structure of attributes like 'time of purchase', 'point of sale' etc. is stored in *dimension tables*. These tables are linked to the central fact table by foreign keys.

The transactions are in the center of the schema. A typical query against these data is for instance the first query from above. This means that aggregations along (some of) the dimensions have to be performed to answer the queries. As transaction data are most often numerical — such as price, cost, or number of units sold —, computations based on these values are well supported by current database systems and allow fast answers. The responsiveness of analysis systems is fundamental for the required interactive use in database marketing. Most current implementations of Data Warehouse systems support therefore only analysis based on those numerical values.

The traditional approach of adding up or counting the transaction values does not help for an analysis of customers and customer groups. The central object that needs to be modeled is the customer himself, not the transaction. The analyst wants to learn about the behavior of a customer, not only about transaction focussed numbers. The straightforward solution seems to be to simply consider the table Customer as center of the snowflake, instead of the table Sales Fact. The problem is that now the foreign key between both tables points in the wrong direction. Therefore the dimensions around the Sales Fact table cannot be used directly for aggregation of customer-centered data. For instance, while it makes perfectly sense to assign customer features (like the gender) to a transaction, it is not clear *a priori* how to assign transaction features (like point in time) to a customer.

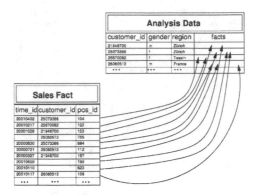

Fig. 2. The Reverse-Pivot from a transaction-oriented to a customer-oriented data model

The standard approach to solve this problem is to add fields derived from the **Sales Fact** table to the **Customer** table. This process is known as *reverse pivoting*. It is sketched in Figure 2. What part of the transaction data is of interest for the analysis depends on the context. In [Py99] it is therefore suggested to ask a domain expert which fields should be added to the customer data table and be present for future examination. For instance, we might be interested in the time of the first purchase, the time of the last purchase, or in the duration of the longest interval between two purchases.

4   Conceptual Information Systems and Reverse Pivoting

A *Conceptual Information System* is a navigation tool that allows dynamic browsing through and zooming into the data based on the visualization of conceptual hierarchies (called *conceptual scales*) by line diagrams. In this paper, we recall only very briefly the basic notions of Conceptual Information Systems. We assume that the reader is familiar with the notions of formal contexts and concept lattices as given, for instance, in [GW99]. For further details we refer to [SSVWW93], [VW95], or [HSWW00].

Definition 1 (Conceptual Information Systems). *A many-valued context is a tuple* $\mathbb{K} = (G, M, W, I)$ *where* G, M, *and* W *are sets, and* $I \subseteq G \times M \times W$ *is the graph of a partial function from* $G \times M$ *to* W *(i. e.,* $(g, m, w_1) \in I$ *and* $(g, m, w_2) \in I$ *always implies* $w_1 = w_2$*). The elements of* G, M, *and* W *are called* objects, attributes, *and* attribute values, *respectively, and* $(g, m, w) \in I$ *is read: "object* g *has attribute value* w *for attribute* m*". An attribute* m *may be considered as a partial mapping from* G *to* W*; therefore,* $m(g) = w$ *is often written instead of* $(g, m, w) \in I$*. A conceptual scale for an attribute* $m \in M$ *is a one-valued context* $\mathbb{S}_m := (G_m, M_m, I_m)$ *with* $m(G) \subseteq G_m$*. The context* $\mathbb{R}_m := (G, M_m, J_m)$ *with* $(g, n) \in J_m : \iff (m(g), n) \in I_m$ *is called the* realized scale for the attribute $m \in M$*. The* derived context of $\mathbb{K}$ with respect *to the conceptual scales* $\mathbb{S}_m := (G_m, M_m, I_m)$, $m \in M$, *is the formal context* $(G, \bigcup_{m \in M}(\{m\} \times M_m), J)$ *with* $(g, (m, n)) \in J : \Longleftrightarrow (m(g), n) \in I_m$. *A many-valued context together with a collection of conceptual scales with line diagrams of their concept lattices is called a* Conceptual Information System.

There is a strong analogy between the conceptual model of Conceptual Information Systems and star schemes. The fact table corresponds to the many-valued context (with the minor difference that the existence of set G is equivalent to a primary key in the fact table, which is not required in star schemes[1]). The conceptual scales correspond to the dimension tables of the star schema. Although the above definition is not directly equivalent to snowflake schemes, it can easily be extended.

Conceptual Information Systems are thus able (as OLAP systems) to handle data which are structured along a star or snowflake schema. The difference is that Conceptual Information Systems focus more on conceptual aspects of the data, while OLAP systems are centered on numerical evaluations. An extension of Conceptual Information Systems by numerical features is presented in [SW98], and the use of Conceptual Information Systems as OLAP tools is discussed in [St00].

Figure 3 shows the snowflake schema from Figure 1 as schema of a Conceptual Information System. Each conceptual scale can be used directly for a fact table (i. e., a many-valued context), as long as there is a path of foreign keys in the appropriate direction. For instance, all conceptual scales for customers can be directly reused for the transactions: we can reuse the scale 'age' of the customer dimension to cluster the transactions according to the age of the customers.

However, the main purpose of the database marketing group is analyzing the customers — and not the transactions — in order to find homogeneous groups of customers for direct mailing actions. Hence the Customer table has now to be considered as fact table. The scales for age, gender, and address can of course still be used, but as the foreign key between the Customer table and the Sales Fact table points in the wrong direction, the scales of the Sales Fact table cannot be reused directly. A reverse pivoting is necessary.

In the project with the database marketing department, the reverse pivoting was performed manually (i. e., implementing SQL queries) as described in the previous section. The aim was to gain experience about the types of queries needed. The result was the Conceptual Information System described in [HSWW00]. We could observe that in fact the system enabled an interactive process of conceptual data analysis leading to useful knowledge.

The system was constructed in tight cooperation with the database marketing department. Therefore the domain experts were involved from the beginning. The fields considered useful were included in the data table. After some time of data analysis however, the analysts got more and more familiar with the data and the information system. This led to a request for more in-depth information and to the need for more sophisticated queries. Therefore the data table had to

---

[1] In our database marketing application, the transaction ID plays the role of the primary key.

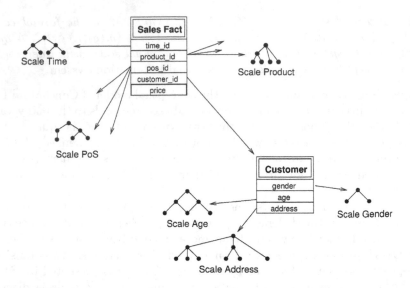

Fig. 3. The schema from Figure 1 as schema of a Conceptual Information System.

be adapted and expanded several times. It became clear that this process had to be supported by a methodology, such that it (or at least parts of it) could be automated.

Very often the new information the analysts were looking for was describable in terms of the knowledge stated in already given scales. Therefore we developed different constructions for deriving new scales from the already given ones. The constructions substitute the explicit coding of the reverse pivoting in the database table. A modification of the fact table as in the previous section need not be performed any longer,[2] and the results can be visualized directly. In the next section, we present the construction and discuss its use.

5   Conceptual Scaling Constructions

How can a scale that was designed for the transactions be reused for the analysis of the customers? We show how to construct new scales which can then be applied for clustering the set of customers, i.e., which can be used for the reverse-pivoted data. We also discuss how their visualization can be supported by *iceberg concept lattices*, which are introduced in [S+01].

There are two more constructions which support other aspects of reverse pivoting: adaptable and parameterized scales. Because of space restriction, we will not present them here. They are described in detail in the diploma thesis [He00]. Instead, we present another construction which is not directly related to reverse

---

[2] In trade-off for a better computational performance it may sometimes still be sensible to do so.

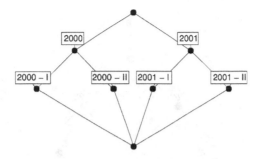

Fig. 4. The scale $\mathbb{S}_{\text{Time}}$.

pivoting, but which has nevertheless proven to be very useful in the database marketing scenario: so-called *scale prisms*, which allow the comparison of the same attributes for different groups, e. g. the comparison of sales in consecutive quarters or in different regions.

The examples given in this section show only the results for a small example database. Even though the inspiration for the constructions stems from a real-world system implemented for the database marketing department, we are not allowed to show results from this system due to company policy.

5.1   Power Scales

As example we use the scale $\mathbb{S}_{\text{Time}} := (G_{\mathbb{S}}, M_{\mathbb{S}}, I_{\mathbb{S}})$ in Figure 4. Applied on the transactions, it clusters them by half year intervals. Now we show how it can be used for the construction of a new scale which can be applied to the customers.

An analyst from the database marketing department often thinks of customers as "customers in the year 2000" or "customers in the first half of year 2001", but the scale in Figure 4 does not allow for all possibilities: Suppose a customer bought something in the second half of year 2000 and something in the first half of year 2001, but nothing in the other intervals. There is no appropriate concept in the scale for containing this customer. The only concept below the attributes "2000 - II" and "2001 - I" is the bottom element – which would imply all other attributes as well.

Obviously the problem is that customers may have *combinations* of attributes where transactions only have one of them. For our first construction, the attributes we want to associate with a customer are all attributes of all his transactions. In other words, the customer should have an attribute assigned if there exists at least one transaction, which is in relation to this attribute. Formally we define:

Definition 2 (Power Scale). *Let* $\mathbb{S} := (G_{\mathbb{S}}, M_{\mathbb{S}}, I_{\mathbb{S}})$ *be a conceptual scale. Then the* power scale *of* $\mathbb{S}$ *is the scale* $\mathbb{S}^{\mathfrak{P}} := \left(\mathfrak{P}(G_{\mathbb{S}}), M_{\mathbb{S}}, I_{\mathbb{S}}^{\mathfrak{P}} \right)$ *with* $(A, m) \in I_{\mathbb{S}}^{\mathfrak{P}} :\Leftrightarrow \exists g \in A$ *with* $(g, m) \in I_{\mathbb{S}}$, *for* $A \subseteq G_{\mathbb{S}}$ *and* $m \in M_{S}$.

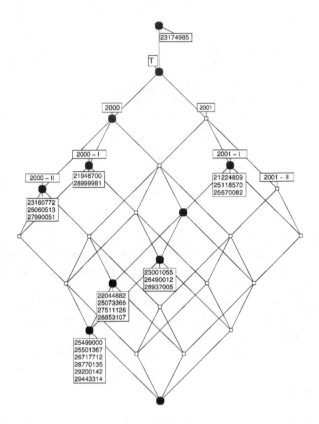

Fig. 5. The power scale $\mathbb{S}_{\text{Time}}^{\mathfrak{P}}$

The line diagram of the power scale in our example is shown in Figure 5. We have applied it to a small example database containing 22 customers. It shows for instance that there are three customers (listed in the lower middle of the diagram) who have bought something in the first half of year 2000 and in the first half of year 2001, but nothing in the two other half year intervals.

The resulting concept lattices can be large relative to the concept lattice of the original scale:

Lemma 1. *If a scale* $\mathbb{S}$ *has* n *concepts, the following equation holds for the number* m *of concepts of* $\mathbb{S}^{\mathfrak{P}}$:

$$n \leq m \leq 2^n \ .$$

The concept lattice of the power scale therefore can be much more complex than the concept lattice of the original scale. A first step to derive a line diagram for the power scale based on the line diagram of the original scale is the following theorem, which asserts that the structure of the original concept lattice is preserved. The proof of this theorem is given in [He00].

Theorem 1. *In the concept lattice* $\mathfrak{B}(\mathbb{S}^{\mathfrak{P}})$ *exists a* $\vee$-*subsemilattice isomorphic to the concept lattice* $\mathfrak{B}(\mathbb{S})$ *of the original scale. The concepts of this subsemilattice are all concepts from* $\mathfrak{B}(\mathbb{S}^{\mathfrak{P}})$ *whose intents are intents of a concept in* $\mathfrak{B}(\mathbb{S})$.

As the order is preserved, the layout of the original scale may be used as a basis for the layout of the power scale. Additionally we know that the concept lattices of power scales are distributive (cf. [He00]), which allows to apply the algorithms presented in [Sk89], yielding readable line diagrams.

5.2   Power Scales of the &–Product

A particular advantage of conceptual scales is that they can easily be combined to give a more detailed view on the data. This possibility allows for more flexibility in the analysis process. Combining scales is usually done by building the semi-product of the scales [GW99]. The new conceptual scale can be represented by a nested line diagram. E. g., if there are two scales $\mathbb{S}_{\mathrm{Year}}$ and $\mathbb{S}_{\mathrm{PoS}}$ showing the year in which, and the point of sale where a transaction was performed, the scale $\mathbb{S}_{\mathrm{Year}} \boxtimes \mathbb{S}_{\mathrm{PoS}}$ shows the information of both scales simultaneously: The user can for instance see which transactions were performed at the department store in the year 2000.

The straightforward approach to study the customer behavior according to both the times and the locations of purchase would be to apply the power scale construction on the semi-product $\mathbb{S}_{\mathrm{Year}} \boxtimes \mathbb{S}_{\mathrm{PoS}}$: The power scale $(\mathbb{S}_{\mathrm{Year}} \boxtimes \mathbb{S}_{\mathrm{PoS}})^{\mathfrak{P}}$ allows to see, for instance, the group of all customers who bought something in the year 2000 and who bought something in the department store. But unfortunately, this is not equal to the group of customers who have effectuated at least one transaction in the department store in the year 2000. If a customer bought something in the department store in 2001 and something else somewhere else in year 2000, he still fulfills the two criteria. Hence we obtain a larger group of customers than expected.

But obviously the information we need exists in the data. The solution is to replace the semi-product by the so-called &-product. The &-product construction is introduced in [GW99], but has not yet been investigated or used much before. The &-product of two scales is defined as follows:

Definition 3 (&-product). *Let* $\mathbb{S}_1 := (G_1, M_1, I_1)$ *and* $\mathbb{S}_2 := (G_2, M_2, I_2)$ *be conceptual scales. The* &-*product of these scales is defined as* $\mathbb{S}_1 \& \mathbb{S}_2 := (G_1 \times G_2, M_1 \times M_2, J)$ *with* $((g,h),(m,n)) \in J :\Leftrightarrow (g,m) \in I_1 \wedge (h,n) \in I_2$.

The concept lattice of the &-product has less concepts (and hence less information) than the one of the semi-product. It is isomorphic to a $\wedge$-subsemilattice of the latter one. On the other hand, its context is considerably larger. Therefore the semi-product has been advantageous in the Conceptual Information Systems built in the past (which had no need of reverse-pivoting).

Applying the power scale construction to these products inverts the size relation. If there is an attribute in each scale which is related to all objects, the

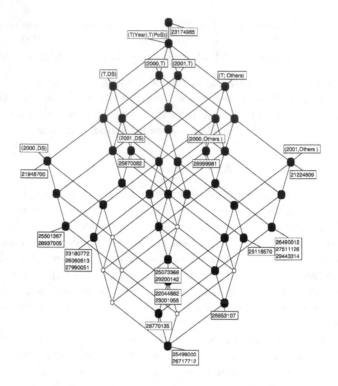

Fig. 6. The power scale $(\mathbb{S}_{\text{Year}} \& \mathbb{S}_{\text{PoS}})^{\mathfrak{P}}$

concept lattice $\underline{\mathfrak{B}}((\mathbb{S}_1 \boxtimes \mathbb{S}_2)^{\mathfrak{P}})$ contains more concepts (and more information) than the concept lattice $\underline{\mathfrak{B}}((\mathbb{S}_1 \& \mathbb{S}_2)^{\mathfrak{P}})$, which is isomorphic to a sublattice of the former one. In Figure 6 we see the power scale $(\mathbb{S}_{\text{Year}} \& \mathbb{S}_{\text{PoS}})^{\mathfrak{P}}$. Its left-most concept shows, for instance, that customer 21948700 and all $2+3+2+1+2 = 10$ customers listed further below bought something in the department store in year 2000. The immediate super-concept of this concept contains all customers who bought something in the department store and something (eventually different) in year 2000.

5.3   Iceberg Concept Lattices

As seen in the last two subsections, concept lattices of power-scales can become very large. But in most cases the user is not interested to see all concepts of the lattice at once. For analyzing overall trends, the concepts containing no or only few customers are not of interest. For those cases it is useful to display only a part of the concept lattice. This can be done by applying *iceberg concept lattices* [S+01], i.e. by hiding all concepts whose extent is below a certain percentage of the total set of customers. This way, the number of concepts to be displayed can be reduced considerably, making the diagram easier to read. In Figure 7 the iceberg lattice version of the diagram in Figure 6 is given. Only those concepts whose extents contain at least 50 % of the 22 customers are shown.

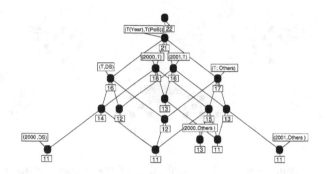

Fig. 7. An iceberg lattice of the power-scale $(\mathbb{S}_{\mathrm{Year}}\&\mathbb{S}_{\mathrm{PoS}})^{\mathfrak{P}}$

There is no problem in labeling iceberg concept lattices as long as the labeling is restricted to the cardinalities (which most often is sufficient for database marketing). Problems may arise when the names of the objects have to be displayed. This is discussed in [He00] for the more general case of a 'Skalenbund'.

5.4   Scale Prisms

An important functionality in database marketing is the comparison of the same attributes for different groups, e. g. the comparison of sales in consecutive quarters or in different regions. A conceptual scale allows a detailed view on the data for one part only. For the comparison of several parts or facets, the construction of a *scale prism* can be useful. Often those facets are (partially) ordered, e. g. when considering time intervals, where one interval contains another.

Definition 4 (Scale Prisms). *A* scale prism *consists of an ordered set* $(F, \leq)$ *(the* facets*) and a set of conceptual scales* $(\mathbb{S}_f)_{f \in F}$ *which are consistent with the context.*

One could argue that the definition of scale prisms should include some common features of the indexed scales. However, our experience shows that the situations in which one wants to apply the construction are so many-fold, that such a constraint would be too restrictive. It is a modeling decision of the user to choose scales which are reasonably related.

The scale prism is displayed like a nested line diagram: A line diagram of the ordered set $(F, \leq)$ is drawn as outer diagram, and each realized scale is drawn in the corresponding node. To emphasize that it is not meant to be read as an ordinary nested line diagram (since different orders are meant in the outer diagram and the inner diagrams), the lines of the outer diagram are drawn as dotted lines.

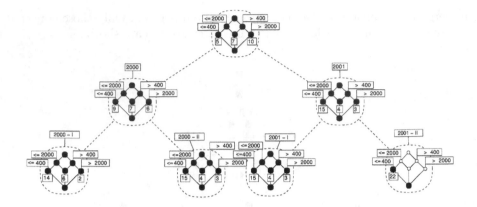

Fig. 8. A scale prism showing the distribution of the amount of money spent per customer in several time intervals

An example is given in Figure 8. If we compare, for instance, the left-most concepts in the four lowest nodes of the outer diagram, we observe that the number of customers who spent relatively few money (i.e., below 400 Swiss francs) has increased from 14 in the first half of year 2000 to 15 in the second half of year 2000. It remained constant for the next half year, and increased again to 22 in the second half of year 2001.

6 Conclusion

In this paper, we have presented the power-scale as a mathematically founded construction for conceptual scales which supports reverse pivoting in Conceptual Information Systems. We have shown that the power-scale and the power-scale of the &-product offer three advantages:

- The constructions can be considered as predefined operations for data preparation which allow to bypass SQL programming every time the data analyst requires a new conceptual scale.
- The constructions go along with the visualization technique of line diagrams of concept lattices. This allows to closely integrate the reverse pivoting operations in the graphical user interface.
- The constructions can be combined with the technique of iceberg concept lattices, i.e., all but the most general concepts can be pruned in order to reduce the complexity of the visualization.

We have further presented scale prisms as means for data analysis with Conceptual Information Systems in an OLAP-style application.

In order to test the usefulness of the constructions, the operations have been first performed manually in the described database marketing scenario. In order to make the techniques available to a larger group of users, it is planned to implement the techniques in the management system TOSCANA for Conceptual Information Systems.

An interesting open research question is the combination of the power-scale constructions with other data mining techniques. For instance, they might be used to improve the performance of algorithms for computing association rules or for discovering subgroups by restricting the search space.

Additionally, the experiences gained with the system show that reverse pivoting is a computationally very intensive task. The queries to be processed yield possibilities for major improvements on the database backend. This should be subject to further research.

References

[GW99]   B. Ganter, R. Wille: *Formal Concept Analysis : Mathematical Foundations*. Springer Verlag, Berlin – Heidelberg – New York, 1999.

[He00]   J. Hereth: *Formale Begriffsanalyse im Data Warehousing*. Diploma thesis, TU Darmstadt, 2000.

[HSWW00] J. Hereth, G. Stumme, U. Wille, R. Wille: Conceptual Knowledge Discovery and Data Analysis. In: B. Ganter, G.W. Mineau (eds.): *Conceptual Structures: Logical, Linguistic, and Computational Issues*. LNAI 1867. Springer, Berlin – Heidelberg – New York 2000, 421-437. Available at http://wwwbib.mathematik.tu-darmstadt.de/Math-Net/Preprints/Listen/shadow/pp2092.html

[Ki96]   R. Kimball: *The Data Warehouse Toolkit*. John Wiley & Sons, New York 1996.

[Py99]   D. Pyle: *Data preparation for data mining*. Morgan Kaufman Publishers, San Francisco 1999.

[SSVWW93] P. Scheich, M. Skorsky, F. Vogt, C. Wachter, R. Wille: Conceptual Data Systems. In: O. Opitz, B. Lausen, R. Klar (eds.): *Information and Classification*. Springer, Berlin–Heidelberg 1993, 72–84

[Sk89]   M. Skorsky: How to draw a concept lattice with parallelograms. In: R. Wille (ed.): *Klassifikation und Ordnung*. Indeks-Verlag, Frankfurt 1989, 191–196.

[St00]   G. Stumme: Conceptual On-Line Analytical Processing. In: K. Tanaka, S. Ghandeharizadeh, Y. Kambayashi (eds.): *Information Organization and Databases*. Chpt. 14. Kluwer, Boston–Dordrecht–London 2000

[S+01]   G. Stumme, R. Taouil, Y. Bastide, N. Pasqier, L. Lakhal: Computing Iceberg Concept Lattices with Titanic. *J. on Knowledge and Data Engineering* (submitted)

[SW98]   G. Stumme, K. E. Wolff: Numerical Aspects in the Data Model of Conceptual Information Systems. In: Y. Kambayashi, Dik Kun Lee, Ee-Peng Lim, M. K. Mohania, Y. Masunaga (eds.): *Advances in Database Technologies. Proc. Intl. Workshop on Data Warehousing and Data Mining, 17th Intl. Conf. on Conceptual Modeling (ER '98)*. LNCS **1552**, Springer, Heidelberg 1999, 117–128

[Th97]   E. Thomsen: *OLAP Solutions: Building Multidimensional Information Systems*. John Wiley & Sons, New York 1997.

[VW95]   F. Vogt, R. Wille: TOSCANA — a graphical tool for analyzing and exploring data. In: R. Tamassia, I. G. Tollis (eds.): *Graph Drawing '94*. LNCS **894**. Springer, Berlin–Heidelberg 1995, 226–233

Refinement of Conceptual Graphs

Juliette Dibie-Barthélemy[1], Ollivier Haemmerlé[1], and Stéphane Loiseau[2]

[1] Institut National Agronomique Paris-Grignon
16, rue Claude Bernard, F-75231 Paris Cedex 05
{Juliette.Dibie,Ollivier.Haemmerle}@inapg.inra.fr
[2] Université d'Angers, UFR Sciences
boulevard Lavoisier, F-49045 Angers Cedex 01
Stephane.Loiseau@info.univ-angers.fr

Abstract. The semantic validation of a knowledge base (KB) consists in checking its quality according to constraints given by an expert. The refinement of a KB consists in correcting the errors that are detected during the validation, in order to restore the KB validity. We propose to perform the semantic validation and refinement of a KB composed of conceptual graphs in two stages. First, we study the coherence of the KB with respect to negative constraints, which represent the knowledge that the KB must not contain. When the KB is not coherent, we propose a solution to correct all the errors of the KB. Second, we study the completeness of the KB with respect to positive constraints, which represent the knowledge that the KB must contain. When the KB is not complete, we propose an assistant, which helps the user to correct the errors of the KB one by one.

1 Introduction

Knowledge validation proposes solutions to ensure the quality of a knowledge base (KB). It provides quality criteria that must be checked by the KB as well as tools to check these criteria automatically. Most of the works on validation deal with the rule basis [1,2,3,4]. The other works essentially study either the validation of knowledge models [5,6] or the validation of description logics [7,8]. The validation of a KB can conclude that the KB is not valid: the KB contains errors. The refinement of a KB takes place after the validation and consists in correcting the errors of the KB in order to restore its validity. Few works on validation study how to correct the errors of a KB that is not valid [9,10].

This paper deals with the semantic refinement of a KB built on the conceptual graph model [11] as formalized in [12]. Our work is based on the semantic validation of a KB presented in [13], which is briefly reminded here for the need of the paper. We suppose that the semantic validation is made on an existing KB, so it is made a posteriori and aims at ensuring that the assertional knowledge of the KB conforms to some expert knowledge, which is given for validation purpose and is assumed reliable. This expert knowledge is expressed in terms of constraints which constitute an extension of the conceptual graph model. When

H. Delugach and G. Stumme (Eds.): ICCS 2001, LNAI 2120, pp. 216–230, 2001.
© Springer-Verlag Berlin Heidelberg 2001

a KB is not semantically valid, it contains errors. The errors are localized in the KB in the form of partial subgraphs called *conflicts*. *Actions* are associated with each conflict in order to remove the errors. A main point with refinement is to find solutions to correct errors with minimal changes on the KB.

We propose to perform the semantic validation and refinement of a KB with respect to two categories of constraints: negative constraints, which allow one to represent the knowledge that the KB must not contain, and, on the contrary, positive constraints, which allow one to represent the knowledge that the KB must contain. These two categories of constraints are assumed reliable and non contradictory. On the one hand, we propose to study the coherence of the KB with respect to negative constraints. When a KB is not coherent, we propose a solution based on revision plans to refine it. A revision plan is a set of actions which allows one to restore the coherence of the KB. Our approach is inspired from that of [10] which concerns the refinement of rule basis. On the other hand, we propose to study the completeness of the KB with respect to positive constraints. When a KB is not complete, we propose an assistant based on actions to help the user to correct the errors which have been detected during the validation one by one. Note that the notion of constraints is a notion that has already been studied in the conceptual graph model [12,14,15]. A complete comparison between these works and our constraints is presented in [16].

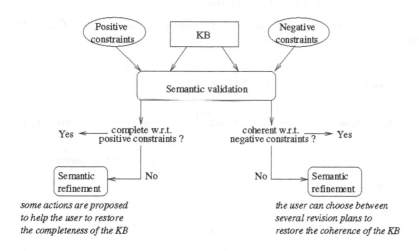

Fig. 1. Refinement process.

This paper is organized as follows. Section 2 deals with the coherence of a KB with respect to negative constraints and presents the associated notions of conflicts and actions. Section 3 deals with the completeness of a KB with respect to positive constraints and presents the associated notions of conflicts and actions. Section 4 studies the refinement of a KB that is not semantically valid with respect to negative and positive constraints.

2   Base Incoherences

In the following, we consider a knowledge base $\mathcal{KB}$ composed of a reliable support S and an irredundant conceptual graph under normal form Γ [17]. A conceptual graph G is *irredundant* if there is no projection from G into one of its strict subgraphs. A conceptual graph is *under normal form* if each of its individual markers appears exactly once.

The coherence of $\mathcal{KB}$ is studied with respect to negative constraints. An intuitive idea of negative constraints is that $\mathcal{KB}$ must not contain the knowledge represented by these constraints. We propose two kinds of negative constraints: the negative existential constraints and the maximal descriptive constraints. The negative existential constraints allow one to represent some knowledge that must not be deduced from $\mathcal{KB}$. The maximal descriptive constraints allow one to represent some knowledge that must conform to restrictions when they appear under a given form in $\mathcal{KB}$. We say that $\mathcal{KB}$ is *coherent* with respect to negative constraints if and only if $\mathcal{KB}$ satisfies each of them.

We remind the definitions of the two kinds of negative constraints, the way $\mathcal{KB}$ satisfies them and, in the case when $\mathcal{KB}$ does not satisfy them, we present the notions of conflicts and actions in order to localize the errors of $\mathcal{KB}$ and to provide a way of removing those errors.

2.1   Negative Existential Constraints

A negative existential constraint is a conceptual graph, noted GE^-, that is *satisfied* by $\mathcal{KB}$ if there does not exist a projection from it into Γ [13].

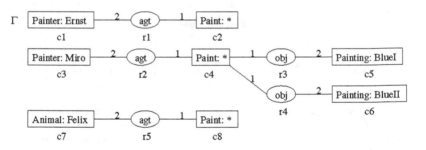

Fig. 2. Γ represents the assertional knowledge of $\mathcal{KB}$.

Fig. 3. $\mathcal{KB}$ does not satisfy the negative existential constraint GE^-, because there exists a projection Π from GE^- into Γ: the animal Felix paints.

We introduce the notion of image graph in order to study the case when $\mathcal{KB}$ does not satisfy a negative existential constraint.

Definition 1 Let h and g be partial subgraphs of respectively the conceptual graphs H and G, and Π a projection from G into H. h is the *image graph of g relatively to* Π, noted $\Pi(g)$, if there exists a surjective projection Π' from g into h such that $\Pi' \subseteq \Pi$.

A set of conflicts is associated with each negative existential constraint not satisfied by $\mathcal{KB}$. These conflicts allow one to explain why $\mathcal{KB}$ does not satisfy the constraint. To be more precise, a conflict of $\mathcal{KB}$ with respect to a negative existential constraint GE^- is an image graph of GE^- into Γ. Such a conflict must be minimal in the sense that there must not exist a projection from GE^- into any of its strict partial subgraphs.

Definition 2 Let GE^- be a negative existential constraint and Π a projection from GE^- into Γ. If there does not exist a projection Π' from GE^- into Γ such that $\Pi'(GE^-) \subset \Pi(GE^-)$, then the partial subgraph $\Pi(GE^-)$ of Γ is a *conflict* of $\mathcal{KB}$ with respect to GE^-, noted $\mathrm{CF}(GE^-, \Pi)$.

Example 1 *According to figure 3, the partial subgraph $\Pi(GE^-)$ of Γ limited to the vertices c_7, r_5 and c_8 is a conflict of $\mathcal{KB}$ with respect to GE^-.*

In order to eliminate a conflict $\mathrm{CF}(GE^-, \Pi)$, the conflict has to be modified so that there is no more projection from GE^- into it. In order to get as close as possible to the initial KB, two different actions are possible: unrestricting[1] a vertex or suppressing a vertex. The first action sets the problem of the choice of the super-type to unrestrict the label of the vertex, a type being able to have several super-types. In order to limit the combinatory, we propose the *action of suppression* $\mathbf{sup}(s)$ where s is a vertex of the conflict: (1) if s is a relation vertex, $\mathbf{sup}(s)$ suppresses s and its adjacent edges; (2) if s is a concept vertex, $\mathbf{sup}(s)$ suppresses s and its neighbour relation vertices with their adjacent edges. As a matter of fact, according to the definition of the conceptual graph model [12], a relation vertex cannot have a pendent adjacent edge.

We say that an action associated with a conflict of $\mathcal{KB}$ *eliminates the conflict* if the KB obtained by its application on $\mathcal{KB}$ does not contain the conflict any more.

Property 1 An action of suppression associated with a conflict $\mathrm{CF}(GE^-, \Pi)$ eliminates the conflict.

Proof 1 The action of suppression suppresses one or several vertices of the conflict. A conflict being minimal, the strict partial subgraph of the conflict obtained by the application of the action on $\mathcal{KB}$ is not a conflict of $\mathcal{KB}$ with respect to GE^- any more.

---

[1] i.e. replacing the type of the vertex by a super-type and/or, if the vertex is an individual concept vertex, replacing its marker by the generic marker.

Example 2 *Three actions of suppression are possible to eliminate the conflict of example 1:* $\mathbf{sup}(c_7)$ *which suppresses the concept vertex c_7 and the relation vertex r_5 with its adjacent edges,* $\mathbf{sup}(r_5)$ *which suppresses r_5 with its adjacent edges, and* $\mathbf{sup}(c_8)$ *which suppresses c_8 and r_5 with its adjacent edges.*

A set of actions of suppression is associated with each conflict $CF(GE^-, \Pi)$, each action allowing one to eliminate the conflict. The cardinal of this set is the number of vertices of the conflict.

Example 3 *The set of actions of suppression to eliminate the conflict of example 1 is: $EA = \{\mathbf{sup}(c_7), \mathbf{sup}(r_5), \mathbf{sup}(c_8)\}$.*

2.2   Maximal Descriptive Constraints

A maximal descriptive constraint is a triplet $(GE, GS, \Pi_s)_{max}$ where GE is an input irredundant conceptual graph, GS an output irredundant conceptual graph and Π_s a projection from GE into GS. It is *satisfied* by $\mathcal{KB}$ if and only if for each projection Π from GE into Γ, there exists at most one projection Π' from GS into Γ such that $\Pi'(head(GS)) = \Pi(GE)$, where $head(GS)$ represents the partial subgraph $\Pi_s(GE)$ of GS [13].

Example 4 *Let $(GE, GS, \Pi_s)_{max}$ be the following maximal descriptive constraint, which means that if a painter paints, then he must paint at most one painting.*

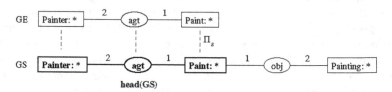

$\mathcal{KB}$ *of figure 2 does not satisfy $(GE, GS, \Pi_s)_{max}$ because the painter Miro paints more than one painting. More precisely, there exists a projection Π'_e from GE into Γ with $f_2 = \Pi'_e(GE)$ the partial subgraph of Γ limited to the vertices c_3, r_2 and c_4, and there exist two projections Π_1 and Π_2 from GS into Γ such that $\Pi_1(head(GS)) = f_2$ and $\Pi_2(head(GS)) = f_2$.*

A set of conflicts is associated with each maximal descriptive constraint not satisfied by $\mathcal{KB}$.

Definition 3 *Let $(GE, GS, \Pi_s)_{max}$ be a maximal descriptive constraint and Π a projection from GE into Γ. If there exist n ($n > 1$) distinct projections $\Pi_1, \ldots, \Pi_n$ from GS into Γ such that $\forall j \in [1,n]$, $\Pi_j(head(GS)) = \Pi(GE)$, then the partial subgraph of Γ limited to the vertices of the subgraphs $\Pi_j(GS)$, $j \in [1,n]$, is a conflict of $\mathcal{KB}$ with respect to the maximal descriptive constraint, noted $CF((GE, GS, \Pi_s)_{max}, \Pi)$.*

Example 5 *According to example 4, the partial subgraph of Γ limited to the vertices c_3, r_2, c_4, r_3, c_5, r_4, c_6 is a conflict of KB with respect to $(GE,GS,\Pi_s)_{max}$.*

In order to eliminate a conflict $\mathrm{CF}((GE, GS, \Pi_s)_{max}, \Pi)$, two strategies are possible: (1) modifying the conflict so that there is not any more projection from GE into it or (2) modifying the conflict so that there exists at most one projection from GS into it, while keeping the projection from GE into it. We propose to use the action of suppression with the following two aims in view: preferring the second strategy and suppressing the minimum vertices of the conflict, in order to get as close as possible to the initial KB. So, the action of suppression can then bear on the vertices of one or several subgraphs of the conflict, depending on the form of $\Pi_j(GS)$, $j \in [1, n]$. There exist two distinct cases:

- if there exist p $(1 < p \le n)$ distinct projections Π_1, ..., Π_p from GS into Γ such that $\forall j \in [1, p]$, $\Pi_j(GS) = \Pi(GE)$, then, in order to eliminate the conflict, we have to suppress one of the vertices of the partial subgraph $\Pi(GE)$ of Γ and the strategy used is the first one;
- else the conflict is composed of m $(m \le n)$ distinct partial subgraphs $\Pi_j(GS)$ such that $\exists i \in [1, m]$ with $\Pi_i(GS) \ne \Pi(GE)$;
  - if $\exists k \in [1, m]$ such that $\Pi_k(GS) = \Pi(GE)$, then, in order to eliminate the conflict, we propose to suppress a vertex of each distinct partial subgraph $\Pi_j(GS)$, $j \in [1, m]$ and $j \ne k$, of Γ;
  - else, in order to eliminate the conflict, we propose to suppress a vertex of each distinct partial subgraph $\Pi_j(GS)$, $j \in [1, m]$, except one if $m > 1$.

In order to determine the subgraphs on which the action of suppression must bear, we introduce the notion of sub-conflict.

Definition 4 Let $H = \mathrm{CF}((GE, GS, \Pi_s)_{max}, \Pi)$ be a conflict and Π_1, ..., Π_n the n $(n > 1)$ distinct projections from GS into Γ such that $\forall j \in [1, n]$, $\Pi_j(head(GS)) = \Pi(GE)$. If there exist Π_1, ..., Π_p, $1 < p \le n$, such that $\forall j \in [1, p]$, $\Pi_j(GS) = \Pi(GE)$, then H is composed of the only one *sub-conflict* $\Pi(GE)$, noted $SCF(GE, \Pi)$; else H is composed of the m $(m \le n)$ distinct subgraphs $\Pi_j(GS)$, called *sub-conflicts* and noted $SCF(GS, \Pi_j)$, $j \in [1, m]$.

Example 6 *The two sub-conflicts of the conflict of example 5 are SCF(GS, Π_1), the partial subgraph of Γ limited to the vertices c_3, r_2, c_4, r_3 and c_5, and SCF(GS, Π_2), the partial subgraph of Γ limited to the vertices c_3, r_2, c_4, r_4, c_6.*

A conflict is composed of at least one sub-conflict. In order to eliminate it, we propose to eliminate either its sub-conflict, or $m - 1$ of its m $(m > 1)$ sub-conflicts. To achieve this, we use the action of suppression $\mathbf{sup}(s)$ where: (1) s is a vertex of $\Pi(GE)$ if the sub-conflict is $SCF(GE, \Pi)$; (2) s is a vertex of $\Pi_j(GS)$ such that it does not belong to $\Pi(GE)$ if the sub-conflict is $SCF(GS, \Pi_j)$.

Property 2 An action of suppression associated with a sub-conflict of a conflict $\mathrm{CF}((GE, GS, \Pi_s)_{max}, \Pi)$ eliminates the sub-conflict.

Proof 2 An action of suppression associated with a sub-conflict SCF(GE, Π) suppresses one or several vertices of the subgraph $\Pi(GE)$. Its application on $\mathcal{KB}$ entails that the projection Π no longer exists from GE into the conceptual graph of the new KB, and the sub-conflict is eliminated. An action of suppression associated with a sub-conflict SCF(GS, Π_j) suppresses one or several vertices of the subgraph $\Pi_j(GS)$ that do not belong to the subgraph $\Pi(GE)$. Its application on $\mathcal{KB}$ entails that the projection Π_j no longer exists from GS into the conceptual graph of the new KB, and the sub-conflict is eliminated.

A set of actions of suppression is associated with each of the m sub-conflicts of a conflict CF((GE, GS, $\Pi_s)_{max}$, Π). A collection of m sets of actions of suppression is thus associated with the conflict. This conflict can be eliminated by the application on $\mathcal{KB}$ of an action of either the only one set of the collection ($m = 1$), or $m - 1$ sets of the collection ($m > 1$). The collection must then check the following property to allow one to eliminate the conflict.

Property 3 Let CF((GE, GS, $\Pi_s)_{max}$, Π) be a conflict composed of m sub-conflicts. If $m > 1$, then at most one of its m sub-conflicts has an empty set of actions of suppression, else the set of actions of suppression associated with its only one sub-conflict is not empty.

Proof 3 The proof comes from definition 4. If the conflict CF((GE, GS, $\Pi_s)_{max}$, Π) is composed of only one sub-conflict, then this sub-conflict is either the sub-conflict SCF(GE, Π) and its set of actions of suppression is not empty, or the sub-conflict SCF(GS, Π_1) which is not equal to $\Pi(GE)$ (otherwise this sub-conflict would be SCF(GE, Π)) and its set of actions of suppression is not empty. If the conflict is composed of m ($m > 1$) sub-conflicts SCF(GS, Π_j) and two of its sub-conflicts have empty sets of actions of suppression, then those sub-conflicts are both equal to $\Pi(GE)$, which is impossible.

Example 7 *According to example 6, the set of actions of suppression associated with SCF(GS, Π_1) is ES_1={sup(r_3), sup(c_5)} and with SCF(GS, Π_2) is ES_2={sup(r_4), sup(c_6)}. The collection of sets of actions of suppression associated with the conflict of example 5 is thus: CES={ES_1, ES_2}.*

The action of suppression checks the following property.

Property 4 An action of suppression associated with a conflict of $\mathcal{KB}$ with respect to a negative constraint is *conservative*: each negative constraint satisfied by $\mathcal{KB}$ remains satisfied after the application of the action on $\mathcal{KB}$.

Proof 4 A negative constraint is satisfied by $\mathcal{KB}$ if $\mathcal{KB}$ does not contain the knowledge described in it. Since an action of suppression suppresses some knowledge from $\mathcal{KB}$, each negative constraint satisfied by $\mathcal{KB}$ remains satisfied after the application of the action on $\mathcal{KB}$.

3   Base Incompleteness

We have studied the coherence of $\mathcal{KB}$ with respect to negative constraints in section 2. We now study the completeness of $\mathcal{KB}$ with respect to positive constraints, and thus deal with the problem of missing knowledge in $\mathcal{KB}$. An intuitive idea of positive constraints is that $\mathcal{KB}$ must contain the knowledge represented by these constraints. We propose two kinds of positive constraints: the positive existential constraints and the minimal descriptive constraints. The positive existential constraints allow one to represent some knowledge that must be deduced from $\mathcal{KB}$. The minimal descriptive constraints allow one to represent some knowledge that must appear in $\mathcal{KB}$ only under some given forms. $\mathcal{KB}$ is said *complete* with respect to positive constraints if and only if $\mathcal{KB}$ satisfies each of them.

We remind the definitions of the two kinds of positive constraints, the way $\mathcal{KB}$ satisfies them and, in the case when $\mathcal{KB}$ does not satisfy them, we present the notions of conflicts and actions in order to localize the errors of $\mathcal{KB}$ and to provide a way of removing those errors.

3.1   Positive Existential Constraints

A positive existential constraint is a conceptual graph, noted GE^+, that is *satisfied* by $\mathcal{KB}$ if there exists a projection from it into Γ [13].

$$GE^+ \quad \boxed{\text{Painter: Picasso}}$$

Fig. 4. $\mathcal{KB}$ of figure 2 does not satisfy the positive existential constraint GE^+, because there is no projection from GE^+ into Γ: there is no painter Picasso in $\mathcal{KB}$.

A positive existential constraint GE^+ is not satisfied by $\mathcal{KB}$ if there is no projection from GE^+ into Γ. We say that there is a *conflict* of $\mathcal{KB}$ with respect to GE^+.

Example 8 *According to figure 4, there is a conflict of $\mathcal{KB}$ with respect to GE^+.*

In order to eliminate a conflict of $\mathcal{KB}$ with respect to a positive existential constraint GE^+, Γ has to be modified so that there exists a projection from GE^+ into it. We propose the *action of addition* **add**(GE^+), which adds the conceptual graph GE^+ into Γ and which is a natural action to eliminate the conflict. One action of addition is associated with each conflict of $\mathcal{KB}$ with respect to a positive existential constraint and allows one to *eliminate the conflict* (the proof is immediate).

Example 9 *The action of addition to eliminate the conflict of example 8 is* add*(GE^+).*

3.2   Minimal Descriptive Constraints

A minimal descriptive constraint is a couple $(GE, \bigvee_{i=1}^{m} (GS_i, \Pi_s^i))_{min}$ where GE is an input irredundant conceptual graph, each GS_i, $i \in [1,m]$, is an output irredundant conceptual graph and each Π_s^i, $i \in [1,m]$, is a projection from GE into GS_i. It is *satisfied* by $\mathcal{KB}$ if and only if for each projection Π from GE into Γ, there exists at least one projection Π' from an output conceptual graph GS_i into Γ such that $\Pi'(head(GS_i)) = \Pi(GE)$, where $head(GS_i)$ represents the partial subgraph $\Pi_s^i(GE)$ of GS_i [13].

Example 10 *Let $(GE, (GS_1, \Pi_s^1) \vee (GS_2, \Pi_s^2))_{min}$ be the following minimal descriptive constraint, which means that a person who makes an action is either a painter who paints a painting or a poet who writes.*

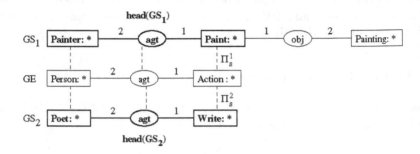

    $\mathcal{KB}$ of figure 2 does not satisfy $(GE, (GS_1, \Pi_s^1) \vee (GS_2, \Pi_s^2))_{min}$ because the painter Ernst who paints does not paint a painting. More precisely, there exists a projection Π_e from GE into Γ with $f_1 = \Pi_e(GE)$ the partial subgraph of Γ limited to the vertices c_1, r_1 and c_2, and there exist neither a projection Π from GS_1 into Γ such that $\Pi(head(GS_1)) = f_1$, nor a projection Π' from GS_2 into Γ such that $\Pi'(head(GS_2)) = f_1$.

    A set of conflicts is associated with each minimal descriptive constraint not satisfied by $\mathcal{KB}$.

Definition 5 Let $(GE, \bigvee_{i=1}^{m} (GS_i, \Pi_s^i))_{min}$ be a minimal descriptive constraint and Π a projection from GE into Γ. If there is no projection Π' from any GS_i, $i \in [1,m]$, into Γ such that $\Pi'(head(GS_i)) = \Pi(GE)$, then $\Pi(GE)$ is a *conflict* of $\mathcal{KB}$ with respect to the minimal descriptive constraint, noted $CF((GE, \bigvee_{i=1}^{m} (GS_i, \Pi_s^i))_{min}, \Pi)$.

Example 11 *According to example 10, the partial subgraph of Γ limited to the vertices c_1, r_1 and c_2 is a conflict of $\mathcal{KB}$ with respect to $(GE, (GS_1, \Pi_s^1) \vee (GS_2, \Pi_s^2))_{min}$.*

In order to eliminate a conflict $CF((GE, \bigvee_{i=1}^{m} (GS_i, \Pi_s^i))_{min}, \Pi)$, we propose an *action of modification* which specializes the conflict so that there exists at least one projection Π' from an output conceptual graph GS_i into Γ with $\Pi'(head(GS_i)) = \Pi(GE)$. Because of the lack of space, we only give, in the following, the intuition of such an action [16]. An action of modification for the conflict according to GS_i consists in *merging* each vertex $y_j = \Pi_s^i(x_j)$ (the image of a vertex x_j of GE by Π_s^i) of the partial subgraph $head(GS_i)$ with $\Pi(x_j)$, a vertex of the conflict. This, in order to obtain a common specialization between the conflict and GS_i. The merging of two concept vertices $c_1 = (t_1, m_1)$ and $c_2 = (t_2, m_2)$ leads to a concept vertex $c = (t, m)$ which is a maximal common specialization of both c_1 and c_2. The resulting concept type t is a maximal common subtype of t_1 and t_2 (not necessarily unique when the set of concept types is not a lattice). The resulting marker m is either the individual marker m_1 or the individual marker m_2 or the generic marker. Note that the concept vertices c_1 and c_2 cannot necessarily be merged. The merging of two relation vertices $r_1 = (t_1)$ and $r_2 = (t_2)$ leads to a relation vertex $r = (t)$ where t is a maximal common subtype of t_1 and t_2. Two relation vertices can only be merged if all their neighbour concept vertices can be merged.

Example 12 *The conflict of example 11 can be specialized according to the output conceptual graph GS_1 into H, which is no longer a conflict of $\mathcal{KB}$ with respect to the minimal descriptive constraint $(GE, (GS_1, \Pi_s^1) \vee (GS_2, \Pi_s^2))_{min}$.*

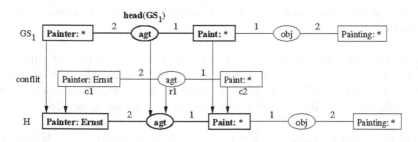

Property 5 An action of modification associated with a conflict $CF((GE, \bigvee_{i=1}^{m} (GS_i, \Pi_s^i))_{min}, \Pi)$ eliminates the conflict.

Proof 5 The action of modification specializes the conflict into H' so that there exists a projection Π_s from an output conceptual graph GS_i into it. Let Π' be the projection from GE into H' that corresponds to the projection Π from GE into H. We have to prove that the projection Π_s from GS_i into H' is such that $\Pi_s(head(GS_i)) = \Pi'(GE)$. As they were built, the vertices of the partial subgraph $\Pi_s(head(GS_i))$ of H' are the same as the vertices of the partial subgraph $\Pi'(GE)$ of H'. So, the projection Π_s is such that $\Pi_s(head(GS_i)) = \Pi'(GE)$. The action of modification thus allows one to eliminate the conflict H.

A set of actions of modification is associated with a conflict CF$((GE, \bigvee_{i=1}^{m}$ $(GS_i, \Pi_s^i))_{min}, \Pi)$. It is the set of actions that propose to specialize the conflict according to each output conceptual graph GS_i. This set can be empty because there may exist no common specialization between the conflict and each GS_i.

4 Base Refinement

$\mathcal{KB}$ is not coherent (resp. complete) with respect to negative (resp. positive) constraints, if it does not satisfy at least one of the constraints. The restoration of the coherence and the completeness of $\mathcal{KB}$ is then based on the notions of conflicts and actions. We propose solutions to restore the coherence of $\mathcal{KB}$, then we discuss the restoration of its completeness.

4.1 Coherence Restoration

$\mathcal{KB}$ is not coherent if it contains knowledge that it must not contain. To restore the coherence of $\mathcal{KB}$ then consists in removing the undesirable knowledge. For that, we propose revision plans that are sets of actions of suppression. In order to get as close as possible to the initial KB, a revision plan must be minimal.

Definition 6 P is a *revision plan* for $\mathcal{KB}$ if and only if P is a minimal set of actions of suppression such that the KB obtained by the application of all its actions on $\mathcal{KB}$ is coherent.

The revision plans are obtained from (1) the set of actions of suppression associated with each conflict of $\mathcal{KB}$ with respect to a negative existential constraint and (2) the collection of sets of actions of suppression associated with each conflict of $\mathcal{KB}$ with respect to a maximal descriptive constraint.

Theorem 1 (see proof in [16]) Let $EA_1, \ldots, EA_n$ be the sets of actions of suppression associated with each conflict of $\mathcal{KB}$ with respect to a negative existential constraint. Let $CES_1, \ldots, CES_p$ be the collections of sets of actions of suppression associated with each conflict of $\mathcal{KB}$ with respect to a maximal descriptive constraint, where $\forall i \in [1, p]$, $CES_i = \{ES_1^i, \ldots, ES_{n_i}^i\}$. P is a revision plan for $\mathcal{KB}$ if and only if P is a minimal set of actions of suppression such that:
(1) $\forall j \in [1, n]$, $EA_j \cap P \neq \emptyset$;
(2) $\forall i \in [1, p]$, (a) if $n_i = 1$, then $ES_1^i \cap P \neq \emptyset$;
    (b) if $\exists\, t \in [1, n_i]$ such that $ES_t^i = \emptyset$, then $\forall k \in [1, n_i]$, $k \neq t$, $ES_k^i \cap P \neq \emptyset$;
    (c) else, given $t_i \in [1, n_i]$, $\forall k \in [1, n_i]$, $k \neq t_i$, $ES_k^i \cap P \neq \emptyset$.

The revision plans of $\mathcal{KB}$ are computed with the algorithm of the calculation of diagnoses of physical systems proposed by Reiter [18]. This kind of algorithm is known to be NP-complete. This lack of efficiency has to be shaded in so far as we assume that the initial KB only contains few conflicts.

Example 13 *Let EA be the set of actions of suppression of example 2, CEA = $\{EA\}$, and CES the collection of sets of actions of suppression of example 7. The revision plans of $\mathcal{KB}$ are calculated from the sets $E_1 = \{CEA, CES \setminus ES_1\} = \{\{\sup(c_7), \sup(r_5), \sup(c_8)\}, \{\sup(r_4), \sup(c_6)\}\}$ and $E_2 = \{CEA, CES \setminus ES_2\} = \{\{\sup(c_7), \sup(r_5), \sup(c_8)\}, \{\sup(r_3), \sup(c_5)\}\}$:*

$P_1 = \{\sup(c_7), \sup(r_4)\}$, $P_2 = \{\sup(c_7), \sup(c_6)\}$, $P_3 = \{\sup(r_5), \sup(r_4)\}$, $P_4 = \{\sup(r_5), \sup(c_6)\}$, $P_5 = \{\sup(c_8), \sup(r_4)\}$, $P_6 = \{\sup(c_8), \sup(c_6)\}$, $P_7 = \{\sup(c_7), \sup(r_3)\}$, $P_8 = \{\sup(c_7), \sup(c_5)\}$, $P_9 = \{\sup(r_5), \sup(r_3)\}$, $P_{10} = \{\sup(r_5), \sup(c_5)\}$, $P_{11} = \{\sup(c_8), \sup(r_3)\}$, $P_{12} = \{\sup(c_8), \sup(c_5)\}$.

4.2   Discussion about the Completeness Restoration

$\mathcal{KB}$ is not complete if it does not contain knowledge that it must contain. Then restoring the completeness of $\mathcal{KB}$ consists in adding the desirable knowledge to $\mathcal{KB}$. This restoration is a problem of knowledge acquisition, which we would rather solve in interaction with the user than manage with an automatic refinement system. Moreover, the actions of addition and modification raise two problems. On the one hand, the set of actions of modification associated with a conflict of $\mathcal{KB}$ with respect to a minimal descriptive constraint can be empty. On the other hand, the actions of addition and modification are not conservative: their application on $\mathcal{KB}$ can change the satisfaction of positive constraints and can then loop. We thus propose an assistant to restore the completeness of $\mathcal{KB}$, which helps the user to eliminate the conflicts of $\mathcal{KB}$ one after the other with respect to positive constraints, by giving him the actions of addition or modification that are associated with the conflict.

5   Conclusion

This paper deals with the refinement of a knowledge base (KB) composed of conceptual graphs that is not coherent with respect to negative constraints and/or that is not complete with respect to positive constraints. The refinement process lays on the notion of conflicts, which provides a way of localizing the errors of the KB, and the notion of actions, which provides a way of correcting those errors. When the KB is not coherent, a solution based on revision plans is proposed to restore the coherence of the KB. When the KB is not complete, an assistant is proposed to help the user to restore the completeness of the KB.

The semantic validation and refinement of conceptual graphs can be compared with works done on knowledge validation. Most of those works provide quality criteria that must be checked by the KB as well as tools to check these criteria automatically. In particular, the coherence and completeness criteria have been studied in many works [19,20,1,2,9,3,4,10] and can be compared with our notions of coherence and completeness. Nevertheless, few works have studied these two criteria together. Moreover, few works have dealt with the refinement of an invalid KB.

The restoration of the completeness of a KB is a problem of knowledge acquisition, since it consists in adding the desirable knowledge described in the positive constraints to the KB. It would be interesting to study it thoroughly with the works on knowledge acquisition and learning [21,5]. One of the reasons why the refinement of a KB is studied in two stages is that the suppression of some knowledge of the KB can question the completeness of the KB and, on the opposite, the adding of some knowledge in the KB can question the coherence of the KB. In the future, an interesting work could be to study how to refine the KB globally with respect to negative and positive constraints.

The work presented in this paper has been implemented on the CoGITo platform [22] which is a tool designed for the implementation of applications based on conceptual graphs. It has been tested on a real application in accidentology [23] which concerns modelling knowledge of multiple experts and was developed by the ACACIA team of the INRIA research institute. The semantic validation and refinement were made on a KB composed of approximately one hundred concept types, fourteen relation types and seven conceptual graphs with respect to five constraints which were built from typical conditions presented in report [23]. This KB appeared to be non semantically valid and some solutions were proposed to restore its validity.

The refinement and the revision which study the consistency of sets of logical formulae [24] propose some comparable approaches. They both study the effect of the addition of a new information in a valid or consistent KB, and, if necessary, solutions to restore the validity or the consistency of the KB. In our work, the new information corresponds to validity constraints. Nevertheless, there exist two main differences between the refinement and the revision. First, the user plays an important role in the refinement process, which is not the case in the revision process. Second, the refinement deals with the validity of a KB (with respect to some given criteria) whereas the revision deals with its logical consistency. Note that many other works than revision have addressed the question of managing inconsistency such as formalisms of non-monotonic reasoning [25,26], truth maintenance systems [27] or works on description logics [8]. We think that the lack of negation and rules in the KB that we study make the comparison between these works and ours, for the moment, non-relevant.

References

1. M.C. Rousset. On the consistency of knowledge bases: the COVADIS system. *European Conference of Artificial Intelligence, ECAI'88*, pages 79–84, 1988.
2. P. Meseguer. Verification of multi-level rule-based expert systems. *National conference of American Association of Artificial Intelligence, AAAI'91*, pages 323–328, 1991.
3. S. Loiseau. Refinement of knowledge bases based on consistency. *European Conference of Artificial Intelligence, ECAI'92*, pages 845–849, 1992.
4. A. Preece and N. P. Zlatareva. A state of the art in automated validation of knowledge-based systems. *Expert Systems with Applications*, 7(2):151–167, 1994.
5. C. Haouche and J. Charlet. KBS validation : a knowledge acquisition perspective. *European Conference of Artificial Intelligence, ECAI'96*, pages 433–437, 1996.

6. F. Sellini. *Contribution à la représentation et à la vérification de modèles de connaissances produits en ingéniérie d'ensembles mécaniques.* PhD thesis, Ecole Centrale de Paris, mars 1999.

7. M.C. Rousset and A.Y. Levy. Verification of knowledge bases based on containment checking. *national conference of American Association of Artificial Intelligence, AAAI'96*, pages 585–591, 1996.

8. P. Hors and M.C. Rousset. Modeling and verifying complex objects : A declarative approach based on description logics. *European Conference of Artificial Intelligence, ECAI'96*, pages 328–332, 1996.

9. A. D. Preece, R. Shinghal, and A. Batarekh. Verifying expert systems: a logical framework and a pratical tool. *Expert Systems with Applications*, 5(3/4):421–436, 1992.

10. F. Bouali, S. Loiseau, and M. C. Rousset. Revision of rule bases. *EUROpean conference on VAlidation and Verification of knowledge based systems, EUROVAV'97*, pages 193–201, 1997.

11. J.F. Sowa. *Conceptual structures: information processing in mind and machine.* Addison Wesley Publishing Company, 1984.

12. M.L. Mugnier and M. Chein. Représenter des connaissances et raisonner avec des graphes. *Revue d'Intelligence Artificielle*, 10(1):7–56, 1996.

13. J. Dibie, O. Haemmerlé, and S. Loiseau. A semantic validation of conceptual graphs. In *Proceedings of the 6th International Conference on Conceptual Structures, ICCS'98, Lecture Notes in Artificial Intelligence*, pages 80–93, Montpellier, France, august 1998. Springer Verlag.

14. P. Kocura. Conceptual graph canonicity and semantic constraints. In Peter W. Eklund, Gerard Ellis, and Graham Mann, editors, *Conceptual Structures: Knowledge Representation as Interlingua - Auxilliary Proceedings of the 4th International Conference on Conceptual Structures*, pages 133–145, Sydney, Australia, August 1996. Springer Verlag.

15. G. W. Mineau and R. Missaoui. The representation of semantic constraints in conceptual graph systems. In *Proceedings of the 5th International Conference on Conceptual Structures, ICCS'97, Lecture Notes in Artificial Intelligence 1257*, pages 138–152, Seattle, U.S.A., 1997. Springer Verlag.

16. J. Dibie-Barthélemy. *Validation et Réparation des Graphes Conceptuels.* PhD thesis, Université PARIS-IX Dauphine, octobre 2000.

17. M. Chein and M.L. Mugnier. Conceptual graphs: fundamental notions. *Revue d'Intelligence Artificielle*, 6(4):365–406, 1992.

18. R. Reiter. A theory of diagnosis from first principles. *Artificial Intelligence*, 32:57–95, 1987.

19. T. A. Nguyen, W. A. Perkins, T. J. Laffrey, and D. Pecora. Checking an expert system knowledge base for consistency and completness. *International Join Conference of Artificial Intelligence, IJCAI'85*, 1:375–378, 1985.

20. A. Ginsberg. Knowledge-base reduction : a new approach to checking knowledge bases for inconsistency and redundancy. *National conference of American Association of Artificial Intelligence, AAAI'88*, pages 585–589, 1988.

21. M. J. Pazzani and C. A. Brunk. Detecting and correcting errors in rule-based expert systems: an integration of empirical and explanation-based learning. *Knowledge Acquisition*, 3:157–173, 1991.

22. O. Haemmerlé. *CoGITo : une plate-forme de développement de logiciels sur les graphes conceptuels.* PhD thesis, Université Montpellier II, Janvier 1995.

23. R. Dieng. Comparison of conceptual graphs for modelling knowledge of multiple experts: application to traffic accident analysis. *Rapport de recherche 3161, INRIA, Sophia Antipolis, Avril 1997.*
24. L. Sombé. Révision de bases de connaissances. *Actes des quatrièmes journées nationales PRC-GDR en Intelligence Artificielle*, 1992.
25. J. McCarthy. Circumscription: a form of non-monotonic reasonning. *Artificial Intelligence*, 13(1–2):27–39, 1980.
26. D. Makinson and P. Gärdenfors. Relations between the logic of theory change and nonmonotonic logic. In A. Fuhrmann and M. MOrreau, editors, *The logic of theory change, Lecture Notes in Artificial Intelligence 465*, pages 185–205, Berlin, 1991. Springer Verlag.
27. J. de Kleer. An assumption-based truth-maintenance system. *Artificial Intelligence*, 28(2):127–224, 1986.

Large-scale cooperatively-built KBs

Philippe Martin and Peter Eklund

Distributed System Technology Centre
KVO Laboratory - Griffith University
PMB 50 Gold Coast MC, QLD 9726 Australia
{philippe.martin,p.eklund}@gu.edu.au

Abstract. We describe a knowledge server that permits Web users to retrieve and add knowledge in a shared knowledge base. The following features distinguish WebKB-2 from other ontology servers or KBMSs: (i) the ontology is large (at present, 69,000 categories and 87,800 links mostly coming from WordNet) and extendible at any time by any user, (ii) asynchronous cooperation between users is supported and encouraged (users are encouraged to reuse, complement or correct the knowledge of other users but do not have to agree with each other and may add new names to categories) while the knowledge base is kept unique to maximize knowledge interconnection, retrieval and inconsistency detection, (iii) the proposed knowledge representation languages are designed to be both expressive and readable to permit and encourage the users to enter all the knowledge they want (though that still requires motivation). WebKB-2 is ultimately intended to permit cooperatively-built Yellow-Page like catalogs, that is, permit Web users to publish their information in a way that is automatically retrievable and comparable with other users' knowledge (as opposed to publishing information in plain text documents or even RDF documents). For example, database developpers or car dealers could describe and compare their products in a precise way, supporting precise queries.

1 Introduction

Current Web search engines can retrieve documents that include some given keywords but cannot extract and therefore retrieve and inter-link precise information (or knowledge) from them. The problem is that *knowledge retrieval and interlinking* is (re)done by each individual using his/her own memory. For example, each person trying to find a good database system for a project has to search and read the documentation of many systems, find comparative criteria and, from what he has read, try to classify each system against these criteria. With some luck, existing up-to-date comparisons may be found and feedback from users of some of these systems compiled. Nevertheless, *each* search is likely to be long, have sub-optimal results and remain unknown to (other) database system seekers or providers. This example is typical of many other searchs: car, insurance, employer, employee, software, hardware, etc. In each case, the solution implies *knowledge representation and centralization.*

H. Delugach and G. Stumme (Eds.): ICCS 2001, LNAI 2120, pp. 231–244, 2001.

A **first step** is to develop languages and systems permitting Web users to create or re-use Web documents containing knowledge representations. To this aim, in 1997, we created WebKB-1 [3], a knowledge-based private annotation tool ("annotation" implies "representation", "indexation" and "retrieval"). Many XML extensions were conceived for this purpose, e.g. RDF[1] in 1998. In [4], we discussed the inadequacies of these XML-based languages for precision-based knowledge indexation and retrieval. Nevertheless, in [5], to permit better knowledge exchange/retrieval via RDF, we proposed extensions to RDF and general conventions for knowledge representation (in RDF, CG, ...). However, even if these conventions were adopted, in this distributed (document based) approach, there would still be a lot of small competing and loosely inter-linked ontologies and hence automatic comparison of representations would remain limited and based on lexical matching.

The **next step** toward centralization - and hence better automatic/manual knowledge retrieval, comparison, inter-linking, cross-checking and cooperation between knowledge providers - is to develop knowledge base servers permitting Web users to add representations into a shared repository while reusing or extending a large ontology. For efficiency and commercial reasons, all Web-users would not use the same knowledge server but rather a few general knowledge servers (e.g. managed by portal companies) and more specialized knowledge servers. By (partly) mirroring one another, general servers would probably share a similar general WordNet-like or CYC-like ontology, and competing specialized knowledge servers would also share some similar content[2]. Thus, it would not really matter where a Web user publishes information first, and this centralized approach would keep the advantages of the current distributed approach.

Early 2000, we stopped adding features to WebKB-1 and began the design and implementation of WebKB-2, a *large-scale knowledge server permitting each users to re-use, complement or annotate the knowledge of other users*[3]. This type of tool is new (yet). Close relatives are "ontology servers" (e.g. Ontolingua[4] and Ontosaurus[5]) but they mostly have small ontologies (a few hundred categories) and do not support/encourage a tight interlinking/annotation of knowledge from various users. Large-scale KBMSs also rarely support a large "dynamic" ontology (users cannot change the ontology interactively, a re-compilation or re-indexation phase is necessary). In this article, we first present a few elements that we think necessary to enable a large-scale cooperatively-built knowledge base, then we describe the cooperation mechanisms in WebKB-2, followed by necessary extensions to classic "searches for specializations of graphs". Finally, we compare our approach to others.

---

[1] http://www.w3.org/RDF

[2] The processes of mirroring and answering queries involving several knowledge bases is by itself permitted by the similarity or interconnection of various used ontologies.

[3] WebKB-2 also inherits the features of WebKB-1, i.e. it can exploit Web documents mixing text and representations and use them as private knowledge bases.

[4] http://WWW-KSL-SVC.stanford.edu:5915/

[5] http://www.isi.edu/isd/ontosaurus.html

2 Elements for a large-scale cooperatively-built KB

2.1 Notations

We believe a general knowledge base server should support a knowledge representation language that is (i) expressive enough to permit and encourage the user to be exact in his/her representations, and (ii) limiting the number of different (and not automatically comparable) ways to express the same information. If the language is not expressive enough, users will either enter incorrect information, develop various incomparable ways to represent it, or simply not use the server. Any of these cases hinder knowledge comparison, cross-checking, inter-linking and reuse. This does not mean that the server should have inference capabilities that exploit all the subtleties of the language. Implementing such an inference engine would actually often involve application-dependant choices. Instead, the server should perform minimal consistency checks and help filter the knowledge relevant to answer a query (i.e. the knowledge relevant for an application).
We designed 2 notations – Frame-CG (FCG) and Formalized English (FE) – to improve on the readability and expressivity of the Conceptual Graph linear notation (see [6] for details). Though we have not yet implemented it, a translator could be built to transform CGLF or CGIF into FE or FCG, and conversely for simple cases [6].

2.2 Lexical/syntactic/semantic/ontological recommendations

To reduce the number of ways, information can be expressed (and therefore increase the chance of automatic comparison), the WebKB-2 user is asked to follow conventions (most of which were described in [6]): lexical recommendations (use English singular nouns as category names), semantic recommendations (be precise, contextualize statements, re-use and complement existing knowledge), syntactic recommendations (e.g. how to represent various kinds of quantifiers, collections, intervals, contexts, 2nd order types/relations), ontological recommendations (e.g. how to represent states and processes, descriptions, indexations, characteristics, measures, numbers, collections, temporal/spatial/logical entities/relations). These recommendations and the associated "patterns" are also a guide. Most are more or less enforced by the use of the notations and the re-use of the existing types in the shared knowledge base.

2.3 Structures for a multi-users KB

The elements of the KB of WebKB-2 are: *categories* (concept/relation types, individuals), *links between categories* (e.g. specialization and exclusion links), category *names* and *links between categories and names* (each category has a unique *identifier* but may have many names; conversely, each name/word may belong/refer to various categories - as many categories as the word has different meanings), *concept* nodes (a graph is itself a concept), *relation* nodes, and *users* (each of the previous elements is connected to (an object representing) its creator, and each user is represented by an individual in the ontology). All connections in WebKB-2 can be traversed in both ways (e.g. each concept with a certain type is accessible from that type) programmatically and from browsing interfaces.

The connections from/to creators are necessary for handling updates by multiple users and permitting each user to filter or focus on knowledge from certain users by referring to them with an identifier, one or several type or supertypes, or even a graph description. The alternative choice of storing knowledge from each creator in a different module would not permit as much flexibility in the management and filtering of knowledge from multiple creators (or it would be harder to implement).

The connections between categories and words are necessary to permit lexical freedom (any user can add new names to existing categories) and the use of words instead of category identifier within graphs. This last feature spares the users the tedious work of looking for the identifiers of each category to build their graph (statements or queries). If the word used in a concept refers to only one category, or if the other categories can be eliminated given the signature of the relations connected to the concept, the category that is (most probably) relevant can be found. Otherwise, the list of candidate categories can be proposed for the user to select. For a query graph, there is no harm in making an automatic choice and let the user refine the query if a wrong category has been selected.

2.4   Reuse of a natural language ontology

A natural language ontology (including the connections between categories and words) is the backbone of a large shared KB. It permits WebKB-2 to relate, compare and retrieve knowledge representations. It provides the user with various meanings for a word or various distinctions for a notion, most of which s/he would have not thought about, thence leading the user to enter more precise and comparable representations. For the same reason, provided semantic constraints are associated to the top level categories of the ontology, it permits some automatic checking on the users' statements and extensions to the ontology.

We initialized the current knowledge base of WebKB-2 with the content of the lexical database WordNet 1.6[6]: 94,500 nouns and 66,000 categories referred by nouns (in accordance to our lexical conventions, we ignored information related to verbs, adverbs and adjectives). Various kinds of links connect these categories: specialization, exclusion, similar, member, part, substance, plus their reverse links. The interpretation of the links other than specialization, exclusion and similar is not always clear nor consistent within Wordnet. For example, a part link from the category airplane to the category wing could mean that "any airplane has for part at least 1 wing" or "all airplanes have for part the same wing", "any wing is part of a plane", etc. We assumed the first interpretation was correct for all kinds of direct links (e.g. part, substance, etc.) and therefore opposite for their inverse links (i.e. part of, substance of, etc.). This interpretation is exploited in our graph comparison/retrieval algorithms.

---

[6] http://www.cogsci.princeton.edu/~wn/

We distinguished **specialization** links into **subtype** links and **instance** links by isolating 2900 individuals. We also made a few other structural corrections (e.g. we removed 3 redundant subtype links, 4 links with same source and destination, and introduced the link **location** to replace some unfortunate usages of the links **subtype** and **part**). Finally, to permit knowledge representation and automatic checking, we inserted WordNet top-level categories into a top-level ontology of about 100 concept types, and complemented it by an ontology of 140 basic relation types signed on these top-level concept types.

To each WordNet category, we associated a unique key name using the first of the category names (in WordNet, the first name is the most common name used for referring to the category). When various categories share a same first name, suffixes such as "/2" and "/3" are used to create unique key names. No suffix is appended for the category that is most likely referred by a name (again according to WordNet). Then, we manually modified some of the key names to simplify the knowledge representation task (WordNet name/category orders are generated based on name frequencies in some manually tagged corpora).

2.5   Flexible ways to refer to a category

Flexible identifiers and multiple interpretation modes permit to take advantage of the connections between users, categories and names.

In WebKB-2, a category identifier is either an URL, an e-mail address or the concatenation of the creator identifier and the key name, e.g. `wn#domestic_dog`, `wn#time`, `wn#time/2`, `wn#time/3`, `pm#IR_system`. (Category identifiers with same key names but different creators refer to different categories and hence, hopefully, represent different objects). A category identifier may also show all the names given by the creator, e.g. `pm#IR_system__information_retrieval_system` and `wn#dog__domestic_dog__Canis_familiaris` (to represent names given by other users, "name" links – abbreviated by the character '_' – must be used). WordNet categories may also be entered without their creator identifier (i.e. without the "wn" prefix), e.g. `#time`. More precisely, this is the case except *within graphs* when a list of default creators has been specified (e.g. with the command "`default creators: pm wn;`" in input files). For instance, if `pm` and `wn` are the default creators, the graph `[a #car]` is accepted if either `pm#car` or `wn#car` have been declared. The order of the creators in the list is important (the first candidate category is preferred).

Words (i.e. category names) are simply entered as such, e.g. `domestic_dog` and `time`. Category names, instead of category identifiers, are accepted within graphs only if the option has been selected (command "`use names;`" in input files). Signatures are used for eliminating candidate categories. If there is more than 1 candidate, the parsing stops or issues a warning depending on an internal ambiguity acceptance level (for our main purpose, ambiguities should not be allowed but an application of WebKB-2 that requires an automated agent to be used as a knowledge provider will probably accept ambiguities). If ambiguities are accepted and a list of default creators specified, WebKB-2 exploits this list to select the best candidate category.

Apart from signatures, type constraints explicitly associated to categories within a graph may be used to guess categories. For instance, the graph
[a transformation \\pm#process] means: there exists an individual instance of a concept type that has "transformation" as one of its names and that is a subtype of pm#process. This permits WebKB-2 to eliminate the two other senses proposed by WordNet for "transformation": the mathematical function and the transmutation. Top-level types such as pm#process are proposed in WebKB-2 menus to help construct graphs.

For better readability, we will often use names instead of category identifiers in the example graphs of this article.

3   Mechanisms for cooperative editing of the KB

The WebKB-2 user is asked to be as precise as possible when making statements (to avoid conflicts and permit to answer queries more adequately). For instance, a user (lets say "user1") should not simply represent that "birds fly" (in FCG: "[user1#birdsFly [any bird, agent of: a flight]]") since this is not always true. If this happens, other users are encouraged to "correct" the information. In WebKB-2, any user can do this by connecting the "faulty" graph to a more precise version using a relation of type pm#corrective_specialization (then, depending on display options, the first version may or may not be filtered by WebKB-2 when answering queries). Similarly, if a user thinks a statement from another user can be generalized, s/he can use a relation of type pm#corrective_generalization. For example, if "user1" stated that "birds fly" and "user2" wants to correct and specialize that by "a study made by Dr Foo found that in 1999, 93% of healthy birds could fly", s/he can write:

```
[user1#birdsFly, corrective_specialization:
   [user2#93pcHealthyBirdsCanFlyAccordingToFoo
     [[93% of (bird, experiencer of: a good health),
          agent of #: a flying  //'#' means ''can''
     ], time: 1999], source: (a study, author: Foo@bird.org)]
  ]]
```

(Note: if a graph is not explicitly named, WebKB-2 generates a name for it).

We believe a scalable approach for cooperation between users of a knowledge base server implies that two seemingly incompatible goals are reached:
(i) each user should be able to represent what s/he considers true, and correct or complement other users' knowledge in a non-destructive manner, use the categories and names s/he wants (providing that lexical recommendations are respected and existing categories reused or specialized), and should not have to discuss and find an agreement with other users each time a conflict arises,
(ii) knowledge from different users should remain consistent and tightly interconnected to permit comparison, search, cross-checking and optimal unification.

We have partly shown how these different points can be achieved and that they are not incompatible, providing users connect their categories and graphs to other existing ones. Removal/modification/addition protocols are also required for semantic conflicts to be managed asynchronously and without person-to-person agreement. The following four points describe our approach.

1) A user may **remove** a category, link or graph only if s/he has created it and unless this induces an inconsistency in the user's knowledge. If the category, link or graph being removed is used by other users or is necessary for their knowledge to remain consistent, it is actually not removed but its creator is changed to one of the users relying on its existence. In WebKB-2, inconsistency detection currently only exploits relation signatures and exclusion links. However, we plan to exploit inconsistencies detected by users and signaled with a relation of type pm#contradiction between two graphs.

2) The creator of a category may **modify** a link connected to this category – so that the link uses an alternate category – unless this modification induces an inconsistency. The creator of a relation type may modify its signature unless such a change induces an inconsistency (in which case, s/he must modify the ontology or related graphs so that the inconsistency disappear). A user may not modify a graph that s/he has not created but s/he can connect it to another graph via a relation of type pm#corrective_specialization, pm#corrective_generalization or pm#correction. This last relation type should only be used when the ontology cannot be modified (or another relation type used) for correcting the first graph. Since graphs can be used for representing links these three relation types may also be used to state alternate links. Depending on display/filtering options, corrected graphs or links are displayed/used for inference or not.

3) A user may **add** a graph or a link (even if s/he is not the creator of the linked categories) unless that induces an inconsistency or a redundancy (WebKB-2 does not accept a graph that already has a specialization or a generalization in the KB). If this happens, the user must either refine his/her graph before re-trying to add it, modify the ontology or use one of the three "corrective" relations cited above.

4) In any of these previous cases, when the knowledge of a user is modified by another user, the change should automatically be e-mailed to the first user or presented to him/her the next time s/he logs on to WebKB-2.

An alternative approach is to always allow the creator of a category to add, modify or remove the categories or links s/he has created even when that induces an inconsistency in other users' knowledge. Then, the inconsistency has to be repaired automatically. Since the update means a change of interpretation of a category (at least from the viewpoint of the other users), a way to repair the inconsistency is to duplicate the categories and links that should not be modified in order to avoid inconsistencies (i.e. the modified category and some of its subtypes from the same user). The duplicates are attributed to other users. Although this approach allows each user to ignore how his/her categories are used by other users, it is far less optimal than manual corrections, reduces cooperation between users and the tight interlinking of their knowledge. It is also complex to implement and cannot be extended to handle graph modifications in the same way.

4   Search mechanisms

4.1   Searching categories and links

Fig. 1 shows a WebKB-2 interface for searching categories or links according to a category identifier or name and/or a link connected to the category(ies) (there may be several categories if a name is provided) and an optional destination. The parameters shown in Fig. 1 specify a display of the category pm#thing (the uppermost concept type) and all its direct or indirect subtypes created by the user rdf or users that are members of the KVO group (M pm#KVO_group) but not from f_modave and any Australian (^ #Aussie). These filtering constraints resolve to the users rdf and pm. Subtype links and categories that do not belong to these users are explored but not shown (though increases in the indentation shows the number of intermediary categories that have not been displayed). Fig. 2 shows the result in the default format. The characters '!', '^' and '>' respectively represent links of type exclusion, instance of and subtype.

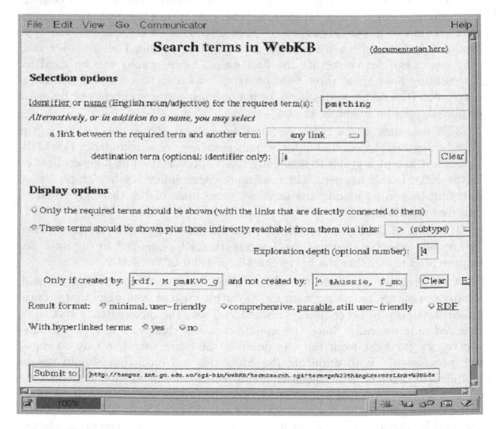

Fig. 1. Query for the subtypes of pm#thing that belong to the user "pm"

Fig. 2. Result of the previous query

4.2 Searching graphs

Classic searches for graph specializations permit searches "by the content" but need to be extended for more flexibility in the formulation of the query graph and to increase the number of relevant answers. WebKB-2 uses four extensions.

1) Let us assume the KB includes the graphs [John, owner of: a car] and [John, owner of: an appartment]. A classic search for graphs specializing the query graph [a man, owner of: a car, owner of: a lodging] would not retrieve the previous graphs since only the union of these specializes the query graph. When WebKB-2 looks for specializations, it also looks for other graphs including coreferent categories: identical individuals, identical types universally quantified or using the same coreference variable. If they permit to answer the query graph, these different graphs are displayed separately – joining them would often not produce a meaningful graph (e.g. their embedding graphs could not be joined). Two other graphs that could be presented in answer to the previous query:

```
[ [[Tom \\IBM_employee, owner of: an apartment], time: 2000], author: Tom]
[[any IBM_employee, owner of: a car], author: IBM]
```

2) Searches should also take into account knowledge represented via links instead of graphs. For instance, let us assume the categories representing the geographical areas "Gold Coast" and "Southport" are connected via a part link and the knowledge base includes the following graph.

```
[philippe.martin@gu.edu.au, agent of: (the renting,
        object: (an apartment, part: 1 bedroom, location: Southport),
        instrument: 140 Australian_dollars, period: a week,
        beneficiary: Spirit_Of_Finance)]
```

WebKB-2 exploits the ontology to present this graph in answer to the query graph `[an apartment, location: (a district, part of: Gold_Coast)]`.

3) Let us assume the graph `[John, owner of: a lodging]` is in the knowledge base and a query graph is `[a man, owner of: an apartment]`. The first graph is not a specialization of the query graph since `wn#housing/2__lodging` is a supertype of `wn#apartment__flat` not the reverse. However, a user may want such a graph to be provided. This is why WebKB-2 provides two graph search commands: "spec" to search specializations of the graph given in parameter, and "?" to search graphs comparable to the one given in parameter. With the second command, supertypes of categories in the query graph are also used. The first graph would not answer the query "? `[a man, owner of: a bike]`" since `wn#housing/2` is neither a subtype nor a supertype of `wn#bicycle__bike`.

4) Structural flexibility should be permitted in query graph specification. We believe the simplest way (both for the user and from an implementation perspective) is to allow the specification of path sequences with common regular expression operators ("*" for "0, 1 or many times", "+" for at "at least 1 time", "?" for "0 or 1 time"). Let us assume the following graph is in the KB.

```
[philippe.martin@gu.edu.au, agent of:(a research, within_group: KVO_group)]
```

Users looking for a person conducting research at "Griffith Uni., Gold Coast campus" are unlikely to find this graph via classic searches for specialization only. However, since `pm#School_of_IT_at_Griffith_Uni_Gold_Coast_Campus` is connected via a part link to `pm#KVO_group` and via a location link to `QLD#GCcGU__Gold_Coast_campus_of_Griffith_Uni`, and since `pm#relation` is the uppermost relation type, it should be possible to find this graph with:

```
      spec [a person, agent of: (a research, relation+: GCcGU)
or:   spec [a research, (relation: a thing)+ location: GCcGU)
or:   spec [a research, relation 3+ (part of: a group)3+ location:GCcGU)
```

("3+" means that at most 3 relations of the specified type should be traversed).

Fig. 3 shows one of WebKB-2's interfaces for searching graphs. Names, instead of category identifiers, have been used and "pm" has been specified as the creator of the graphs to retrieve. Fig. 4 shows the result. It first indicates that 2 categories share the name "Gold_Coast" and that the first has been selected. Then, a graph ("with hyperlinked categories") answering the query is presented.

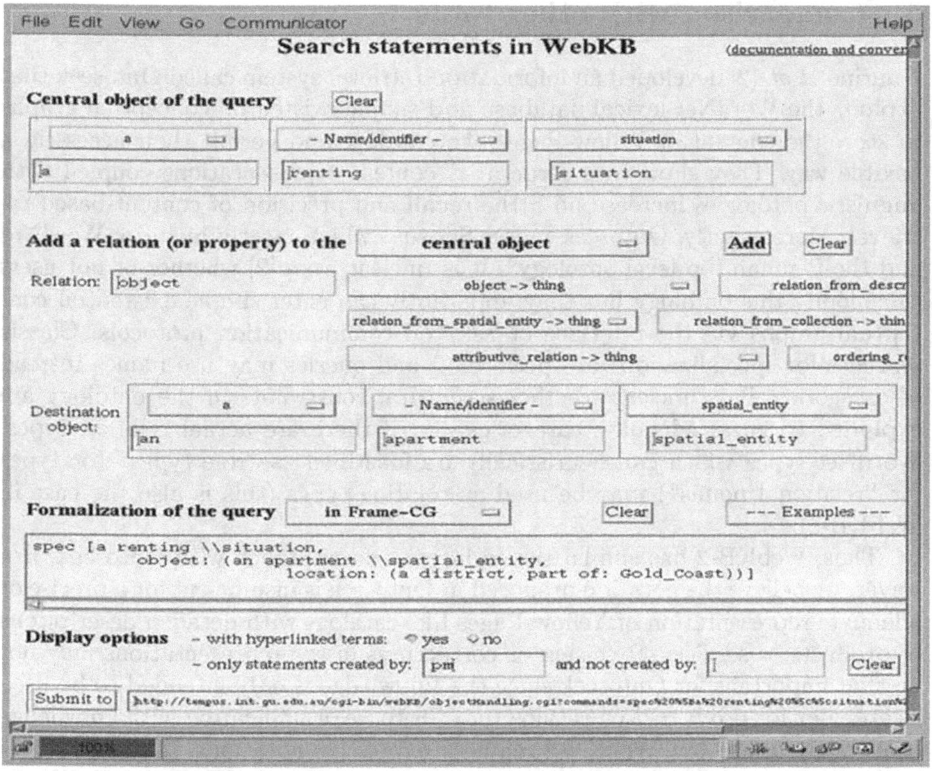

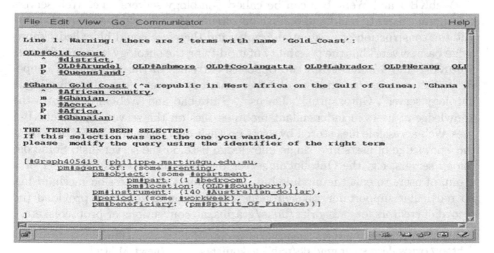

Fig. 3. Query for the specializations of a graph

Fig. 4. Result of the previous query

5   Comparison with other tools

Guarino *et al.* [2] developed an information retrieval system called Ontoseek that exploits the WordNet lexical database and simple existential conceptual graphs to store the content of Yellow-Pages like catalogs and permit their access in a flexible way. They show that structured content representations coupled with linguistic ontologies increase both the recall and precision of content-based retrieval. More exactly, Ontoseek reuses Sensus[7] which mostly includes WordNet and the Penman top-level ontology[8]. It is unclear from [2] whether or not users can modify this ontology but they apparently can enter simple existential conceptual graphs via the interface or ask/tell communication protocols. Classic searches for specializations are performed and queries may use names instead of categories. It is unclear whether structural constraints in the ontology are exploited to guess adequate categories and if there are actual relation types. WordNet types which can heuristically be identified as "role types" (or types for "relational nouns") may be used as relation types (this is also the case in WebKB-2).

Thus, WebKB-2 has similarities in intent and approach with Ontoseek. However, we believe the notation proposed in Ontoseek is insufficient for a precise or adequate representation of Yellow-Pages like catalogs with detailed descriptions of products or services. Precision or correctness in the representations may not be that important for Ontoseek since the knowledge is only intended to be used as an index for products in a catalog (not for reuse or unification with knowledge from many users) but WebKB-2 requires expressive notations, the handling of multiple users, and knowledge representation conventions. We have also shown in the previous section the insufficiency of classic searches for specializations.

WebKB-1 and WebKB-2 can be called "ontology servers", i.e. Web servers that permit users to build and publish ontologies. Most ontology servers also permit the construction of existential graphs and therefore could be called "knowledge base servers" but the possibility of modifying the ontology is a rarer feature. WebKB-1 and WebKB-2 are two opposite extremes in the handling of cooperation between users: while most other ontology servers (e.g. the Ontolingua ontology server[9], Ontosaurus[10], Ikarus[11], Tadzebao and WebOnto[12]) store the knowledge of users in independant modules/files on the server disk, WebKB-1 uses Web-accessible files stored by users on their own disks and WebKB-2 stores the knowledge of users in a single knowledge base on the server disk. Some ontology servers, e.g. the Ontolingua server or Ontosaurus permit any user or a group of users to edit the module but, apart from locking/session mechanisms, no particular support for asynchronous cooperation is generally provided (no record of creators for categories/links/graphs, no conventions or protocols, etc.).

---

[7] http://www.isi.edu/natural-language/projects/SENSUS-demo.html
[8] http://www.darmstadt.gmd.de/publish/komet/gen-um/newUM.html
[9] http://WWW-KSL-SVC.stanford.edu:5915/
[10] http://www.isi.edu/isd/ontosaurus.html
[11] http://www.csi.uottawa.ca/ kavanagh/Ikarus/IkarusInfo.html
[12] http://ksi.cpsc.ucalgary.ca:80/KAW/KAW98/domingue/

An exception we know of is the Co4 system[13] which has protocols modeled on submission procedures for academic journals, i.e. on peer-reviewing, resulting in a hierarchy of knowledge bases, the uppermost containing the most consensual pieces of knowledge while the lowermost ones are the knowledge bases of each user. This approach certainly leverages some problems of module-based approaches but would doubtly scale to large knowledge bases or a large number of users. The Ontoloom/Powerloom authors mainly rely on knowledge comparison procedures and the pre-existence on a large ontology to guide and check users in their extension of a unique knowledge base.

Modules are an easy way to delimit knowledge about a particular subject and handle competing formalizations, but since categories between modules are generally not inter-connected, automatic comparisons of knowledge representations from/re-using different modules is unlikely to succeed. For the same reason, even when general descriptions of the content of modules are made using graphs, the selection of adequate modules to reuse or search is a difficult task. From a knowledge retrieval point of view, the indexation of knowledge according to some knowledge domains or other characteristics is a coarse-grained approach. In WebKB-2, this selection problem does not exist: categories are tightly interlinked, and each link or relation in the knowledge base may be used as an index for retrieving a relevant piece of knowledge, thus permitting to take into account any combination of characteristics specified in a query not just combinations given by users in general indexations.

Compared to other large scale KBMSs, a notable feature of WebKB-2 is that the ontology is large and can be dynamically/interactively modified by the users (no lengthy re-compilation phase or graph re-indexation is necessary). This feature is shared by the Parka-DB system[14] which we considered to re-use for implementing WebKB-2. We also considered the SHORE deductive database[15] as well as standard relational databases. However, we found more flexibility and programming ease was provided by the free-to-use object-oriented main-memory database system called FastDB[16] (in case the database grows larger than 4Gb (on a 32 bit system), FastDB can be replaced with a disk-based version called GigaBASE[17]). In Parka-DB the ontology is also entirely loaded in memory but the graphs remain on disk. Large-scale CG systems also begin to appear. Santiago[18] is based on the MG database system[19] and requires a re-indexation phase each time a type is added. In Bernd Groh's relational database based CG system[1], no re-indexation phase is necessary.

[13] http://ksi.cpsc.ucalgary.ca/KAW/KAW96/euzenat/euzenat96b.htm
[14] http://www.cs.umd.edu/projects/plus/Parka/parka-db.html
[15] http://www.cs.wisc.edu/shore/
[16] http://www.ispras.ru/ knizhnik/fastdb.html
[17] http://www.ispras.ru/ knizhnik/gigabase.html
[18] Gerard Ellis' system. See CG mailing list.
[19] http://www.mds.rmit.edu.au/mg/

6 Conclusion

We have presented an approach permitting Web users to search and coopera-
tively build a shared knowledge base, and engineered a system supporting this
approach[20]. The approach permits and relies on knowledge reuse and intercon-
nections at a local level, e.g. categories are connected to names, creators and
other categories, while concepts and graphs are interconnected via relations or
the categories they reuse. In coarser-grained approaches, these connections are
often not represented (and, we believe, more difficult to represent in a manageable way) and therefore cannot be automatically combined to permit knowledge
comparison or more relevant and complete knowledge retrieval. We also described
our approach to permit asynchronous cooperation, and necessary extensions to
classic searches for specializations.

Entering information in WebKB-2 is more difficult than entering sentences
in a document, but information from documents cannot be interconnected to
respond to precise queries and is therefore lost to most people. We believe that
entering information in WebKB-2 is easier than in most other systems thanks
to the adopted notations, the initialisation of the knowledge base with WordNet
and our top-level ontology, and the possibility of using everyday words instead
of category identifiers. Some information remain difficult to represent precisely
but we think that WebKB-2, or an extension of it with nicer interfaces, could
be used by Yellow-Pages-like-services or community servers to permit people to
advertize products and services or publish information.

References

1. Groh, B., P. Eklund.: Algorithms for creating relational power context families
   from conceptual graphs, In ICCS'99, 7th International Conference on Conceptual
   Graphs, Springer Verlag, LNAI 1640, pp. 389–400, 1999.
2. Guarino, N., Masolo, C., Vetere, G.: Ontoseek: Content-based Access to the Web.
   In: IEEE Intelligent Systems, Vol. 14, No. 3 (1999) 70–80
3. Martin, Ph.: The WebKB set of tools: a common scheme for
   shared WWW Annotations, shared knowledge bases and information retrieval. In: ICCS'97, 5th International Conference on Conceptual Structures, Springer Verlag, LNAI 1257 (1997), 585–588. URL
   http://meganesia.int.gu.edu.au/~phmartin/webKB/doc/papers/cgtools97/
4. Martin, Ph., Eklund, P.: Embedding Knowledge in Web Documents: CGs versus XML-based Metadata Languages. In: ICCS'99, 7th International Conference
   on Conceptual Structures, Springer Verlag, LNAI 1640 (1999) 230–246. URL
   http://meganesia.int.gu.edu.au/~phmartin/WebKB/doc/papers/iccs99/iccs99.ps
5. Martin, Ph., Eklund, P.: Conventions for Knowledge Representation via RDF. In: WebNet2000, ACCE press 378–383. URL
   http://meganesia.int.gu.edu.au/~phmartin/WebKB/doc/papers/webnet00/
6. Martin, Ph.: Conventions and Notations for Knowledge Representation and Retrieval. In: ICCS'00, 8th International Conference on Conceptual Structures, Springer Verlag, LNAI 1867 (2000) 41–54. URL
   http://meganesia.int.gu.edu.au/~phmartin/WebKB/doc/papers/iccs00/iccs00.ps

---

[20] Accessible at http://meganesia.int.gu.edu.au/~phmartin/WebKB/shared.html

Conceptual Graphs and Metamodeling

Olivier Gerbé[1], Guy W. Mineau[2], and Rudolf K. Keller[3]

[1] HEC Montreal.
3000, chemin de la Côte-Sainte-Catherine, Montréal, Québec, Canada H3T 2A7
Olivier.Gerbe@hec.ca
[2] Université Laval
Québec, Québec, Canada G1K 7P4
mineau@ift.ulaval.ca
[3] Université de Montréal
C.P. 6128 Succursale Centre-Ville, Montréal, Québec, Canada H3C 3J7
keller@IRO.UMontreal.ca

Abstract. Metamodeling is often identified as a key layer in the development of an information system because it formally defines the modeling primitives that will be used in subsequent modeling activities. We use the Conceptual Graph (CG) theory for illustration purposes. The simplicity of the CG notation and its flexibility to represent metalevel knowledge through the use of contexts makes it a serious contender for the representation of a metamodeling theory. Therefore, this paper presents a CG based metamodeling framework for the modeling of information systems.

1 Introduction

Metamodeling is often identified as a key layer in the development of an information system [2,3] because it formally defines the modeling primitives that will be used in subsequent modeling activities. By defining the modeling language, the semantic constraints of the domain can be embedded into it, restricting the expressivity of the modeling language accordingly, ensuring greater consistency throughout the modeling of the domain. Also, queries concerning the modeling language itself can be answered. Using its formal definitions, the modeling language can be explained, which provides the essentials for the establishment of an on-line task support system. A reduction in the number of work hours spent to understand the modeling language, and therefore a gain in productivity, is sought. When many people act as knowledge modelers throughout some organization (as with consultant firms for instance), or when employee turn over is high, this gain in productivity is considerable [11].

Furthermore, through the use of these formal definitions, the modeling language can be validated. A valid modeling language, one in which all definitions are together compatible, i.e., do not produce any inconsistencies, improves the ability of the knowledge modeler(s) to produce a set of object definitions which are consistent with one another. Finally, with a metamodeling approach the modeling language is defined in a declarative formalism. This allows partial

H. Delugach and G. Stumme (Eds.): ICCS 2001, LNAI 2120, pp. 245–259, 2001.
© Springer-Verlag Berlin Heidelberg 2001

or full mapping between different modeling languages, leading to systems integration. Therefore, system interoperability issues can be approached from a metamodeling point-of-view. For all these reasons, we too advocate the use of a metamodeling layer in the development of an information system. We use the Conceptual Graph (CG) theory for illustration purposes. The simplicity of the CG notation and its flexibility to represent metalevel knowledge through the use of contexts makes it a serious contender for the representation of a metamodeling theory. Therefore, this paper[1] presents a CG based metamodeling framework for the modeling of information systems. Section 3 introduces the basic ontology required to develop a CG based metamodeling language. Section 4 presents a mapping function from the metalevel to the data level, allowing the definitions stated at the former level to be used at the latter level. Section 5 presents how this framework can be used to create an arbitrary number of metalevels. Section 6 introduces metarules that state restrictions and properties of metalevel definition primitives. Section 7 concludes and presents future directions for our research.

2   Literature Review

Metamodeling and conceptual graphs have not been extensively investigated. John Esch in [7] introduces metamodeling through two predefined relations: Kind that links a concept to its type and Subt that links two concept types that are in a subtype relationship. He defines, using the relationship Subt, a type hierarchy for each higher order level, and links, using relation Kind, types and concepts from different levels. But Esch does not deal with conceptual relations and relation types. In [26] Michel Wermelinguer defines more formally higher order types and proposes a translation to first order logic. He defines one hierarchy for all the concept types and one hierarchy for all the relation types and organizes them in regard of their nature and their order. Pavel Kocura in [12] deals with the semantics of attribute relations in conceptual graphs and introduces some second order concept types like: TYPE, REL_TYPE and relation types like ATTR and VALUE_TYPE. He also presents some mapping rules from higher level to lower level using Attribute (ATTR) relations. But none proposes a complete metamodel of the conceptual graph language itself.

3   Modeling Constructs in the CG Formalism

Through a mapping to first-order logic (FOL), the CG theory is recognized as a general knowledge representation language. The simple CGs which are fully mappable to FOL formulae are called first-order CGs. Additional features such as sets [21,8], contexts [5,6,7,16], and various quantifiers [1] provide higher reasoning capabilities by allowing modal [10], temporal [4,19,20] and fuzzy [1] reasoning systems to be devised based on the CG representation language. First-order CGs

---

[1] This work is part of a research project supported by HEC Montreal.

are composed of concepts and relation nodes. Concepts represent objects (either physical or not) of some type. They are composed of the type of the object followed by a reference to the object that they represent, called a referent. Relations represent semantic links between objects. Relations are typed. The distinction between a concept and a relation is rather arbitrary. At times relations may be seen as objects. For simplicity purposes pertaining to both the modeling activities and the subsequent efficiency of the knowledge handling operators, the knowledge engineer must decide on a domain ontology that reflects the concepts of the domain and their possible relationships, all seen as primitive elements of the modeling language [14]. Section 3.1 introduces the representation primitives needed to describe concept types; Section 3.2 does the same with the definition of relation types. Together these sections provide the basic constructs needed to set up a metamodeling layer in a CG system.

3.1   Defining Concept Types

The definition of a concept type is about an object that is being defined (and specified) at the metalevel so that it can be used as a concept type at the data level. For example, let us introduce the object [ConceptType: Driver] which states that Driver is an object of type ConceptType (which is predefined). And let us use Driver in a concept [Driver: *x]. The former is useful to describe the properties of concept type Driver, therefore providing it with a formal definition. The latter is then permitted and can be used to describe individual drivers who will comply with the definition of concept type Driver as given by the former definition. Therefore we first need a concept type ConceptType that is used to represent and define concept types. Then, when a concept type t is to be defined, we need to attach the corresponding concept type concept [ConceptType: t] to some definition graphs, represented by (embedded graphs) concepts, that will provide it with different roles and restrictions. These roles are indicated by the type of relation that links concept [ConceptType: t] to its definition graphs. The different relation types that are needed to define a concept type are: csubt, def, rstrct, and sntx; each links concept type concept [ConceptType: t] to a concept type concept, a definition graph, a restriction graph (or a rule graph), and a syntax graph, respectively. Each of these relation types is defined in the subsections below. For example, concept type Driver could be defined using the CG of Figure 1 asserted at the metalevel.

Subtyping: csubt

From Figure 1, the subtype relation Driver < Person can be extracted from the csubt relation. The concept type hierarchy is therefore built from all such relations extracted from all concept type definitions. This creates an inheritance network among concept types where all linked pairs of concept types are part of a partial order of generality defined by the csubt relation.

Using a Concept Type at the Data Level: sntx

The syntax graph presents how the concept defined at the metalevel, a concept

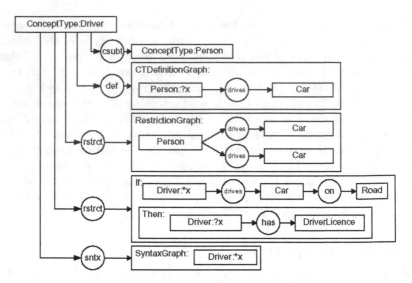

Fig. 1. An example of a concept type definition.

type in the case of Figure 1, must be used at the data level[2]. Here again, rules on how to compose a concept based on a concept type definition can be stated at the metalevel (see Section 6). These rules ensure that the syntactical forms are used according to the metalevel definitions on which the acquired objects are based.

Genus and Differentia: def

In Figure 1, it is stated that a driver is a person who drives a car. The genus of the definition is thus the concept of type Person; its differentia is the statement that s/he must drive a car to be recognized as a driver. This statement is equivalent to the lambda expression of Figure 2. Additional validation rules can be

$$\text{Driver}(x_1) = [\text{Person: } \lambda x_1] \rightarrow (\text{drives}) \rightarrow [\text{Car}]$$

Fig. 2. The lambda expression extracted from the definition graph of Figure 1.

stated (as shown in Section 6) on how to ensure that the definition graph has a concept whose referent is ?x, and that the concept type of this concept is in accordance with the concept type which is the destination concept of the csubt relation.

---

[2] The use of *x in the syntax graph and ?x in the definition graph is a lexical convention (see /refCGIG for semantics of *x and ?x). That does not mean they refer to the same individual except if there are in the same graph

Constraints on the Use of Types: rstrct

There are two types of restriction graphs, each one imposing a constraint on the use of the concept type being defined: restriction graphs and rule graphs. Restriction graphs introduce graphs that must never project themselves onto any other graph in a CG system[3]. Therefore, they are graphs that represent situations that must never occur. In our example of Figure 1, a driver cannot drive two cars at the same time. Rule graphs introduce complementary definitions to the main definition of a type, but only under certain conditions. With our example, when a driver drives a car that is on the road, then s/he must have a driver's license. That is, when the if-graph projects itself onto some graph in the CG system, then the then-graph must also project itself onto the same graph (providing that the coreferenced variables are bound to the same concepts). Restriction and rule graphs permit the representation of a large subset of the semantic constraints found in database literature. They were introduced under a slightly different representation in [18]. For a more complete introduction on restriction and rule graphs see [17] and [13]; for a formal definition of their associated extensional semantics see [15].

3.2   Defining Relation Types

As before, a predefined RelationType concept type is required to express that some object is a relation type. The primitive relations defined above either hold for the definition of relation types or have counter-parts. Figure 3 gives some example of the definition of a relation type.

Subtyping: rsubt

In Figure 3, relation type goingto is defined as a subtype of the Link[4] relation type. A relation type hierarchy can be built from the rsubt relations found in the definitions of all relation types. Here we chose to specialize the type inheritance relation (subt) for concept types (csubt) and relation types (rsubt). This choice, rather than using the subt relation directly, is justified for the following reasons. First, linked elements concept types and relation types are different. Second, the way to verify the validity of the relation is also different (See metarules in Figure 16 and Figure 18.)

Using a Relation type at the Data Level: sntx

Finally, the use of the relation type r at the data level must be represented. The syntax graph associated with the definition of r does that. As mentioned before, syntactical formation rules can be expressed at the metalevel in order to validate the use of a relation type at the data level (see Section 6).

---

[3] The projection that we consider here is injective. See [9] for the appropriate motivation. We believe that under certain simplifying assumptions (see [14]) an injective projection should be sought.

[4] The relation type Link is primitive and states a relationship between two concepts. Link is at the top of the relation type hierarchy.

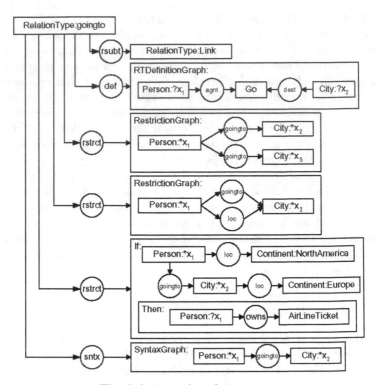

Fig. 3. The definition of a relation type..

Genus and Differentia: def

From Figure 3, one can see that the relation type goingto between two parameters x1 and x2, is defined as a person x1 who is the agent of a verb Go, for which the destination is a city x2. From this (relation type) definition graph, the lambda expression of Figure 4 could be extracted. Again, a metalevel rule can be expressed to verify that the definition of a relation type r conforms to the relation type of its supertype r' (according to the rsubt relation) (see Section 6). Using the lambda expression produced from the definition of a relation type r,

$$goingto(x_1, x_2) = [Person: \lambda x_1] \rightarrow (agnt) \rightarrow [Go] \leftarrow (dest) \leftarrow [Person: \lambda x_2]$$

Fig. 4. The lambda expression extracted from the definition graph of Figure 3.

the signature of r is then known. Therefore the canonical basis of the CG system, B, can be built from the analysis of the definition graph of each relation type. Also, a metarule enforcing the signature of each relation type can be expressed, as will also be presented in Section 6.

Constraints on the Use of Types: rstrct

As with concept type definitions, restrictions can be defined on the use of a relation type at the data level. For instance, Figure 3 states that a person cannot go to two different cities that s/he cannot go to a city where s/he is already located, and that when a person is located in North America and goes to a city in Europe, then s/he must own an airline ticket.

4   Mapping Metalevel to Data Level

In [9] we defined a function ω[5] that maps higher level to a lower level objects. Let us recall that definition.

Definition 1. *Function ω is defined over $C \rightarrow \mathcal{E}$ where C is the set of concepts that represent entities of the system and $\mathcal{E}$ is the set of all referenced elements (internal and external elements[6]).*

Applied on a concept, function ω returns the entity represented by the concept. Obviously, the function is defined on the set of concepts that represent entities of the system, i.e., internal elements. Figure 5 shows the way the function ω may be used. Let us have (a) graph `[City:Ottawa]->(cap)->[Country:Canada]` identified by the internal referent `#4387`, (b) two different ways to speak about this conceptual graph, and (c) the application of ω on the first concept of (b). Figure 6 presents how the type of a concept could be accessed using a meta-level

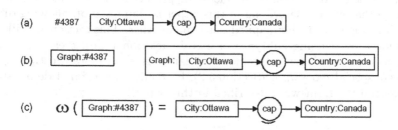

Fig. 5. Function ω.

CG describing the concept. Let us consider `[Concept: [City:Ottawa]]` that is the concept that represents the concept `[City:Ottawa]`. Using the predefined `type` and `ref` relations and `Concept` and `Referent` concept types, let us have the graph of Figure 6. Figure 7 shows the mapping from a higher level to its immediate lower level. Applying function ω on the concept of type `Concept` in the meta-level CG of Figure 6, we obtain the (data level) concept representing the city of Ottawa.

---

[5] The function ω is a generalization of the Sowa's functions τ and ρ[24]. ω is different of Sowa's function *referent* that returns the lexical of the referent field[25].

[6] In the metamodel we distinguish two types of element, the external elements, external with the language, and the internal elements which are the components of the language. The external elements represent the objects of the universe of the speech which is outside the system and which can be referred by internal elements.

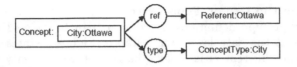

Fig. 6. From a lower level to a higher level.

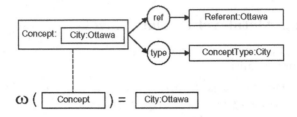

Fig. 7. From a higher level to a lower level.

5   Defining Subtypes of the Primitives of Section 3

The definition primitives of Section 3 and the mapping operator of Section 4 allow the objects of any level to be described by definitions found one level up, at their metalevel. This provides for many layers of modeling levels. One use for such layers is the definition of the modeling primitives from which the modeling language that we are describing in this paper is composed of, allowing different modeling languages to be mapped onto one another. This section will illustrate these ideas by defining subtypes of certain relation types, creating classes of relations, thus specializing the modeling language even more (Section 5.1) and classes of specialized graphs (Section 5.2). By doing so, we aim at demonstrating how general the framework described in this paper is.

5.1   Creating Classes of Relations

The RelationType concept type, used as a primitive element in our modeling language ontology so far, is itself a concept type. Therefore, it could be defined using the definition primitives introduced in Section 3.1. Doing so will allow specializations of it to be defined, refining further the modeling language that will be handed out to the knowledge acquisition modules in charge of modeling the actual application domain. In this section, we intend to show the expressivity of the simple representation tools introduced in Sections 3 and 4. As a first example, let us say that a relation type is a type that is subtype of another relation type. Figure 8 below illustrates this definition. Notice that there is no syntax graph associated with it since a syntax graph represents a precise syntactical form, therefore, a predetermined and fixed number of parameters (the arity of relations of that type) would be required. Therefore, syntactic considerations lead us to define fixed-arity relation types. Figure 9 shows a specialization of concept type RelationType, BinaryRelationType, which will be useful for defining

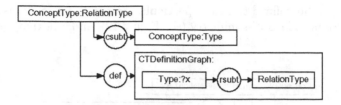

Fig. 8. The definition of a relation type.

binary relations[7]. It imposes a particular syntax graph to all of its elements, providing a fixed arity of two for all relations of that type.

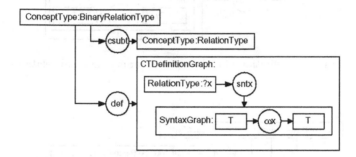

Fig. 9. The definition of a relation type for binary relations.

Other subclasses of relation types can be defined in the same way. For instance, it is possible to define a class of transitive relations through the definition of a subclass of BinaryRelationType. Figure 10 gives such a definition. Other classes

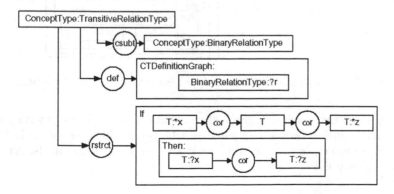

Fig. 10. The definition of a class (type) for all transitive relations.

---

[7] As introduced in [9] in order to simplify the notation we replace $\omega(\texttt{[RelationType:*r]})$ by $\omega\texttt{r}$

of relations can be defined to match particular properties of relations, like symmetry (see Figure 11), anti-symmetry (see Figure 12), and reflexivity (see Figure 13).

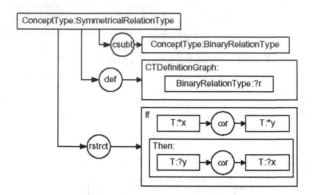

Fig. 11. The definition of a class (type) for symmetrical relations.

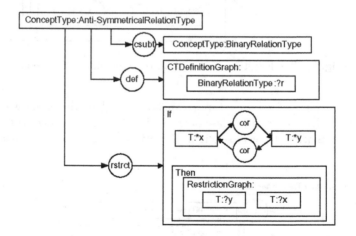

Fig. 12. The definition of a class (type) for anti-symmetrical relations.

In Figure 12 the restriction graph states that if x is in relation r with y and y is in relation r with x then x and y may not be two distinct concepts. In Figure 13 we need to use the syntactic graph to identify in the rule graph the type that the relation may link.

5.2   Creating Classes of Graphs

The CTDefinitionGraph concept type, used as a primitive element so far, may be defined as a concept type. A Concept Type Definition Graph (CTDefinition-

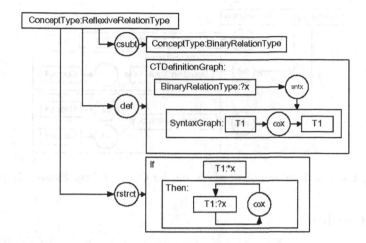

Fig. 13. The definition of a class (type) for reflexive relations.

Graph) is a specialization of DefinitionGraph. Figure 14 presents the definition and restriction graphs of CTDefinitionGraph. The definition graph states that a CTDefinitionGraph has at least one concept with a question mark and the restriction graph states that a CTDefinitionGraph may not have two distinct concepts with question marks. As CTDefinitionGraph above, the CTSyntaxGraph concept type

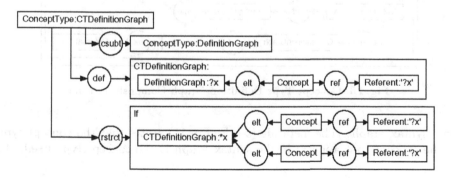

Fig. 14. The definition of a class (type) for Concept Type Definition Graph.

may be defined as a concept type. A Concept Type Syntax Graph (CTSyntaxGraph) is a specialization of DefinitionGraph. Figure 15 presents its definition and restriction graphs. The definition graph states that a CTSyntaxGraph has at least one concept and the restriction graph states that a CTSyntaxGraph may not have two distinct concepts. In this section we demonstrated how the definition primitives of Section 3 and the mapping operator of Section 4 can be used to describe objects of any level by definitions at their metalevel. In the next section we will show how we can complete specifications by adding metarules.

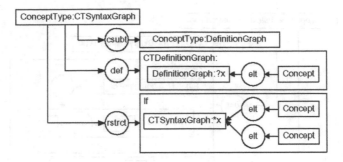

Fig. 15. The definition of a class (type) for Concept Type Syntax Graph.

6   Metarules

This section presents metarules we introduced in earlier sections. The first metarule, as illustrated in Figure 16, states that the definition graph of a concept type has a concept whose referent is ?x, and whose concept type is in accordance with the concept type which is the destination concept of the csubt relation. The second metarule presented here, illustrates the composition rule for concept

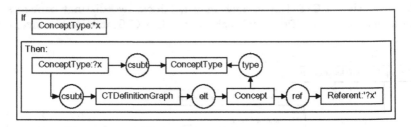

Fig. 16. Concept Type Definition Graph Composition Rule.

type syntax graphs. The metarule (see Figure 17) states that the concept type of the concept that appears in the syntax graph is the concept type itself. In

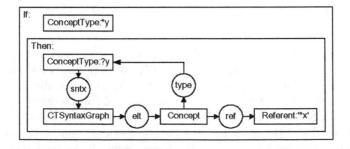

Fig. 17. Concept Type Syntax Graph Composition Rule.

section 3.2 we argued that a metalevel rule can be expressed to verify that the definition of a relation type r conforms to the relation type of its supertype r' (according to the rsubt relation). Figure 18 presents this metarule. If two con-

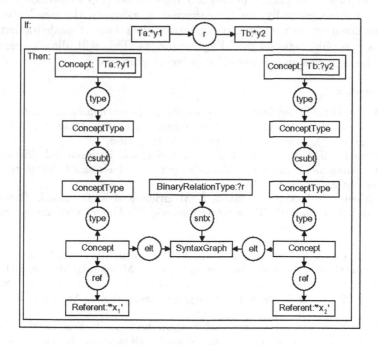

Fig. 18. Signature Compliance Rule.

cepts [Ta:*y1] and [Tb:*y2] are linked by a relation whose relation type is r then the two types Ta and Tb are in csubt relation with the two types of the signature of r as expressed in its syntax graph. Concept [Ta:*y1] identified as the source of the relation has a type that is a specialization of the concept type of the concept (element of the syntax graph) whose referent is *x_1. Respectively, Concept [Tb:*y2] identified as the destination of the relation has a type that is a specialization of the concept type of the concept whose referent is *x_2. These few examples demonstrate that conceptual graphs can easily be used to state restriction or define rule at a metalevel.

7   Conclusion and Future Work

Metamodeling and therefore metamodels are important because they formally define the modeling primitives used in modeling activities. In this paper we introduced basic building blocks in order to use Conceptual Graphs in metamodeling activities. We have seen a metamodeling approach is important because it allows declarative and formal definition of modeling constructs (Section 3 and 5). It authorizes validation of acquired knowledge through formal definition and metarules(Section 6). In this paper we demonstrated that CGs are powerful enough to be used as an universal metamodeling language but a lot of work remains to be done to define a complete metamodeling framework based on CGs. In [9] we demonstrated that CGs may be used to model the main model component of KADS [22,23] (a methodology to develop knowledge-based systems), and more generally, we are currently working on the development of a meta-metalevel where we could, using formal definition of modeling languages, specify a mapping between modeling languages. This will allow integration of information systems even if based on different paradigms.

References

[1] T. Cao and P. Creasy. Fuzzy order-sorted logic programming in conceptual graphs with a sound and complete proof procedure. In *Lecture Notes in Artificial Intelligence #1453*, pages 270–284. Springer-Verlag, 1998.

[2] S. Crawley, S. Davis, J. Indulska, S. McBride, and K. Raymond. Meta information management. In *Formal Methods for Open Object-based Distributed Systems Conference*, Canterbury, UK, July 1997.

[3] S. Crawley, S. Davis, J. Indulska, S. McBride, and K. Raymond. Meta-meta is better-better! In *IFIP WG 6.1 International Working Conference on Distributed Systems*, October 1997.

[4] J. Esch. Temporal intervals. In T. Nage, J. Nagle, L. Gerhloz, and Eklund P., editors, *Conceptual Structures*, pages 363–380. Ellis Horwood, 1992.

[5] J. Esch. Contexts as white box concepts. In G. Mineau, B. Moulin, and J. Sowa, editors, *Proceedings of the 1st International Conference on Conceptual Structures (ICCS'93)*, pages 17–29, Quebec City, Quebec, Canada, August 1993. Springer-Verlag.

[6] J. Esch. Contexts and concepts, abstraction duals. In W. Tepfenhart, J. Dick, and J. Sowa, editors, *Proceedings of the Second International Conference on Conceptual Structures (ICCS'94)*, pages 175–184, College Park, Maryland, USA, August 1994. Springer-Verlag.

[7] J. Esch. Contexts, canons and coreferent types. In W. Tepfenhart, J. Dick, and J. Sowa, editors, *Proceedings of the Second International Conference on Conceptual Structures (ICCS'94)*, pages 185–195, College Park, Maryland, USA, August 1994. Springer-Verlag.

[8] D. Gardiner, B. Tjan, and J. Slagle. Extended conceptual structures notation. In *Proceedings of the 4th Annual Workshop on Conceptual Structures. IJCAI-89*, 1989. Section 3.05.

[9] O. Gerbé. *Un modèle uniforme pour la modélisation et la métamodélisation d'une mémoire d'entreprise.* PhD thesis, Université de Montréal, 2000.

[10] B. Ghosh and V. Wuwongse. Computational situation theory in the conceptual graph language. In *Lecture Notes in Artificial Intelligence #1115*, pages 188–202. Springer-Verlag, 1996.

[11] C. Havens. Enter, the chief knowledge officer. *CIO Canada*, 4(10):36–42, 1996.

[12] P. Kocura. Semantics of attribute relations in conceptual graphs. In G. Ganter, B. Mineau, editor, *Proceedings of 8th International Conference on Conceptual Structures (ICCS2000)*, pages 235–248, Darmstadt, Germany, August 2000. Springer.

[13] G. Mineau. Constraints and goals under the conceptual graph formalism: One way to solve the scg-1 problem. In W. Tepfenhart and W. Cyre, editors, *Proceedings of the 7th International Conference on Conceptual Structures*, pages 334–354, Blacksburg, VA, USA, July 1999. Springer.

[14] G. Mineau. The engineering of a cg-based system: Fundamental issues. In
B. Ganter and Mineau G., editors, *Proceedings of the 8th International Conference on Conceptual Structures*, pages 140–156, Darmstadt, Germany, August
2000. Springer-Verlag.

[15] G. Mineau. The extensional semantics of the conceptual graph formalism. In
B. Ganter and G. Mineau, editors, *Proceedings of the 8th International Conference on Conceptual Structures (ICCS'2000)*, pages 221–234, Darmstadt, Germany,
August 2000. Springer.

[16] G. Mineau and O. Gerbé. Contexts: A formal definition of worlds of assertions. In
Proceedings of 5th International Conference on Conceptual Structures (ICCS'97),
pages 80–94, Seattle, Washington, USA, August 1997.

[17] G. Mineau and R. Missaoui. *Semantic Constraints in Conceptual Graph Systems.*
DMR Consulting Group Inc., Montreal, Quebec, Canada, June 1996. Internal
Research Report #960611A. 39 pages.

[18] G. Mineau and R. Missaoui. The representation of semantic constraints in conceptual graph systems. In D. Lukose, H. Delugach, M. Keeler, L. Searle, and
J. Sowa, editors, *Proceedings of 5th International Conference on Conceptual Structures (ICCS'97)*, pages 138–152. Springer-Verlag, 1997. LNAI #1257.

[19] B. Moulin. The representation of linguistic information in an approach used
for modeling temporal knowledge in discourses. In G. Mineau, B Moulin, and
J. Sowa, editors, *Proceedings of 1st International Conference on Conceptual Structures (ICCS'93)*, pages 182–204, Quebec City, Quebec, Canada, August 1993.
Springer.

[20] B. Moulin and S. Dumas. The temporal structure of a discourse and verb tense
determination. In J. Tepfenhart, J. Dick, and J. Sowa, editors, *Proceedings of
Fourth International Conference on Conceptual Structures (ICCS'94)*, pages 45–
68, College Park, Maryland, USA, August 1994. Springer-Verlag.

[21] H. Pfeiffer and R. Hartley. Additions for set representation and processing to conceptual programming. In *Proceedings of the 5th Annual Workshop on Conceptual
Structures. AAAI-90*, 1990. Section A.15.

[22] A. Schreiber, B. Wielenga, H. Akkermans, W. Van de Velde, and A. Anjewierden. CML: The CommonKADS conceptual modelling language. In L. Steels,
A. Schreiber, and W. Van de Velde, editors, *Proceedings of the 8th European
Knowledge Acquisition Workshop (EKAW'94)*, pages 1–24, Hoegaarden, Belgium,
1994. Springer-Verlag.

[23] G. Schreiber, B. Wielenga, R. de Hoog, H. Akkermans, and W. Van de Velde.
CommonKADS: A comprehensive methodology for KBS development. *IEEE Expert*, pages 28–36, December 1994.

[24] J. Sowa. Relating diagrams to logic. In John F. Sowa Guy W. Mineau,
Bernard Moulin, editor, *Proceedings of the First International Conference on Conceptual Graphs (ICCS'93)*, volume 699, pages 1–35, Quebec City, Quebec, Canada,
August 1993. Springer-Verlag.

[25] J. F. Sowa. *Conceptual Structures: Information Processing in Mind and Machine.*
Addison-Wesley, 1984.

[26] M. Wermelinger. Conceptual graphs and first-order logic. In G. Ellis, R. Levinson,
W. Rich, and J. Sowa, editors, *Proceedings of the Third International Conference
on Conceptual Structures (ICCS'95)*, pages 323–337, Santa Cruz, CA, USA, August 1995. Springler-Verlag.

Making Virtual Communities Work: Matching Their Functionalities

Aldo de Moor and Willem-Jan van den Heuvel

Infolab
Tilburg University
P.O. Box 90153, 5000 LE Tilburg, The Netherlands
ademoor/wjheuvel@kub.nl

Abstract. Virtual professional communities increasingly make use of standard information tools, like mailers and groupware applications, to support their collaborative activities. However, the requirements of these communities and the technologies in use change rapidly, so that requirements and available functionalities continuously need to be recalibrated. Changing their mappings is not trivial, because of the many dependencies between the business processes and tool components. To increase the efficiency of the specification process, functionality matching approaches need to be developed that are sensitive to the socio-technical semantics of the community. In this way, the technical feasibility of a proposed change can be more easily determined.

In this paper, we propose a concrete matching approach based on the RENISYS method for legitimate user-driven system specification. The approach consists of a series of matching process steps which are based on a functionality matching meta-model. We illustrate how such an approach could be used in practice by applying it to a proposed system change process in the case of an electronic journal.

1 Introduction

Virtual professional communities and their information systems are good examples of complex socio-technical systems. There is significant pressure on these systems to change, because of change drivers of many different kinds. Technological, economic, political and many other factors contribute to a continuous need for evolution of the requirements and supporting information technologies. However, change processes are costly, and effects of changes are often unclear. Therefore, often considerable resistance to change exists. To reduce this resistance, it must be clear to users what are the consequences of a proposed change in the socio-technical system. An important barrier is taken away if changes are legitimate, in the sense that they are both meaningful and acceptable to the community. One approach increasing this legitimacy is the RENISYS method [5]. Other effects of change, such as those on non-functional constraints like quality and usability aspects, need to be taken into account as well. Yet another very important category of change aspects is ensuring a good match between functional requirements and the available IT resources. The question "do we still have adequate technological support after implementing the proposed change?" needs to be answered positively, especially since in virtual communities work processes are complely or mostly enabled

H. Delugach and G. Stumme (Eds.): ICCS 2001, LNAI 2120, pp. 260–274, 2001.

by information technologies. Otherwise, disruption of the socio-technical infrastructure will interrupt the evolution of the community. Furthermore, when the technical complexity of specification changes can be reduced, then the efficiency of the change process can be increased so that more attention can be paid to other, non-functional aspects. This will lead to information systems that are better tailored to the specific needs of the community.

In Sect. 2, we first give an overview of existing theory and practice concerning functionality matching, and introduce a case to illustrate the ideas. In Sect. 3, we then introduce a meta-model specifically developed for matching required and enabled functionalities in virtual professional communities. This meta-model is based on the RENISYS method. A concrete matching process, grounded in this meta-model is introduced in Sect. 4. Some conclusions and directions for future research are given in Sect. 5.

2 Functionality Matching: Theory and Practice

In this section, we first define our view on functionality matching. After reviewing related work, we introduce a case that is used to explain the ideas introduced in this paper.

2.1 Theory: What Is Functionality Matching?

Information systems for virtual professional communities are generally not constructed from scratch. Instead, applications supporting collaboration are developed by experimenting with widely available *information tools*, which originally were often developed for other purposes [7]. We define an information tool as a unit of software that completely or partially enables some information and communication processes. An *information process* allows a single user to produce a new *information object* out of already existing objects. An example is a researcher writing a review of a paper. The focus of a *communication process*, which involves multiple communicating entities, is on the transfer rather than on the production of information objects. Examples of information tools range from mailers, list servers, and chat tools to numerous kinds of web applications.

In order to understand the role that an information tool plays as part of the socio-technical network information system, we need to look at both the *functionality* that the tool provides and its *usability*, which concerns the extent to which the functions provided by the tool are understood and applied by its users to their particular tasks. Together, these notions determine the *effective functionality* [9], which we define as that part of the available functionality used to support the activities of the community. This needs to be known to assess the effect of changes in the information system.

Usability is not a property of the tool itself, but rather of the tool in its context of use. Therefore, usability has been defined as the evaluation of the extent to which users can translate their intentions into effective actions to access the functionality [8]. We have decomposed this definition to focus on two aspects: (1) who can *access* a particular information tool in which capacity and (2) how to *represent* the user requirements, the tool functionality and their linkages. To deal with the first aspect, some framework is needed to model the functionality of tool types and the access rights of particular users to particular instances of these tools. The second aspect requires some ontology describing

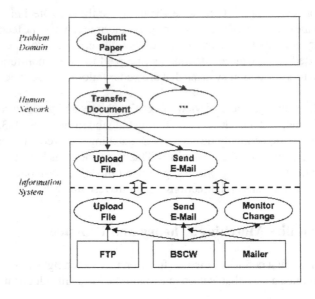

Fig. 1. Functionality Specification Example: The Paper Submission Process

the key entities of the complete socio-technical system, in order for users to propose and discuss about functionality specifications. In the RENISYS method, both aspects are addressed by the *reference framework* that is used to represent specifications that define the socio-technical system. The reference framework consists of a *problem domain* ontology modelling goals and activities, a *human network* ontology representing the organizational structures in which these tasks are carried out, and an *information system* domain ontology in which the technical functionality used by the virtual community for its work is defined. An example of the dependencies between elements from these domains is shown in Fig. 1. The relationships between the particular elements of the figure are explained in Sect. 2.3.

2.2   Related Theory and Approaches

To some extent, research into matching algorithms that compare the functionality of two software components on the basis of some kind of component specification, has been done in the areas of information retrieval [15], cooperating (or interoperable) information systems ([13], [3]) and software reusability ([14], [12]). These solutions assume that the functionality of components can be represented as a collection of signatures.

A component signature explicitly separates the definition of the services of the component from the actual implementation. The services are defined as methods (functions) with input parameters, input types and the output type. This separation is critical for interoperability across programming languages, operating systems and even networks. The Interface Definition Language (IDL) is a prominent example of a interface specification language, that has been proposed by the Object Management Group (OMG) and constitutes the foundation for their object request broker (middleware) architecture. IDL specifications can be used to specify component attributes, parent classes, typed events, methods (including input and output parameters and their data types), and exceptions.

In the following excerpt we give a simple example of an IDL specification:

```
module MyCommunity {
  interface Administrator : Person {
    attribute integer ID;
    void registerNewMember (in short MemberID, in integer ID,
          in String Community) raises (NotAuthorized);
    void deregisterMember (in short MemberID, in integer ID,
          in String Community) raises (NotAuthorized);}
} /* End MyCommunity
```

The excerpt specifies an interface of a Administrator component. This class inherits the characteristics and behavior from the parent class Person. Administrator has an attribute ID with integer as its datatype. Moreover, this class exhibits three methods to other interested classes: registerNewMember, deregisterMember and NotAuthorized. The exception NotAuthorized occurs whenever the person that tries to invoke one of these methods is not authorized.

Most interface matching approaches now compare the methods, and pre and postconditions of a collection of interface specifications, that are stored in some kind of interface repository, with a given specification. The solutions generally have some mechanism to deal with partial matches, and result in the best matching interface specification(s).

Although these ideas are applicable for acquiring potentially reusable component definitions for example from a component repository, they do not deal with the specific functionality evolution characteristics of virtual professional communities. More particularly, such socio-technical systems require efficient mechanisms to deal with changes to configurations of tools, requirements, and users. Besides that, current matching approaches only match functionality in the narrow sense, omitting the usability aspect.

In our view, mapping tool functionality to the requirements of virtual communities, requires firstly a functionality specification language that adds more social-technical system semantics to the rather low-level interface definitions, and secondly, a mapping procedure that is based on a more sophisticated process that makes use of these semantics.

This does not mean that component mapping is unnecessary. On the contrary, these approaches are essential to *construct* the support information tool components, e.g., mailing component, chat enabling components and registration components for composing virtual community applications. However, they are not capable of dealing with the more complex, and high level information tool requirement specifications specified by the (mostly non-technical) community members themselves. Questions like "do we still have enabling components if we change the community structure?" can not be answered with interface mapping approaches as the specification languages can not capture the semantics.

Thus, what is needed are approaches that can deal with the specific functionality matching problems of virtual communities, so that changes in functionality can be analyzed in their broader usage context.

2.3   Case: The Electronic Journal on Comparative Law

IWI, a Dutch organization stimulating new ways of distributing scientific information, funded a project to create an Electronic Journal on Comparative Law (EJCL)[1]. The project group included participants from various academic law institutes, university libraries, and computer centers. The goal was to have all publishing activities, ranging from paper submission to editing, peer review, and publication, being done completely via the Web. The initial basic set of requirements defined by the project team members gradually grew in scope and complexity. Furthermore, the set of simple information tools over time included more advanced groupware applications.

One interesting observation from a functionality matching perspective concerns the definition of the technological support for the paper submission process (Fig. 1). The submission of papers was considered as a document transferring process, which consisted of two required communication processes: first, an author has to upload a file, then he sends an e-mail to the editor with the submission details. The technical committee responsible for the selection of the right tools proposed to enable the file uploading process using a standard FTP tool. This tool enables basic file transfer. However, the project coordinator then proposed to use a BSCW-server instead. This tool has been optimized for file distribution processes, as it enables advanced, userfriendly, and secure file transfer. Furthermore, it can be used to send e-mails as well as monitor changes in file updates and accesses. The effects of *replacing* the FTP-server with a BSCW-server are not clear. Both tools enable their own sets of information and communication (IC) processes. Their effective functionality needs to be known before this change is technically feasible. The approach we introduce next is capable of dealing with such change complexities.

3   A Functionality Matching Meta-model

To develop an approach that can facilitate the functionality change process, we first need to define a functionality matching meta-model. This metamodel can be used to model the exact relations between tools, users, and the functionalities that are required and enabled. We use this static model to define the actual functionality matching *process* in Sect. 4.

Before presenting the meta-model, we first operationalize the concept of effective functionality by listing a number of axioms.

Effective Tool Functionality Axioms. These axioms form the foundation of the functionality matching meta-model. In any change process, their validity must be guaranteed.

- An information tool can enable one or more information and communication processes.
  Example: a mailer allows a user to compose a mail (information process) and send or receive a mail (communication processes).

---

[1] http://law.kub.nl/ejcl

- Different information tools may have partially *overlapping* functionality, i.e. each enabling the same information or communication process, while also enabling different such processes at the same time.

  Example: Both a mailer and a web browser allow one to send a mail. However, only with a mailer can a user also organize sent and received messages, whereas sophisticated HTML document access is just possible with a web browser.

- All network participants involved in a required information/communication process must have at least one *enabling information tool* at their disposal.

  Example: a participant may have a required communication process of sending a mail. Thus, the participant must have access to, for instance, a mailer or a web browser.

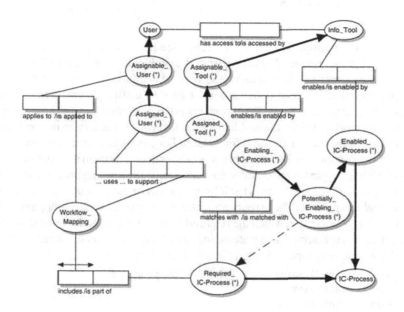

Fig. 2. A Functionality Matching Meta-Model

The Meta-model. In the meta-model, we describe how in RENISYS the following elements are specified: (1) the enabled functionality (which tools enable which IC-processes), (2) the required functionality (which IC-processes are required), (3) the enabling functionality (which required IC-processes can be enabled by the tools), (4) functionality access (which users have access to which tool instances), and (5) functionality assignment (which users use which tool instances for what workflow mappings). Fig. 2 shows the relations between the different entities necessary in the functionality specification process[2]. The semantics of this figure are explained in the remainder of this section.

Enabled Functionality. Any IC-process enabled by some information tool is called an *enabled IC-process*. Such a process is represented as a a state definition which conforms to a specialization of the following type definition of the enable-relation:

[Type : [Enable : *x] → (Def) → [T : *x]−
    (Inst) → [Info_Tool]
    (Obj) → [IC_Proc]].

Example

The following state definition says that uploading a file is enabled by an FTP tool:

[State : [Enable : #267]−
    (Inst) → [FTP]
    (Obj) → [Upload_File]].

□

Required Functionality. The RENISYS reference framework distinguishes three domains, as mentioned before. Workflows from the problem domain are called *activities*, from the human network *interactions* and from the information system domain *IC-processes*. Functionality requirements consist of information or communication (IC) processes in their usage context. Requirements are represented by workflow mappings, which relate a workflow from the problem domain, via one in the human network domain to a workflow in the information system domain. For example, a workflow mapping can say that a (problem domain) editorial process is a form of a (human network) discussion process which is supported by an (information system) file sending process, among others. The *required IC-process* then is the IC-process part of the workflow mapping, in this case the *send file*-process. The activity and interaction part of such a mapping together identify the usage context in which the required IC-process operates. A particular workflow mapping is represented as a state definition which conforms to a specialization of the workflow mapping type definition. This definition is:

[Type : [Workflow_Mapping : *x] → (Def) → [Mapping : *x]−
    (Part) → [Activity]
    (Part) → [Interaction]
    (Part) → [IC_Proc]].

Example

[State : [Workflow_Mapping : #123]−
    (Part) → [Submit_Paper]
    (Part) → [Transfer_Document]
    (Part) → [Comm_Process]].

---

[2] The diagram is a variant of NIAM-notation [6]. Bold arrows indicate subtype relations, the predicates represent other relations. Only the entity types User, Info_Tool, IC_Process, and Workflow_Mapping are basic concept types. The other entities distinguished in the functionality specification process are roles that these types play. They are denoted by an asterix.

This workflow mapping specifies that 'the *paper submission* process is a *document transfer* process that is supported by some *communication* process'. The latter process is thus a required IC-process. Note that this process is defined as a generic communication process, because the specifier does not either know, or care, by which particular type of communication process the paper submission process is to be supported. This means that many degrees of freedom are left in the choice of the tools that are to support this particular workflow mapping.

□

Enabling Functionality. For the required IC-process of each workflow mapping, a set of *potentially enabling IC-processes* exists. These are those IC-processes that are (1) enabled by some tool and (2) are a subtype of the required IC-process. This makes sense, because the specifiers of a workflow mapping would define the required IC-process to be generic if they are indifferent or do not know yet which particular enabling IC-process should satisfy it, as in the previous example. So, the more generic the required IC-process, the more enabled IC-processes can match with it, thus the more potentially enabling IC-processes for a particular workflow mapping there are. Out of this set of potentially enabling IC-processes, at least one *enabling IC-process* must be selected for the workflow mapping to become operational.

Example

Assume the workflow mapping #123 defined earlier, and assume that the set of enabled IC-processes equals
{[Send_Mail], [Receive_Mail], [Send_File], [Edit_Textfile]}, of which all but the edit-textfile process (which is an information process) are communication processes. The set of potentially enabling IC-processes is
{[Send_Mail], [Receive_Mail], [Send_File]}, since these are all subtypes of the *communication*-process. Out of this set of potentially enabling processes, the specifier selects the *send mail*-process as the (actually) *enabling IC-process*.

□

Functionality Access. Each user has *access* to a certain set of *tool instances*, represented in the form of state definitions that conform to this type definition of the access-relation:

[Type : [Access : *x] → (Def) → [T : *x]−
  (Poss) ← [User]
  (Obj) → [Info_Tool]].

Example

The following state definition (representing a state-of-affairs in the domain) indicates that John has access to mailer #4 at the Infolab.

[State : [Access : #213]−
  (Poss) ← [User : #John]
  (Obj) → [Mailer : #4@Infolab]].

□

Some types of information tools are *complex*, in the sense that users can access only part of the functionality of the tool. A typical example of such a complex information tool is a web server that consists of many different pages, each enabling different functionality.

The meaning of a complex information tool is the following:

[Type : [Complex_Info_Tool : *x] → (Def) → [Info_Tool : *x]−
    (Part) → [Entity]].

Example

This definition of a complex information tool indicates that user John only has access to the home page of the BCFOR-web server:

[State : [Access : #215]−
    (Poss) ← [User : #John]
    (Obj) → [Web_Server : #BCFOR] → (Part) → [Web_Page : #home.html]].

□

Functionality Assignment. For each workflow mapping, it should be determined for all users in the community whether the workflow mapping applies to them. If so, out of the tools accessible to a particular user, one or more should be selected. This selected tool is to support him in the required IC-process that is part of the workflow mapping.

Users are in the set of *assignable users*, a subset of all community members, for a workflow mapping if he or she is permitted to be involved in it. Such permissions can in principle be calculated from the action norms that define the workflow behaviour of users (see [5]), however, for simplicity, we allow users to be assigned to a workflow mapping manually here.

An information tool is in the set of *assignable tools* for a workflow mapping if it enables the enabling IC-process, i.e., the particular IC-process type chosen to match with the required IC-process. Thus, assignable tools for the *send mail* required IC-process of the previous example could be, for instance, mailers and BSCW, since both tools offer some form of mail-sending functionality.

The actual assignment of the tool that is to support a particular assignable user in a specific workflow mapping is not automated in our approach. The main reason is that the choice of which tool to use for a work process depends on many circumstances beyond functionality matching, such as the non-functional requirements mentioned in the introduction. For example, the users themselves could be intensively involved in this assignment process, since they can best assess their own requirements and preferences.

The functionality assignment is represented by a so-called *support*-relation. This definition assigns some assignable user and an assignable tool to the workflow mapping. This user is referred to as the *assigned user*, the tool is called the *assigned tool*. The type definition of the support-relation is:

[Type : [Support : *x] → (Def) → [T : *x]−
    (Poss) ← [User]
    (Inst) → [Info_Tool]
    (Obj) → [Workflow_Mapping]].

Example

The requirement that user John is to use (possibly among other tools) BSCW
server #3 to enable him to submit papers is represented by:

[State : [Support : #167]−
  (Poss) ← [User : #John]
  (Inst) → [BSCW : #3]
  (Obj) → [Workflow_Mapping : #123]].

☐

Often, it may be necessary to specify that a particular required IC-process is to
be supported by a particular type of tool, without knowing yet who are its users. For
example, a team leader wants all of his staff to use the same tool for a particular workflow.
The representation of such a *required implementation* is:

[Type : [Req_Impl : *x] → (Def) → [Entity : *x]−
  (Inst) → [Info_Tool]
  (Obj) → [Workflow_Mapping]].

Example

The following state definition concisely represents that all users should be able
to access the BSCW-server for submitting papers:

[State : [Req_Impl : #165]−
  (Inst) → [Web_Server : #BSCW]
  (Obj) → [Workflow_Mapping : #123]].
☐

4 The Functionality Matching Process

In the previous section, we introduced the static meta-model in which the matching
relations between requirements (i.e. workflow mappings), tools, and users were specified.
Next, we use this model to construct a process that can be used to assess the effects
of changes in the system specifications on the match between required and enabled
functionalities. We briefly introduce the matching steps in Sect. 4.1, and we illustrate
the process using material from the case described earlier in Sect. 2.3.

4.1 Matching Process Steps

The matching process consists of 5 stages: (1) creating a base of system specifications,
(2) proposing some change to the specifications, (3) formulating a set of functionality
matching criteria, (4) calculating the match, and (5) interpreting the results.

1. Define System Specifications. Virtual professional communities are continuously changing socio-technical systems. At a time t=0, before the change is proposed, we assume that the current system specifications are properly matched with respect to the required and enabled functionalities.

Example A set of enabled-functionality state definitions declares that FTP enables the upload file-process; BSCW enables the upload file, send e-mail and monitor change-process; mailers enable the send e-mail process. To model the required functionality, two workflow mapping definitions WM1 and WM2 represent that the submit paper-process is a transfer document-process which is enabled by the upload file-process (WM1) and the send e-mail-process(WM2), respectively. For WM1, the only potentially enabling IC-process is the upload file-process (enabled by FTP and BSCW), for WM2 the only potentially enabling IC-process is send e-mail (enabled by BSCW and mailers). These are also selected as the enabling IC-processes for the workflow mappings. The selection process is trivial in this case, since there is only one potentially enabling IC-process here. In other cases, however, there may be more options to choose from, if there are deeper IC-process type hierarchies. Two functionality access definitions declare that John has access to mailer #4 and FTP-server #EJCL, three other definitions say that Mary has access to mailer #22, FTP-server #EJCL and BSCW-server #EJCL. Finally, to assign the functionality, two support definitions declare that John is supported in WM1 by FTP-server #EJCL and in WM2 by mailer #4. Two other definitions say that Mary is supported in WM1 by FTP-server #EJCL and in WM2 by mailer #22.

These definitions are represented in the Peirce[3] conceptual graph workbench. To illustrate, one of these definitions is given here:

```
[State: [Support:#300] -
    <- (Poss) <- [User:#John]
    -> (Inst) -> [FTP:#EJCL]
    -> (Obj) -> [Workflow_Mapping:#WM1]].
```

2. Propose Specification Change. At t=1, some specification change is proposed by one of the users. Such a change concerns the creation, modification or deletion of one or more specification knowledge definitions like the ones presented above. Note that the legitimacy of the user being involved in such a change process is guaranteed in the RENISYS method by performing the proper calculations on the set of applicable composition norms (see [5]). These norm calculations say which users may, must, or may not be involved in these knowledge definition change processes. For instance, there may be a norm that says that all system administrators must be involved in the creation of new access-definitions. A change proposal can be in any part of the socio-technical system.

Example Instead of using FTP to upload files in the paper submission process, the project coordinator proposes to use BSCW. This means that the support for workflow mapping WM1 needs to be changed.

3. Formulate Matching Criteria. Many different kind of functionality matches are conceivable. *Matching criteria* (or constraints) need to be specified on which the match

---

[3] http://www.cs.adelaide.edu.au/users/peirce

is to be performed. Such criteria are expressed in terms of the elements of the meta-model. For instance, one criterion could be that when upgrading a tool by installing a new version (i.e., changing its enabled functionality), all existing workflow mappings that the old version supports must still be supported after the change. Once formulated, each criterion needs to be expressed in one or more *matching criteria graphs*. These graphs are the CG-queries necessary for retrieving the knowledge definitions that satisfy the matching criteria.

Example The change process concerns the replacing of tool instances in support-definitions (i.e., definitions that say which users use what tool instances to enable a particular workflow mapping). The matching criteria are (1) all tool instances of FTP in support-definitions of WM1 need to be replaced by tool instances of BSCW. Before this can be done, however, (2) all users that are part of the support definitions selected in (1) need to have an access-relation to at least one instance of the BSCW-tool. In this way, their requirements continue to be enabled. The accompanying matching criteria graphs are:

```
(1) [State: [Support] -
        (Inst) -> [FTP]
        (Obj) -> [Workflow_Mapping: #WM1]]
(2) [State: [Access] -
        (Poss) <- [User:#x]
        (Obj) -> [BSCW]]
```

4. Calculate the Match. Using the functionality specifications of step 1 and the matching criteria graphs of step 3, the actual match is calculated. In general, such a match can be calculated by projecting the matching criteria graphs on the knowledge base of functionality specification graphs.

Example Matching criteria graph (1) is first projected on the specification knowledge base. Using the specialisations function of the Peirce workbench, the following result is returned:

```
> (Specialisations) -> [[State: [Support] -
    -> (Inst) -> [FTP]
    -> (Obj) -> [Workflow_Mapping:#WM1]]]?
[State: [User: #Mary]->(Poss)->[Support: #302]-
    (Inst)->[FTP: #EJCL]
    (Obj)->[Workflow_Mapping: #WM1],].
[State: [User: #John]->(Poss)->[Support: #300]-
    (Inst)->[FTP: #EJCL]
    (Obj)->[Workflow_Mapping: #WM1],].
true
>
```

This means that two users, Mary and John, currently make use of an FTP server.

Next, the matching criteria graph (2) is projected in similar fashion on the knowledge base, with *x replaced by #Mary and #John, respectively. Only for Mary, a specialization is returned. This means that she already has access to the BSCW tool, but John not yet.

5. Interpret the Matching Results. Based on the criteria of step 3, the matching results of step 4 can be interpreted in different ways. Different *courses of action* can be taken to deal with functionality mismatches. For example, if one criterion says that no users should have access to a particular type of tool, then nothing needs to be done if no results are returned in step 4, whereas otherwise one or more functionality specifications may need to be redefined.

Example Since for John no access-relation has been returned, there first must be a specification process that gives him access to the BSCW-tool. To do so, an e-mail could automatically be sent to the system administrator. After access has been granted by means of an access-definition, the now superfluous definitions that described the FTP-support for the upload-file process can be removed.

4.2   Discussion

The functionality matching meta-model was based on the semantics introduced in the RENISYS specification method, which was explained in detail in [5]. In the literature, such a meta-model plus approach for supporting virtual communities in the specification of their network information systems was lacking at the time. Extensions are needed in various directions to realize a practical methodology. For example, we now assume that semantic mismatches between the required and the enabled functionality specification have already been resolved. In reality, much middleware consists of functionality components, such as information services, that are much more complex than the heavily simplified information and communication processes described in this paper. Furthermore, for implementation purposes links to low-level technical functionality specifications need to be established.

Another required extension is to expand the functionality matching metamodel with roles. In the current approach, users (e.g. John and Mary) are directly coupled to information tools. However, roles are an important construct for functionality specification to become more efficient. Roles can be loosely defined as collection of information and communication processes that can be performed by an *actor*. An *actor role*, such as an editor, can be played by various users at the same time. In our view, this concept enhances the matching process by limiting the necessity to determine for each individual user its workflow mappings and tool assignments.

Another limitation of the current approach is that only a few dependencies between specifications have been modelled so far. For example, besides the basic assignment dependencies, there are many others conceivable. One issue concerns the relations between client and server tools: installing a BSCW server also means that users need to have a BSCW client (i.e. Web browser). This dependency has not been modelled yet.

We do not claim that from a theoretical perspective, this RENISYS-based approach is the only or even the best possible one. However, we do claim that the issues raised and elements of the functionality matching approach introduced are relevant in all matching approaches.

Once implemented and sufficiently extended, we also think that the functionality matching approach could become a true application, in the sense of [4]. Such an application should aid in the solution of actual problems, and be more than just a tool. Generic CG tools already exist and can be used to provide the basic functionality of the application. An important application area of our approach could be in developing testbeds [2], such as envisioned in the PORT project [1], in which many members of the CG-community are involved and which aims to develop a testbed methodology: "...The testbed methodology in a collaboratory research program provides a *virtual observational context* in which to study the needs of collaborators (in remote interaction with instruments, colleagues, and data) and to develop technology in response to those needs, for testing in that context. In testbeds, those collaborating must be able to monitor themselves in the process of examining how a proposed technology might augment their work."[4]. We feel that our approach, including its meta-model, could help to provide such a virtual observational context. One tool we are currently experimenting with is WebKB[5]. This tool seems to be well suited to construct such testbed applications, since it combines relatively advanced graph operations with a user-friendly, web-based interface. In this way, for example pulldown-lists can be easily generated with options for users to choose from, i.e. the list values are derived from graph operations on the knowledge base.

5   Conclusions

In this paper, we introduced a concrete functionality matching approach that aims to support virtual professional communities in order to achieve a more adequate evolution of their network information systems. The approach is based on a meta-model containing a detailed high-level, socio-technical semantics of the relations between requirements in the form of information and communication processes, users, and information tools. The approach was illustrated by a real-world case: an electronic law journal.

The functionality matching approach proposed here bridges two theoretical worlds. It is on the one hand related to work on component interface matching, which currently dominates middleware research. A major drawback of existing approaches is that they are defined at a very low level and do not contain any semantics of the evolution of the socio-technical system of virtual professional communities. On the other hand, our approach makes use of the power of conceptual graph theory, notably the availability of graph generalization hierarchies for efficient specification representation and easy calculation of graph matches by means of basic projection operations. Of course, the proposed approach is only a very simple one. The most important contribution currently is that the approach (1) makes explicit use of a functionality matching meta-model to describe high level socio-technical semantics; it recognizes that different communities (2) may apply different matching criteria, so that they can define their own, customized constraints on the evolution of their socio-technical system and (3) interpret the results in their own way by taking potentially different courses of action in case of violation of matching constraints. This tailored approach to defining the implementation of network information systems does justice to the unique and volatile nature of many virtual communities.

---

[4] Proposal for Workshop on the Semantic Web for ICCS 2001, PORT-mailing list, 24 December 2000

[5] http://www.webkb.org

References

1. M.A. Keeler, C. Kloesel and L. Searle. PORT: A Testbed Paradigm for Knowledge Processing in the Humanities, *in: Lecture Notes in Artificial Intelligence*, vol. 1257, Springer-Verlag, pp. 100-113, 1997.
2. J. Lederberg and K. Uncapher. Towards a National Collaboratory: [NSF] Report of an Invitational Workshop. *Rockefeller University, New York City, 13P15, March 1989.*
3. W.J. van den Heuvel, M.P. Papazoglou and M. Jeusfeld. "Connecting Business Objects to Legacy Systems", Proceedings of the CAiSE Conference, Springer, 1999.
4. M. Chein, D. Genest. CG Applications: Where are We 7 Years after the First ICCS? In *Proceedings of the Eighth International Conference on Conceptual Structures, ICCS2000, Darmstadt, Germany, August 14–18, 2000*, 2000, pages 127–139.
5. A. De Moor. Composition norm dynamics calculation with conceptual graphs. In *Proceedings of the Eighth International Conference on Conceptual Structures, ICCS2000, Darmstadt, Germany, August 14–18, 2000*, 2000, pages 525–539.
6. O. de Troyer, R. Meersman, and P. Verlinden. RIDL on the CRIS case: A workbench for NIAM. In T.W. Olle, A.A. Verrijn-Stuart, and L.. Bhabuta, editors, *Computerized Assistance During the Information Systems Life Cycle*, pages 375–459. Elsevier Science Publishers, B.V., 1988.
7. T.A. Finholt and G.M. Olson. From laboratories to collaboratories: A new organizational form for scientific collaboration. *Psychological Science*, 8(1):28–36, 1997.
8. B.R. Gaines, L.J. C. Lee, and M.L.G. Shaw. Modeling the human factors of scholarly communities supported through the Internet and the World Wide Web. *Journal of the American Society for Information Science*, 48(11):987–1003, 1997.
9. N.C. Goodwin. Functionality and usability. *Communications of the ACM*, 30(3):229–233, 1987.
10. E. Bertino. "Integration of Heterogenuous data repositories by using object-oriented views", ACM Transactions on Database Systems, Vol. 17(3):385-422, 1992.
11. S. Chen. Retrieval of Reusable Components in a Deductive, Object-Oriented Environment. PhD thesis, RWTH Aachen, Information Systems Institute, 1993.
12. Scott Henninger. An Evolutionary Approach to Constructing Effective Software Reuse Repositories", ACM Transactions on Software Engineering and Methodology", vol. 6, nr. 2, pp. 111–140, 1997.
13. L. Kalichenko. Workflow Reuse and Semantic Interoperability Issues, *Worklfow Management Systems and Interoperability*, A. Doğaç, L. Kalichenko, M.T. Özsu and A. Seth, NATO ASI Series, Springer, 1998.
14. G. Spanoudakis and P. Constantopoulos. Similarity for Analogical Software Reuse: A Conceptual Modelling Approach, in: Proceedings of the 5th International Conference on Advanced Information Systems Engineering, (CAiSE '93), (eds) Rolland C., Cauvet C., LNCS 685, Springer -Verlag, 1993
15. Am.M. Zaremski and J.M. Wing. Specification Matching of Software Components. ACM Transactions on Software Engineering and Methodology, Vol. 6, No. 4, pp. 333-369, Oct. 1997.

Extension of RDFS Based on the CGs Formalisms

Alexandre Delteil, Catherine Faron-Zucker, and Rose Dieng

INRIA, ACACIA project,
2004 route des Lucioles, BP93, 06902 Sophia Antipolis cedex, France
{Alexandre.Delteil,Catherine.Faron,Rose.Dieng}@sophia.inria.fr

Abstract. RDF(S) is the emerging standard for knowledge representation on the Web. In the CoMMA European IST project dedicated to ontology guided Information Retrieval in a corporate memory, the semantic annotations describing the Intranet documents are represented in RDF(S). In this context, the RDF(S) expressivity appears to be too much limited for efficient IR. In this paper, we present DRDF(S), an extension of RDF(S) to express class and property definitions and axioms as ontological knowledge, and more generally contextual knowledge on the Semantic Web. Our approach is underlain by the existing mapping between RDF(S) and CGs: DRDFS -the extension of RDF(S) we propose- is based on features of the Conceptual Graphs model. We argue that CGs meet the needs of the Semantic Web and we hope that DRDF(S) will contribute to the ongoing work of the W3C committee for improving RDF(S).

1 Introduction

The need of a Semantic Web is now well recognized and always more emphasized [1]. The huge amount of information available on the web has become overwhelming, and knowledge based reasoning now is the key to lead the Web to its full potential. In the last few years, a new generation of knowledge based search engines has arisen, among which the most famous are SHOE [9] and Ontobroker [6]. They rely on extensions of HTML to annotate Web documents with semantic metadata, thus enabling semantic content guided search. For interoperability on the Web, the importance of widely accepted standards is emphasized. The Resource Description Framework (RDF) is the emerging standard proposed by the W3C for the representation and exchange of metadata on the Semantic Web [11]. RDF Schema (RDFS) is the standard dedicated to the representation of ontological knowledge used in RDF statements [12].

In the CoMMA European IST project dedicated to ontology guided Information Retrieval (IR) in a corporate memory, RDF(S) is the knowledge representation language used to annotate the documents of an organization's intranet. These annotations are translated into conceptual graphs which are then exploited for knowledge based IR on the intranet by using the CORESE inference engine implemented in our team [2]. However the expressivity of RDFS appears too much

H. Delugach and G. Stumme (Eds.): ICCS 2001, LNAI 2120, pp. 275–289, 2001.

limited to represent the ontological knowledge of the corporate memory. Axiomatic knowledge -concept definitions, algebraic properties of relations, domain axioms- is crucial for intelligent IR on the Web and the need for inference rules is well-known since the first IR systems on the Semantic Web: they are the key to discover implicit knowledge in Web page annotations for IR to be independent of the point of view adopted when annotating [7]. The need for additional features and conventions in RDF is claimed in [10].

When compared to Object-Oriented Languages(OOL), Description Logics(DL), or Conceptual Graphs(CG), RDF(S) does not enable to declare class or property definitions, nor axioms [4] [5]. In this paper, we propose an extension of RDF(S) with class and property definitions and axioms. We call it DRDF(S) for *Defined Resource Description Framework*. DRDF(S) more generally enables to express contextual knowledge on the Web. Because the RDF philosophy consists in letting anybody free to declare anything about any resource, the knowledge of by who and in which context a special annotation has been stated is therefore crucial: DRDF(S) enables to assign a context to any cluster of annotations. The representation of class and property definitions and axioms is based on this general notion of context.

Our approach of RDF(S) is underlain by the existing mapping between RDF(S) and the CG model: DRDF(S) is based on features of the CG model that provide further representation capabilities. The existing knowledge representation languages have a crucial role to play in providing the grounds of a standard language for the Semantic Web. We argue that the CG model meets the needs of the Semantic Web and we hope that DRDF(S) will contribute to the ongoing work of the W3C committee for improving RDF(S).

Sections 2 and 3 are dedicated to the presentation of the RDF(S) model and its comparison with the CGs model. Section 4 presents the extension of RDF(S) with contexts and existential quantification. The extensions of RDF(S) with class and property definitions and axioms are then presented in sections 5 and 6. The metamodel of DRDF(S) is described in section 7. Finally, we compare DRDF(S) with other Web languages in section 8.

2   RDF(S)

2.1   RDF and RDFS

RDF is the emerging Web standard for annotating resources, such as images or documents, with semantic metadata [11]. These Web resources are identified by their URIs. In addition, anonymous resources provide a limited way of existential quantification. An RDF description consists in a set of statements, each one specifying a value of a property of a resource. A statement is thus a triple (*resource, property, value*), a value being either a resource or a literal. The RDF data model is close to semantic nets. A set of statements is viewed as a directed labeled graph: a vertex is either a resource or a literal and an arc between two vertices is labeled by a property. RDF is provided with an XML syntax. Figure 1 presents an example of RDF graph and its XML serialization. It is

the annotation of the Web page of T-Nova which is a subdivision of Deutsche
Telekom. The examples highlighting our paper are all based on the CoMMA
ontology.

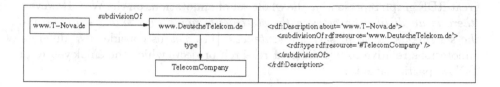

Fig. 1. An RDF annotation

RDF Schema (RDFS) is dedicated to the specification of schemas representing the ontological knowledge used in RDF statements [12]. A schema consists
in a set of class and property declarations. Multi-inheritance is allowed for both
classes and properties. A property is declared with a signature allowing several
domains and one single range: the domains of a property constraint the classes
this property can be applied to, and its range the class the value of this property belongs to. The RDFS metamodel is presented in Figure 2. It is a recursive
definition, the RDFS metamodel being defined as a set of RDFS statements by
using the two core RDFS properties: *subclassOf* and *type* which denote respectively the subsumption relation between classes and the instantiation relation
between an instance and a class.

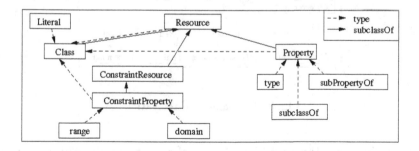

Fig. 2. The RDFS metamodel

To represent domain specific knowledge, a schema is defined by refining the
core RDFS. Domain specific classes are declared as instances of the *Class* resource, and domain specific properties as instances of the *Property* resource.
The *subclassOf* and *subPropertyOf* properties enable to define class hierarchies
and property hierarchies.

2.2 RDF Limitations

• *A Triple Model.* The RDF data model is a triple model: an RDF statement is a triple *(resource, property, value)*. When asserted, RDF triples are clustered inside annotations. An annotation can thus be viewed as a graph, subgraph of the great RDF graph representing the whole set of annotations on the Web. However, *'there is no distinction between the statements made in a single sentence and the statements made in separate sentences'* [11]. Let us consider two different annotations relative to two different research projects which the employee 46 of T-Nova participates to:

- $\{(E_{46}, worksIn, TNova), (E_{46}, project, CoMMA), (E_{46}, activity, endUser)\}$
- $\{(E_{46}, worksIn, TNova), (E_{46}, project, projectX), (E_{46}, activity, developer)\}$.

The whole RDF graph does not distinguish between these two clusters of statements. Employee 46 is both *endUser* and *developer*: the knowledge of which activity inside of a project he is implicated in is lost.

• *RDF Reification.* The RDF model is provided with a reification mechanism dedicated to higher order statements about statements. A statement (r, p, v) is reified into a resource s described by the four following properties: the *subject* property identifies the resource r, the *predicate* property identifies the original property p, the *object* property identifies the value v of p, the *type* property describes the type of s; all reified statements are instances of the *Statement* class. Figure 3 presents the following reification: *'Observer 3002 says that the rating of Newsletter 425 is seminal'.*

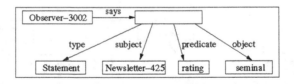

Fig. 3. An example of reification

Let us consider now the reification of a set of statements. It requires the use of a container to refer to the collection of the resources reifying these statements. This leads to quite complicate graphs (see [11], Figure 10). Moreover a statement containing an anonymous resource can not always be reified: the values of the *subject* and *object* properties must have an identifier.

• *Existential quantification.* The RDF model focuses on the description of identified resources but allows a limited form of existential quantification through the anonymous resource feature. Let us consider the RDF graph describing an anonymous resource in Figure 4. It is handled by automatically generating an ID for the anonymous resource. However, such a handling of existential knowledge through constants is a limited solution and a graph containing a cycle with more than one anonymous resource can not be represented in RDF.

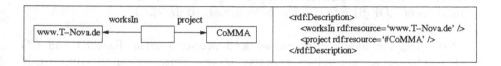

Fig. 4. An example of anonymous resource

• *Classes and properties.* An RDF Schema is made of atomic classes and properties. The RDFS model does not enable the definition of classes or properties. More generally, axioms cannot be represented in the model.

3   RDF(S) and Conceptual Graphs

3.1   Mapping of the RDF(S) and CG Models

Here we present the common features the RDF(S) and CG models share.

• *Both models distinguish between ontological knowledge and assertional knowledge.* This distinction is common to most knowledge representation languages. First, the class (resp. property) hierarchy in an RDF Schema corresponds to the concept (resp. relation) type hierarchy in a CG support; second, and more important, RDFS properties are declared as first class entities like RDFS classes, in just the same way that relation types are declared independently of concept types. This common handling of properties makes particularly relevant a mapping between RDF(S) and CG models. In particular, it can be opposed to OOL approaches, where properties are defined inside of classes.

• *In both models, assertional knowledge is positive, conjunctive and existential.*

• *Both models allow a way of reification.*

• *In both models, assertional knowledge is represented by directed labeled graphs.* An RDF graph G may be translated into a conceptual graph CG as follows:

  – Each arc labeled with a property p in G is translated into a relation node of type p in CG.
  – Each node labeled with an identified resource in G is translated into an individual concept in CG whose marker is the resource identifier. Its type corresponds to the class the identified resource is linked to by a *type* property in G.

- Each node labeled with an anonymous resource in G is translated into a generic concept in CG. Its type corresponds to the class the anonymous resource is linked to by a *type* property in G.

A mapping holds between RDFS and a large subset of the CG model and a translation of RDF(S) into CG has been presented in [2].

Regarding the handling of classes and properties, the RDF(S) and CG models differ on several points. However these differences can be quite easily handled when mapping RDF and CG models.

- *RDF binary properties versus CG n-ary relation types.* The RDF data model intrinsically only supports binary relations, whereas the CG model authorizes n-ary relations. However n-ary relations can be represented by binary properties by using an intermediate resource with additional properties giving the remaining relations (see [11], section 7.3).

- *RDF multi-instantiation versus CG mono-instantiation.* The RDF data model supports multi-instantiation whereas the CG model does not. However the declaration of a resource as instance of several classes in RDF can be translated in the CG model by generating the concept type corresponding to the most general specialization of the concept types translating these classes.

- *Property and relation type signatures.* In the RDF data model, a property may have several domains whereas in the CG model, a relation type is constrained by a single domain. However the multiple domains of an RDF property may be translated into a single domain of a CG relation type by generating the concept type corresponding to the least common generalization of the concept types translating the domains of the property.

3.2   Additional Expressivity of the CG Model

The CG model is provided with additional features insuring a greater expressivity. Regarding the existing mapping between the RDF(S) and CG models, these features will be the key to an extension of RDF(S) based on the CG model.

- *A graph model.* A conceptual graph represents a piece of knowledge separate from the other conceptual graphs of the base it belongs to. Let us consider again the two projects of T-Nova which employee 46 participates in. The statements relative to one project are clustered in one conceptual graph and then separated from the statements relative to the other projects. The two conceptual graphs are the following:

- $[Project : CoMMA] \leftarrow (project) \leftarrow [\top : E_{46}] \rightarrow (activity) \rightarrow [EndUser : *]$
- $[Project : P_X] \leftarrow (project) \leftarrow [\top : E_{46}] \rightarrow (activity) \rightarrow [Developer : *]$.

- *Reification.* A conceptual graph G is reified into a marker whose value is G. Let us consider again the following reification: *'Observer-3002 says that the rating of Newsletter-425 is seminal'.* It is represented by the following conceptual graph:

$[\top : Obs3002] \rightarrow (say) \rightarrow [Proposition : [\top : News425] \rightarrow (rating) \rightarrow [\top : seminal]]$.

In the RDF model, the reification of a set of statements requires the use of a container to refer to the collection of the resources reifying these statements. In the CG model, since the notion of graph is intrinsic to the model, the equivalent reification remains based on the initial basic mechanism.

- *Existential Quantification.* The CG model allows to represent every existential, positive and conjunctive proposition without any restriction.

- *Type definitions and axioms.* In the CG model, concept types and relation types are either atomic or defined [14][8] and graph rules allow the representation of axiomatic knowledge [13].

Starting from the correspondences between the RDF(S) and CG models and the higher expressivity of CGs, we propose an extension of RDF(S) based on CGs features to provide RDF(S) with an expressivity equivalent to the CGs one. We call it DRDFS for *Defined Resource Description Framework.*

4   Extending RDF(S) with Contexts and Existential Quantification

4.1   Extending RDF(S) with Contexts

We introduce a notion of *context* in the RDF model to express independant pieces of knowledge through a general mechanism much more elegant than containers. A context identifies a sub-graph of the whole RDF graph: RDF triples can be clustered by being stated inside of a context. This extension is based on the similarities between the RDF and CG models: a context is just the translation of a conceptual graph. Indeed, a conceptual graph implicitly defines a context and the CG model enables a 'direct' representation of contextual knowledge: quotations, viewpoint, etc. This general notion of context will be the keystone of further extensions of RDF(S), among which class and property definitions and axioms.

To extend RDFS with contexts, we introduce the following new RDF primitives:

- *Context.* A context is a resource of type *Context. Context* is a subclass of *Class.*
- *isContextOf.* A resource is linked by an *isContextOf* property to the context it belongs to.

– *referent*: An anonymous resource is linked by a *referent* property to the identified resource it refers to.

The rules for constructing RDF contexts are based on the translation of conceptual graphs into RDF:

– An individual concept $[C : r]$ of a conceptual graph G is represented by three RDF triples: $(a, type, C)$, $(a, referent, r)$ and $(G, isContextOf, a)$, where a is an anonymous resource (whose ID is automatically generated by RDF parsers)[1].
– A generic concept $[C : *]$ of a conceptual graph G is represented by two RDF triples: $(a, type, C)$ and $(G, isContextOf, a)$.
– A generic concept $[C : *x]$ of a graph G is represented by three RDF triples: $(a, type, C)$, $(a, referent, x)$ and $(G, isContextOf, a)$, where x is an instance of the class *Variable* (this class will be further described in next section).
– A relation R between two concepts $[C_1 : r_1]$ and $[C_2 : r_2]$ of a conceptual graph G is represented by an RDF property P between the two anonymous resources a_1 and a_2.
– The resource G is an instance of the *Context* class; this is represented by the triple $(G, type, Context)$.

Let us consider again the two projects of T-Nova which employee 46 participates in. As shown in Figure 5, the statements relative to one project can now be clustered in a context and then separated from the statements relative to the other projects.

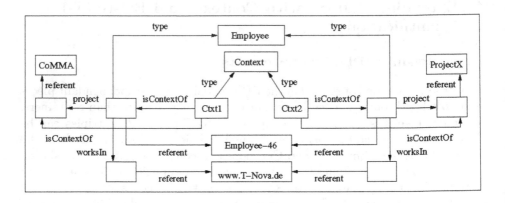

Fig. 5. Two contexts about employee 46

The rules for extracting the set S of the triples belonging to a context from the whole RDF graph are the following:

---

[1] Note that it could be sufficient to link a single anonymous resource of the context to G by the *isContextOf* property

- Select a resource G of type *Context*; $S \leftarrow \{(G, type, Context)\}$.
- Select all the anonymous resources a_i for which the value of the *isContextOf* property is G; for each i, $S \leftarrow S \cup \{(G, isContextOf, a_i)\}$.
- Select all the identified resources r_j values of a *referent* property of a resource a_i; for each i, $S \leftarrow S \cup \{(a_i, referent, r_j)\}$.
- Select all the properties p_{ik} between two resources a_i and a_k; for each ik, $S \leftarrow S \cup \{(a_i, p_{ik}, a_k)\}$.

A context is defined from a resource G of type *Context* as the largest subgraph of the whole RDF graph whose all internal nodes excepted G are anonymous resources a_i. A context is thus an abstraction that enables to talk about representations of resources (through anonymous resources) rather than directly about resources. For instance, in Figure 5, the resource E_{46} representing the employee 46 is referred to by two distinct anonymous resources in two different contexts. Anonymous resources are *externally identified* by the *referent* property.

4.2   Extending RDF(S) with Existential Quantification

The RDF model allows a limited form of existential quantification through the anonymous resource feature. The introduction of the *referent* property provides the RDF model with a general mechanism for existential quantification handling. To extend RDFS with existential quantification, we introduce the following new RDF primitives:

- *Variable*: A variable is a resource of type *Variable. Variable* is a subclass of *Class*.
- *parameter*: A variable is linked by a *parameter* property to the context it belongs to.

An existential quantification is represented by an anonymous resource described by a referent property whose value is an instance of *Variable*. The scope of a variable is the context it belongs to, just like in first-order logic, where the scope of a variable is the formula it belongs to.

In an RDF graph, an anonymous resource can be duplicated into several anonymous resources coreferencing a same variable; the new graph remains semantically equivalent to the initial graph. This enables the XML serialisation of RDF graphs embedding a cycle with anonymous resources. Figure 6 presents an RDF graph that could not be represented in the XML syntax. The cycle is resolved by introducing a second anonymous resource and two *referent* properties sharing the same value x.

The extension of RDF(S) with class and property definitions and axioms will rely on the notion of existentially quantified context.

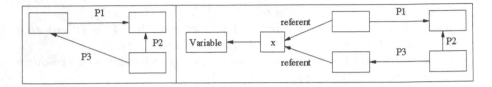

Fig. 6. An example of existential quantification

5   Extending RDF(S) with Class and Property Definitions

5.1   Type Definitions in the CG Model

A concept type definition is a monadic abstraction, i.e. a conceptual graph whose one generic concept is considered as formal parameter. It is noted $t_c(x) \Leftrightarrow D(x)$. The formal parameter concept node of $D(x)$ is called the head of t_c, its type the genus of t_c, and $D(x)$ the differentia of t_c from its genus [15] [8]. For instance, let us consider the following definition of the *WebPage* concept type:

$$WebPage(x) \Leftrightarrow [Document : x] \rightarrow (hasForReprSystem) \rightarrow [System : Html]$$

This defines a web page as a document having HTML for representation system. A defined concept type t_c is a subtype of its genus: *WebPage* is a subtype of *Document*. The classification of defined concept types in the concept types hierarchy depends on the generalization relations between their differentiae.

A relation type definition is a n-ary abstraction, i.e. a conceptual graph with n generic concepts considered as formal parameters. It is noted $t_r(x_1, ..., x_n) \Leftrightarrow D(x_1, ..., x_n)$. For instance, let us consider the following definition of the *colleague* relation type:

$$colleague(x, y) \Leftrightarrow [Person : x] \rightarrow (workIn) \rightarrow [Institute] \leftarrow (workIn) \leftarrow [Person : y]$$

This defines a relation of type *colleague* as holding between two persons working in the same institute. Since the definition of relation types is not based on the Aristotelian principle of genus and differentia, the classification of defined relation types takes into account the matching of formal parameters when projecting a graph definition against another.

5.2   Class and Property Definitions in DRDF(S)

DRDF(S) class and property definition is descended from type definition in the CG model. A class definition is a monadic abstraction, i.e. a context whose one resource of type *Variable* is considered as formal parameter. A property definition is a diadic abstraction, i.e. a context whose two resources of type *Variable* are considered as formal parameters. To extend RDFS with class and property definitions, we introduce the following new RDF primitives:

- *DefinedClass*: A defined class is of type *DefinedClass*. *DefinedClass* is a subclass of *Class*.
- *DefinedProperty*: A defined property is of type *DefinedProperty*. *DefinedProperty* is a subclass of *rdf:Property*.
- *hasDefinition*: A defined class is linked by a *hasDefinition* property to its definitional context.
- *formalParameter*: The variable linked to the definitional context by a *formalParameter* property corresponds to the formal parameter of a monadic lambda abstraction.
- *firstFormalParameter* and *secondFormalParameter*: The variables linked to the definitional context by these properties correspond to the formal parameters of a diadic lambda abstraction.

The *WebPage* class and the *colleague* property are defined in Figure 7.

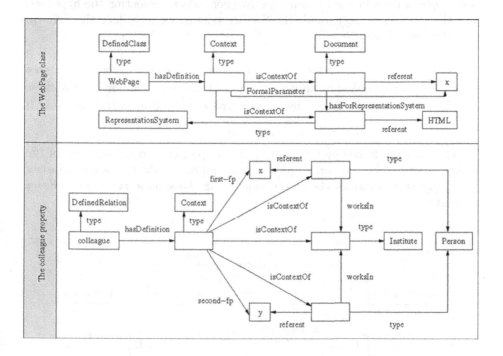

Fig. 7. Definition of the *WebPage* class and the *colleague* property

6 Extending RDF(S) with Axioms

6.1 Graph Rules in the CG Model

Graph rules are an extension of the core CG model introduced in [13]. A rule $R : G_1 \Rightarrow G_2$ is a couple of lambda-abstractions $((\lambda x_1, ... x_n G_1), (\lambda x_1, ... x_n G_2))$, where G_1 and G_2 are two conceptual graphs called hypothesis and conclusion, and $x_1, ... x_n$ are 'connection point's corresponding to n co-reference links between concepts of G_1 and G_2. The following graph rule represents the symmetry of the relation of type *colleague*:

$$G_1 : [Person : x] \rightarrow (colleague) \rightarrow [Person : y]$$
$$\Rightarrow$$
$$G_2 : [Person : y] \rightarrow (colleague) \rightarrow [Person : x]$$

6.2   Axioms in DRDF(S)

DRDF(S) axioms are descended from graph rules of the CG model. An axiom is a couple of lambda abstractions, i.e. two contexts representing the hypothesis and the conclusion. To extend RDFS with axioms, we introduce the following new RDF primitives:

- *Axiom*: An axiom is a resource of type *Axiom*. *Axiom* is a subclass of *Context*.
- *if*: An axiom is linked by an *if* property to the context defining its hypothesis.
- *then*: An axiom is linked by a *then* property to the context defining its conclusion.

The variables linked by a *formalParameter* property to the resource of type *Axiom* correspond to the formal parameters common to the two lambda abstractions. Figure 8 describes the axiom expressing the symmetry of the *colleague* property.

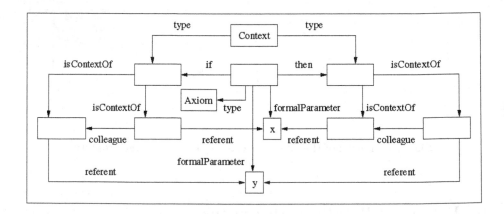

Fig. 8. An axiom representing the symmetry of the *colleague* property

7 The Defined Resource Description Schema

DRDF(S) is made of the set of all the RDF(S) primitives augmented with the ones we have introduced in the previous sections. DRDF(S) is thus a refinement of the core RDFS and remain totally compliant with the RDF triple model. The DRDF(S) metamodel is presented in Figure 9.

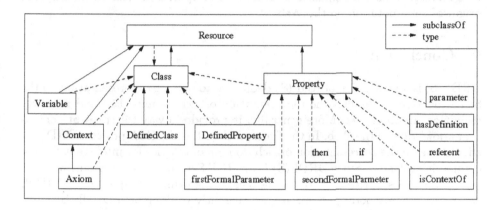

Fig. 9. The DRDFS metamodel

The semantics of DRDFS relies on its translation into the CG formalism. Conceptual graphs are themselves translated into first order logic formulae thanks to the Φ operator defined in [14].

8 Related Work

Several languages for ontology representation and exchange are existing[3], among which RDF(S), OIL [5] and DAML [4] are dedicated to the Semantic Web. Like DRDF(S), OIL and DAML are tentatives of improvement of RDF(S); they are defined as an RDF Schema.

OIL enables to define classes and restrict property ranges and domains through boolean combinations of classes. In particular, it enables negation in class definitions, which is not provided in DRDFS. OIL is based on a DL. When compared to it, what DRDFS provides with its CG's expressivity is the possibility to express any positive, conjunctive and existential graph in a definition. The absence of variables in DLs does not enable to express RDF graphs embedding cycles; the class definitions in OIL are then limited to 'serializable' graphs. Contrary to OIL, DRDFS stays in the spirit of RDF(S), namely the representation of positive, conjunctive and existential knowledge. In our opinion, this better meets the needs of the Semantic Web.

DAML provides primitives to express relations between classes (disjonction, intersection, union, complementarity, ...) and enrich properties (minimal and maximal cardinality, transitivity, inverse, ...). DAML is provided with OOL features. It provides no mechanism for class or property definitions. It is therefore orthogonal to both OIL and DRDFS. As the merge of DAML and OIL led to DAML+OIL, it should be interesting to integrate the DAML features into DRDF(S).

In addition, DRDFS addresses the problem of the representation of contextual knowledge on the semantic web. This is of special interest to identify the origin of an annotation on the Web.

9   Conclusion

DRDF(S) is an extension of RDF(S) dedicated to ontology representation on the Semantic Web. It enables the representation of axioms and class and property definitions in ontologies. More generally, it provides a way to represent contextual knowledge on the Web. In the framework of the CoMMA project, DRDF(S) should enable the representation of rich domain ontologies for intelligent IR in a companie's intranet. Since DRDF(S) is an RDF Schema, it is compliant with existing RDF parser. However, the semantics of the primitives specific to DRDFS can not be understood by them. We are currently working on a DRDF(S) interpreter for the existing platform CORESE.

The grounds of DRDF(S) rely on the existing mapping between RDF(S) and CGs; it is an extension of RDF(S) guided by the CG features. Regarding the similarities the RDF(S) and CG models share, it is a real challenge for the CG community to contribute to the elaboration of a standard language for knowledge representation, interoperability and reasoning on the Semantic Web. We hope that DRDF(S) will contribute to the ongoing work of the W3C committee for improving RDFS.

References

1. Berners Lee, T.: Weaving the Web, Harper San Francisco, 1999.
2. Corby, O., Dieng, R., Hebert, C.: A Conceptual Graph Model for W3C Resource Description Framework. In proc. of ICCS'00, Darmstadt, Germany, LNAI 1867, Springer-Verlag, 2000.
3. Corcho, O., Gomez-Perez, A.: A Roadmap to Ontology Specification Languages. In proc. of EKAW 2000, France, LNAI 1937, p. 80-96, Springer-Verlag, 2000.
4. DAML: Darpa Agent Markup Language. http://www.daml.org.
5. Fensel, D., Horrocks, I., Harmelen, F., Decker, S., Erdmann, M., Klein, M.: Oil in a Nutshell. In proc. of EKAW 2000, France, LNAI 1937, Springer-Verlag, 2000.
6. Fensel, D., Decker, S., Erdmann, M., Studer, R.: Ontobroker: Or How to Enable Intelligent Access to the WWW. In proc. of KAW 1998, Banff, Canada, 1998.
7. Heflin, J., Hendler, J., Luke, S.: Reading between the Lines: Using SHOE to Discover Implicit Knowledge from the Web. In proc. of the AAAI Workshop on AI and Information Integration. WS-98-14. AAAI Press, p. 51-57, 1998.
8. Leclere, M. Reasoning with type definitions. In proc. of ICCS'97, Seattle, WA, LNAI 1257, Springer-Verlag, 1997.

9. Luke, S., Spector, L., Rager, D., Hendler, J.: Ontology based Web Agents, In proc. of the 1st International Conference on Autonomous Agent, 1997.
10. Martin, P., Eklund, P.: Conventions for Knowledge Representation via RDF, In proc. of WebNet 2000, San Antonio, Texas, ACCE press, p. 378-383, 2000.
11. RDF: http://www.w3.org/TR/REC-rdf-syntax/, 1999.
12. RDFS: http://www.w3.org/TR/2000/CR-rdf-schema-20000327/, 2000.
13. Salvat, E., Mugnier, M.L.: Sound and Complete Forward and Backward Chainings of Graph Rules. In proc. of ICCS'96, LNCS 1115, Springer-Verlag, p. 248-262, 1996.
14. Sowa, J.F.: Conceptual Graphs, Conceptual Structures: Information Processing in Mind and Machine, Addison-Wesley, Reading, MA, 1984.
15. Sowa, J.F.: Conceptual Graphs: DpANS. In proc. of ICCS'99, Blacksburg, VA, USA, LNAI 1640, p.1-65, Springer-Verlag, 1999.

Building Concept (Galois) Lattices from Parts: Generalizing the Incremental Methods

Petko Valtchev[1,2] and Rokia Missaoui[1]

[1] Département d'Informatique, UQAM, C.P. 8888, succ. "Centre Ville",
Montréal, Québec, Canada, H3C 3P8
[2] INRIA, France

Abstract. Formal concept analysis is increasingly used as a data mining technique, whence the need of efficient algorithms for handling large sets of volatile data. Recently, we designed a general framework for constructing concept (Galois) lattices from fragmented and/or evolving data based on a lattice assembly operation. In this paper, the framework is adapted to the maintenance of concept lattices upon the insertion of a set of objects into the context, a problem which generalizes the insertion of individual objects considered by the existing incremental methods. The paper provides a set of structural results for the case of single object insertions which underlie a new incremental algorithm. Our method is shown to improve a key flaw of the major incremental technique.

1 Introduction

Formal concept analysis (FCA) [14] and Galois lattices are increasingly used in the resolution of practical problems from software engineering [6] and data mining [7,11].

Recently, we have been investigating the design of flexible algorithmic tools to meet the increasing need for analysis of large and volatile datasets, typically arising within the data mining process. The present paper focuses on the incremental update of an existing lattice upon the extension of the target dataset by a (set of) previously unseen individual(s). Incrementation by a single new individual has been studied by Godin *et al.* [8][1]. Our own approach relies on a complete framework for lattice assembly from parts (see [13]) which is based on operators on data tables, *apposition* and *supposition* [5]. First, the theoretical results are provided which allow the set-wise increment to be dealt with as an assembly of two lattices. In the case of single object incrementation, we adapt parts of the assembly method to the procedure of Godin *et al.*, to design a new incremental algorithm whose worst-case complexity compares to that of the best batch methods [10].

The paper starts with a recall of the FCA basics (Section 2), followed by a short presentation of the reference incremental algorithm (Section 3). It then shifts to context assembly and nested line diagrams (Sections 4) before describing

---

[1] A sketch of an incremental algorithm could be found in [4] as well.

H. Delugach and G. Stumme (Eds.): ICCS 2001, LNAI 2120, pp. 290–303, 2001.

a procedure for assembling partial lattices (Section 5). Finally, the new incremental algorithm is presented and its worst-case complexity is compared to that of the reference procedure (Section 6).

2 Formal Concept Analysis Basics

Formal concept analysis [5] studies the partially ordered structure[2], known under the names of *Galois lattice* [1] or *concept lattice* [14], which is induced by a binary relation over a pair of sets O (*objects*) and A (*attributes*).

Definition 1. *A formal context is a triple* $K = (O, A, I)$ *where* O *and* A *are sets and* I *is a binary (incidence) relation, i.e.,* $I \subseteq O \times A$.

Within a context (see Figure 1 on the left), objects are denoted by numbers and attribute by small letters. Two functions, f and g, summarize the context-related links between objects and attributes.

Definition 2. *The function* f *maps a set of objects into the set of common attributes, whereas* g *is the dual for attribute sets:*

- $f : \mathcal{P}(O) \to \mathcal{P}(A),\ f(X) = \{a \in A | \forall o \in X, oIa\}$
- $g : \mathcal{P}(A) \to \mathcal{P}(O),\ g(Y) = \{o \in O | \forall a \in Y, oIa\}$

For example, w.r.t. the context in Figure 1, $f(134) = fgh$ and $g(abc) = 127$[3]. Furthermore, the compound operators $g \circ f(X)$ and $f \circ g(Y)$ are closure operators over $\mathcal{P}(O)$ and $\mathcal{P}(A)$ respectively. Thus, each of them induces a family of *closed* subsets with f and g[4] as bijective mappings between both families. A couple (X, Y), of mutually corresponding closed subsets is called a *(formal) concept*.

Definition 3. *A formal concept is a couple* (X, Y) *where* $X \in \mathcal{P}(O)$, $Y \in \mathcal{P}(A)$, $X = Y'$ *and* $Y = X'$. X *is called the extent and* Y *the intent of the concept.*

For example, $(134, fgh)$ is a concept, but $(16, efh)$ is not. Furthermore, the set C_K of all concepts of the context $K = (O, A, I)$ is partially ordered by intent/extent inclusion:

$$(X_1, Y_1) \leq_K (X_2, Y_2) \Leftrightarrow X_1 \subseteq X_2 (Y_2 \subseteq Y_1).$$

Theorem 1. *The partial order* $\mathcal{L} = \langle C_K, \leq_K \rangle$ *is a complete lattice with joins and meets computed as follows:*

- $\bigvee_{i=1}^{k}(X_i, Y_i) = ((\bigcup_{i=1}^{k} X_i)'', \bigcap_{i=1}^{k} Y_i)$,
- $\bigwedge_{i=1}^{k}(X_i, Y_i) = (\bigcap_{i=1}^{k} X_i, (\bigcup_{i=1}^{k} Y_i)'')$.

For example, the join and the meet of the concepts $c_1 = (123, cf)$ and $c_2 = (1246, ef)$ (see Figure 1 on the right) are $(12346, f)$ and $(12, abcef)$ respectively.

---

[2] An excellent introduction to partial orders and lattices may be found in [3].
[3] Standard FCA notations use a separator-free form for sets, e.g., 127 stands for $\{1, 2, 7\}$, and ab for $\{a, b\}$.
[4] Hereafter, both f and g are denoted by $'$ and compound operators by $''$.

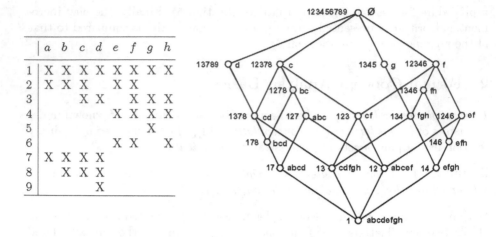

	a	b	c	d	e	f	g	h
1	X	X	X	X	X	X	X	X
2	X	X	X			X	X	
3			X	X		X	X	X
4					X	X	X	X
5						X		
6			X	X			X	
7	X	X	X	X				
8		X	X	X				
9						X		

Fig. 1. Left: Context K (adapted from [5]) with $O = \{1, 2, ..., 9\}$ and $A = \{a, b, ..., h\}$. **Right:** The Hasse diagram of the concept (Galois) lattice derived from K.

3   Building the Lattice Incrementally

A variety of efficient algorithms exist for constructing the concept lattice of a binary table [2,8,4,10][5], most of them being designed to work on small tables that can easily be stored and managed in main memory. However, current databases and data warehouses are known to be volatile and very large in size and therefore require incremental and highly scalable analysis algorithms.

In what follows, the first algorithm constructing Galois lattices incrementally [8] is described and a possible generalization is suggested.

3.1   Principles of the Incremental Approach

Incremental methods construct the lattice $\mathcal{L}$ starting from a single object o_1 and gradually incorporating any new object o_i (on its arrival) into the lattice $\mathcal{L}_{i-1}$ (over a context $\mathcal{K} = (\{o_1, ..., o_{i-1}\}, A, I)$), each time carrying out a set of structural updates [12].

The basic approach in [8] follows a fundamental property of the *Galois connection* established by f and g on $(\mathcal{P}(O), \mathcal{P}(A))$: both families of closed subsets are themselves closed under set intersection [1]. Thus, the whole insertion process is aimed at the integration into $\mathcal{L}_{i-1}$ of all concepts (called *new*) whose intents correspond to intersections of $\{o_i\}'$ with intents of existing concepts, which are not themselves the intent of an existing concept. Hence, three categories of concepts in $\mathcal{L}_{i-1}$ are distinguished: *generator* concepts ($\mathbf{G}(o)$) give rise to new concepts; *modified* concepts ($\mathbf{M}(o)$) evolve by integrating o_i into their extents; *old* concepts ($\mathbf{U}(o)$) remain completely unchanged. The delimitation of

---

[5] See [8,9] for comparative studies.

the three sets together with the creation of the new concepts and their subsequent integration in the existing lattice structure constitutes the main part of the algorithm's task.

3.2   Description of the Algorithm

The original incremental algorithm (see Algorithm 1[6]) could be split into two parts. The first part (lines $5-13$) is in charge of any specific treatment required by the presence of some attributes in the description of o_i which are not in $A_i = \bigcup_{j<i}\{o_j\}'$. The second part (lines $14-29$) consists of a top-down traversal of the lattice $\mathcal{L}_{i-1}$ with a recognition of the current concept's category (lines 20–26). As for the updates, modified concepts merely receive o_i in their extent, whereas the detection of a generator c leads to the creation of a new concept $\bar{c}$ with an intent $Intent(\bar{c}) = Intent(c) \cap \{o_i\}'$ and its extent $Extent(\bar{c}) = Extent(c) \cup \{o_i\}$. The new concept is then integrated into the (partially) completed lattice $\mathcal{L}_i$, a task which requires the detection of all its upper covers, as described by Algorithm 2. This procedure looks through a superset of the actual covers and picks those nodes up: the candidate set is made up of all concepts whose intent is strictly smaller in size than $Int(\bar{c})$. A candidate qualifies only if it is a super-concept of $\bar{c}$, but none of its lower covers (Cov^l) is a super-concept of $\bar{c}$ on its own. An upper cover is linked to $\bar{c}$ and the obsolete links, i.e., those between the generator and an upper cover of $\bar{c}$, are removed. Finally, $\bar{c}$ is linked to its generator (line 27).

Example 1 (Insertion of object 3). Assume $\mathcal{L}$ is the lattice induced by the object set 12456789 (see Figure 3 on the left) and consider 3 as the new object. The three categories are $\mathbf{U}(o) = \{c_{\#7}, c_{\#9}\}$, $\mathbf{M}(o) = \{c_{\#1}, c_{\#2}, c_{\#4}\}$, $\mathbf{G}(o) = \{c_{\#3}, c_{\#5}, c_{\#6}, c_{\#8}, c_{\#10}, c_{\#11}, c_{\#12}\}$. The set of new concepts (identified by their extents) is: $\{13, 123, 134, 1378, 1346, 12378, 12346\}$. For example, the computation of the upper covers of the concept $(134, fgh)$ considers the following set of candidate concepts (identified by their intents): $\{\emptyset, f, g, c, d, cd, bc, fh, ef\}^{[7]}$. The actual upper covers are $\{g, fh\}$, a result that requires 15 to 20 set operations. The result of the whole operation is the lattice $\mathcal{L}$ on the right of Figure 1.

3.3   Requirement Evolution

The one-increment, i.e., adding a single object at a time, as described above, is only a partial solution to the problem of volatile data. As a matter of fact, in most databases and data warehouses, the updates are not object-wise, but rather group-wise, meaning that a whole subset of objects $\delta O = \{o_{i+1}, ..., o_{i+l}\}$ are to be added at a time. Instead of inserting them one by one into $\mathcal{L}_i$, one may think of first extracting the lattice $\delta \mathcal{L}$ corresponding to δO and then construct $\mathcal{L}_{i+l}$ from $\mathcal{L}_i$ and $\delta \mathcal{L}$. The challenging problem is then the *merge* or *assembly* of both lattices, a task which generalizes the conventional, single-object incrementation.

---

[6] The original version is preserved except for early termination tests.

[7] *cf* may as well qualify depending on the order in *Classes[4]*.

```
 1: procedure ADD-OBJECT(In: L a lattice, o an object; Out: L⁺ a lattice)
 2:
 3: Local   : Classes, Classes-New : array [0..‖A ∪ {o}'‖] of concept sets
 4:
 5: if L = ∅ then
 6:    L⁺ ← {NEW-CONCEPT({o}, {o}')}
 7: else
 8:    if {o}' ⊄ Intent(⊥(L)) then
 9:       if Extent(⊥(L)) = ∅ then
10:          Intent(⊥(L)) ← Intent(⊥(L)) ∪ {o}'
11:       else
12:          c ← NEW-CONCEPT(∅, Intent(⊥(L)) ∪ {o}') ; NEW-LINK(c, ⊥(L))
13:          ADD(L,c) ; ⊥(L) ← c
14:    for i from 0 to ‖A ∪ {o}'‖ do
15:       Classes[i] ← ∅ ; Classes-New[i] ← ∅
16:    for all c̄ in L do
17:       ADD(Classes[‖ Intent(c̄) ‖], c̄)
18:    for i from 0 to ‖A ∪ {o}'‖ do
19:       for all c̄ in Classes[i] do
20:          if Intent(c̄) ⊆ {o}' then
21:             ADD(Extent(c̄),o)      {(c̄) is modified}
22:          else
23:             Int ← Intent(c̄) ∩ {o}'    {(c̄) is old}
24:             if not (Int', Int) ∈ Classes-New[‖Int‖] then
25:                c ← NEW-CONCEPT(Extent(c̄) ∪ {o}, Int)    {(c̄) is generator}
26:                ADD(Classes-New[‖Int‖], c)
27:                FIND-UPPER-COVERS(c,c̄) ; NEW-LINK(c̄, c)
28:             ADD(Classes-New[i],c̄)
29:    L⁺ ← ⋃_{i=0}^{‖A∪{o}'‖} Classes-New[i]
```

Algorithm 1: Insertion of a new object into a concept (Galois) lattice.

```
 1: procedure FIND-UPPER-COVERS(In: new, gen concepts)
 2:
 3: Int ← Intent(new)
 4: for j from 0 to ‖Int‖ − 1 do
 5:    for all c̃ in Classes-New[j] do
 6:       if Intent(c̃) ⊆ Int then
 7:          Cover ← true
 8:          for all ĉ in Covˡ(c̃) do
 9:             if Intent(ĉ) ⊆ Int then
10:                Cover ← false
11:                exit-for
12:          if Cover then
13:             if gen ∈ Covˡ(c̃) then
14:                DROP-LINK(gen, c̃)
15:             NEW-LINK(new, c̃)
```

Algorithm 2: Insertion of a new object into a concept (Galois) lattice.

In a recent paper [13], we have proposed an approach for the resolution of the dual problem, i.e., assembly of lattices over contexts sharing the same objects but with different attributes. The assembly procedure relies on a theoretical framework whose keystone is the context apposition operation initially intended to support lattice visualization.

4 Visualization Framework

The research in FCA has yield an original method for visualizing complex concept lattices (see [5]). The method relies on two dual operators for assembling contexts called *subposition* and *apposition* respectively. To facilitate the understanding of complex structures, the lattice corresponding to a context is drawn as a nested structure reflecting the direct product of two or more smaller lattices built on top of some parts of the initial context.

4.1 Subposition of Contexts and Partial Lattices

Subposition is an assembly of contexts sharing the same attributes [5].

Definition 4. *Let $\mathcal{K}_1 = (O_1, A, I_1)$ and $\mathcal{K}_2 = (O_2, A, I_2)$ be two contexts with the same set of attributes A, then the context $\mathcal{K} = (O_1 \dot\cup O_2, A, I_1 \dot\cup I_2)$ is called the subposition of $\mathcal{K}_1$ and $\mathcal{K}_2$: $\mathcal{K} = \frac{\mathcal{K}_1}{\mathcal{K}_2}$.*

For example, with a *global* context $\mathcal{K} = (O, A, I)$ as given in Table 1, where $O = \{1, 2, 3, 4, 5, 6, 7, 8, 9\}$ and $A = \{a, b, c, d, e, f, g, h\}$, let $O_1 = \{1, 2, 3, 4\}$ and $O_2 = \{5, 6, 7, 8, 9\}$. The two lattices corresponding to $\mathcal{K}_1$ and $\mathcal{K}_2$, which will further be referred to as *partial* lattices[8] $\mathcal{L}_1$ and $\mathcal{L}_2$, are given in Figure 2. We consider the direct product of $\mathcal{L}_1$ and $\mathcal{L}_2$, $\mathcal{L}_\times = \mathcal{L}_1 \times \mathcal{L}_2$, which is itself a lattice. The nodes of $\mathcal{L}_\times$ are pairs of concepts (c_1, c_2), where c_i appears in $\mathcal{L}_i$ for $i = 1, 2$, and its order relation $\leq_\times$ is the product of both lattice orders. The concrete concepts will be further identified by an index in the respective lattice, i.e., a unique number ranging between 1 and $\|\mathcal{C}_{\mathcal{K}_i}\|$, denoted $\#i$. Thus, $(c_{\#7}, c_{\#3})$ denotes the product of $(14, efgh)$ from $\mathcal{L}_1$, and $(6, efh)$ from $\mathcal{L}_2$.

4.2 Nested Line Diagrams

Nested line diagrams (NLD) [5] are a visualization means which allows a lattice $\mathcal{L}$ to be drawn as a sub-structure of the *direct* product $\mathcal{L}_\times$ of a set of partial lattices $\mathcal{L}_i$. A NLD basically represents the product lattice $\mathcal{L}_\times$ by combining the respective line diagrams of the lattices $\mathcal{L}_i$ into a unique complex structure. However, neither the nodes of $\mathcal{L}_\times$ nor their precedence links are directly represented. Instead, the information about them is spread over the various levels of nesting. Figure 2 on the right presents the NLD of the product lattice $\mathcal{L}_\times = \mathcal{L}_1 \times \mathcal{L}_2$. As it can be seen, the line diagram of the lattice $\mathcal{L}_1$ is used as an outer frame

---

[8] As $\mathcal{K}_1$ and $\mathcal{K}_2$ are partial contexts for $\mathcal{K}$.

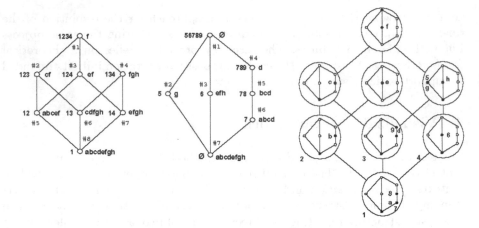

Fig. 2. Left: Partial lattices $\mathcal{L}_1$ and $\mathcal{L}_2$ built from a horizontal decomposition of the context in Figure 1. **Right**: The nested line diagram of the lattice $\mathcal{L}_1 \times \mathcal{L}_2$.

in which the diagram of $\mathcal{L}_2$ is embedded. The lattice $\mathcal{L}$ is represented by an isomorphic sub-structure of $\mathcal{L}_\times$ which is exactly the image of $\mathcal{L}$ by φ (see next section). Within a NLD, the product nodes belonging to the sub-structure, further referred to as *full* nodes, are marked in black, whereas the remaining nodes (called *void*) are left unfilled. Moreover, a node (c_i, c_j) of the product is located by first finding the node for c_i in the outer diagram and then, finding the respective node c_j within the local $\mathcal{L}_2$ diagram. For example, the node $(c_{\#1}, c_{\#3})$ (see the numbering in Figure 2) of the product lattice is located within the NLD on the right side of Figure 2 at the node $c_{\#3}$ of the inner lattice inside the node $c_{\#1}$ of the outer lattice and corresponds to the concept $(12346, f)$.

4.3   Linking the Global Lattice to the Partial Ones

Any concept of $\mathcal{L}$ can be "projected" upon the lattice $\mathcal{L}_\times$ by subsequently restricting its extent to the set of "visible" objects in O_1 and O_2. The resulting mapping constitutes an order homomorphism between $\mathcal{L}$ and the direct product.

Definition 5. *The function $\varphi : C \to C_\times$, maps a concept from the global lattice into a pair of concepts of the partial lattices by splitting its extent over the partial context object sets O_1 and O_2:*

$$\varphi((X,Y)) = ((X \cap O_1, (X \cap O_1)') , (X \cap O_2, (X \cap O_2)')).$$

For example, φ maps the concept $(1278, bc)$ (see Figure 1) into $((12, abcef), (78, bcd))$ from $\mathcal{L}_1$ and $\mathcal{L}_2$ respectively (see Figure 2). The mapping constitutes an *order embedding* of $\mathcal{L}$ into $\mathcal{L}_\times$, $\varphi : \mathcal{L} \hookrightarrow \mathcal{L}_\times$, which preserves lattice meets.

The homomorphism ψ maps $\mathcal{L}_1 \times \mathcal{L}_2$ into $\mathcal{L}$.

Definition 6. *The function* $\psi : C_\times \rightarrow C$ *maps a pair of concepts over partial contexts into a global concept by intersecting their respective intents:*

$$\psi(((X_1, Y_1) , (X_2, Y_2))) = ((Y_1 \cap Y_2)' , Y_1 \cap Y_2).$$

For example, the image of the pair $(c_{\#7}, c_{\#3})$ (see Figure 2) by ψ is the concept $(146, efh)$. The mapping ψ is an *order-preserving* homomorphism whereby the equivalence classes of its **kernel** relation are convex subsets of $C_\times$.

The above properties have a direct impact on visualization: the fact that $\mathcal{L}$ is isomorphic to its image by φ on $\mathcal{L}_\times$ allows the global lattice to be represented by a proper sub-lattice of the direct product (the one induced by the full nodes).

5   Assembling Partial Lattices

In [13] we described a lattice assembly procedure based on structural results about apposition of contexts. In this section, an appropriate procedure is provided for the dual subposition operation.

5.1   Characterizing the Global Lattice

Three problems need to be tackled for concept lattice building from parts: identification of concepts (filtering full nodes of the product lattice), computation of concept intent/extent, and detection of the precedence relations between concepts (upper and/or lower covers).

The relevant structural results could be summarized as follows. First, given a global concept c, its image by φ in the product lattice is the minimum of all its antecedents by ψ.

Proposition 1. $\forall c \in C, \varphi(c) = \max(\psi^{-1}(c))$.

The set of antecedents are exactly the nodes of the product for which the intersection of the respective intents is equal to the intent of c.

Proposition 2. $\forall c \in C, \psi^{-1}(c) = \{(c_i, c_j) | Int(c_i) \cap Int(c_j) = Int(c)\}$.

To ease the effective detection of $\psi^{-1}(c)$ members, we define the function Q with the obvious property that $[n]_Q = \psi^{-1}(\psi(n))$ where n is a node in $\mathcal{L}_\times$:

Definition 7. *The function* $Q : C_\times \rightarrow 2^A$ *computes intent intersections for concept pairs:* $Q(\bar{c}, \underline{c}) = Int(\bar{c}) \cap Int(\underline{c})$.

Intents/extents of global concepts can be computed in a direct way, i.e., without looking at the context, from the characteristics of their images by φ:

Proposition 3. *For any concept* $c = (X, Y)$ *with* $\varphi(c) = ((X_1, Y_1), (X_2, Y_2))$, $X = X_1 \cup X_2$ *and* $Y = Y_1 \cap Y_2$.

Finally, the immediate successors, or upper covers, of c in $\mathcal{L}$, $Cov^u(c)$, could be determined by only examining the upper covers of its image $\varphi(c)$ in the product lattice. Actually, the set $Cov^u(c)$ is made up of the minima among the images by ψ of the upper covers of $\varphi(c)$.

Proposition 4. $Cov^u(c) = min(\{\psi(\bar{c})|\bar{c} \in Cov^u(\varphi(c))\})$.

The above stated properties underly an algorithmic procedure that builds the global lattice by only looking at the structure of the partial ones.

5.2   Description of the Algorithm

The algorithm represents a top-down traversal of the product lattice. Each node (c_i, c_j) is tested for being void or full. For that purpose, the images by ψ of the upper covers of (c_i, c_j) in the product (*PsiImages*) are compared to the intent of the image of (c_i, c_j), i.e., the intersection of both intents. A full node generates an intersection which is not the intent of an already generated global concept. In this case only, a new global concept c is created whose intent and extent are computed according to Proposition 3. Finally, the upper covers of c are computed as the minima of the previously examined set of global concepts (*PsiImages*). The traversal follows a *linear extension* of the product lattice order

```
 1:  procedure ASSEMBLY(In: L₁, L₂ partial lattices; Out: L the global lattice)
 2:
 3:    L ← ∅
 4:    for all cᵢ, cⱼ in C₁ × C₂ do
 5:        I ← Intent(cᵢ) ∩ Intent(cⱼ); Psilmages ← ψ(UPPERCOVERS(cᵢ, cⱼ))
 6:        if not FIND-PSI(I, Psilmages) then
 7:            c ← NEW-CONCEPT(Extent(cᵢ) ∪ Extent(cⱼ), I)
 8:            for all c̄ in MIN(Psilmages) do
 9:                NEW-LINK(c, c̄)
10:            L ← L ∪ {c}
```

Algorithm 3: Assembling the global Galois lattice from a pair of partial ones.

which is achieved through a preliminary sorting of both sets C_1 and C_2. Most of the primitives admit very efficient implementations which take advantage of the lattice properties (see [13] for implementation details).

In what follows, a special case of the assembly framework is used in the design of a more powerful incremental algorithm.

6   Improved Incremental Algorithm

Although suitable, Algorithm 3 may prove inefficient in the case of single-object incrementation because the exhaustive examination of the nodes in $\mathcal{L} \times \mathcal{L}(o)$[9] may lead to many void nodes. In contrast, Algorithm 1 examines only actual concepts of the target lattice, which warns at using it as a "short-cut" for lattice traversal in our own incremental method.

---
[9] $\mathcal{L}(o)$ is the lattice built for the object set $\{o\}$.

6.1 Mapping Concept Categories into Partial Lattices

A preliminary step is the mapping of *modified* ($\mathbf{M}(o)$), ordinary *old* ($\mathbf{U}(o)$) and *generator* ($\mathbf{G}(o)$) concepts of $\mathcal{L}$ into *void* and *full* nodes in $\mathcal{L} \times \mathcal{L}(o)$. This requires a deeper insight into the structure of $\mathcal{L}(o)$.

A single-object lattice $\mathcal{L}(o)$ has at most two concepts: a top one, $\top_o = (\{o\}, \{o\}')$ and a bottom one, $\perp_o$ which only exists if some attributes in the context corresponding to $\mathcal{L}$ are not in $\{o\}'$. In this case, $\perp_o = (\emptyset, A \cup \{o\}')$ (see the lattice $\mathcal{L}(3)$ shown in the middle of Figure 3). The lattice resulting from the

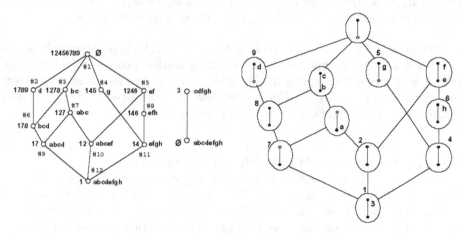

Fig. 3. Left: Lattices $\mathcal{L}$ (objects $1, 2, 4, ..., 8$) and $\mathcal{L}(3)$. **Right**: The nested line diagram of the lattice $\mathcal{L} \times \mathcal{L}(3)$.

assembly of $\mathcal{L}$ and $\mathcal{L}(o)$ is isomorphic, in the sense of FCA, to $\mathcal{L}^+$, obtained by the "insertion" of o into $\mathcal{L}$, so it makes sense to consider the embedding of $\mathcal{L}^+$ into the product $\mathcal{L}_\times = \mathcal{L} \times \mathcal{L}(o)$, and thus to look at its image by φ.

If $\mathcal{L}(o)$ is a singleton, $\mathcal{L}^+$ is easily obtained from $\mathcal{L}$ since all existing concepts are in $\mathbf{M}(o)$ (so they merely integrate o).

Let $\exists \perp_o \neq \top_o$ and consider the order intervals of product nodes that correspond to a concept c in $\mathcal{L}^{10}$:

Definition 8. *The function* $\theta : \mathcal{C} \to \mathcal{C}_\times$ *maps each concept* c *from* $\mathcal{L}$ *into the part of* $\mathcal{L}_\times$ *incident to* c: $\theta(c) = [(c, \perp_o), (c, \top_o)]$.

A chain of the above kind will be referred to as a θ-chain, and the nodes $(c, \perp_o)$ and $(c, \top_o)$ will be denoted as $\theta^l(c)$ and $\theta^u(c)$ respectively. It is noteworthy, that both $c = (X, Y)$ and the image by ψ of $\theta^l(c)$ have the same intent: $Intent(c) = Intent(\psi(\theta^l(c)))$, i.e., the node $\theta^l(c)$ is in $\psi^{-1}(Y', Y)$, where (Y', Y) is the concept in $\mathcal{L}^+$ that corresponds to c (by intent equality).

Proposition 5. *Given a concept* c *in* $\mathcal{L}$, *at least one node in* $\theta(c)$ *is full.*

---

[10] These are actually two-element chains.

Proof. Recall a *full* node is the unique maximum of its class in $\mathbf{kernel}(\psi)$ whereby all class members share the same value for $\mathcal{Q}$. Thus, given $c = (X, Y)$, $\theta^l(c) \in \psi^{-1}(Y', Y)^{11}$. Next, $\mathcal{Q}$ is *non-increasing* upon the product order $\leq_\times$. Consequently, $\max(\psi^{-1}(Y', Y)) \in \{\theta^l(c), \theta^h(c)\}$.

In other words, only three out of the four possible combinations for nodes in $\theta(c)$ being full or void, do occur. Moreover, these are in bijection with concept categories in $\mathcal{L}$ ($\bullet$ denotes a *full* node and $\circ$ a *void* one):

Proposition 6. *Given c in $\mathcal{L}$, $c \in \mathbf{U}(o)$ if and only if $\theta(c) = [\bullet, \circ]$, $c \in \mathbf{M}(o)$ if and only if $\theta(c) = [\circ, \bullet]$, and $c \in \mathbf{G}(o)$ if and only if $\theta(c) = [\bullet, \bullet]$.*

The proposition states that an old concept $c = (X, Y)$ is represented[12] in the product lattice by the full node $\theta^l(c)$, which is further mapped to the concept (X, Y) in $\mathcal{L}^+$ (see Proposition 3). Modified concepts are represented by $\theta^u(c)$ and mapped to $(X \cup \{o\}, Y)$. Generators, like old, are represented by $\theta^l(c)$ and mapped to (X, Y). However, the second full node in the respective θ-chain, $\theta^u(c)$, represents the generated new concept, $(X \cup \{o\}, Y \cap \{o\}')$. For example, on Figure 3, the concept $c_{\#3}$ is in $\mathbf{G}(o)$ (see Example 1) and $\theta(c_{\#3})$ is of the type $[\bullet, \bullet]$, whereas $c_{\#4} \in \mathbf{M}(o)$ and $\theta(c_{\#4})$ is of the type $[\circ, \bullet]$.

The *upper covers* of a full node are invariably computed according to Proposition 4 and this is to be matched to the various cases distinguished by Godin *et al.*. The correspondence is immediate in the case of a *new* concept, but the existing concepts deserve further considerations. The following proposition presents the way the upper cover sets for a concept $\bar{c} = (Y', Y)$ in $\mathcal{L}^+$ are computed with respect to the category of its antecedent in $\mathcal{L}$, $c = (X, Y)$.

Proposition 7. *Given a concept c in $\mathcal{L}$:*

- *if $c \in \mathbf{U}(o)$ then $Cov^u(\psi(\theta^l(c))) = \psi(\theta^l(Cov^u(c)))$,*
- *if $c \in \mathbf{M}(o)$ then $Cov^u(\psi(\theta^u(c))) = \psi(\theta^u(Cov^u(c)))$,*
- *if $c \in \mathbf{G}(o)$ then $Cov^u(\psi(\theta^l(c))) = \psi(\theta^l(Cov^u(c) - \mathbf{M}(o))) \cup \{\psi(\theta^u(c))\}$.*

In other words, for both *old* and *modified* concepts, the *upper covers* of the ψ image of the full nodes in their respective θ-chains, correspond to their own upper covers. In contrast, for a generator concept, $Cov^u(\psi(\theta^l(c)))$ differs significantly from the pure set concepts corresponding to the generator upper covers, $\psi(\theta^l(Cov^u(c)$: it includes the generated new concept whereas all modified upper covers are dropped out.

In sum, we have confirmed the findings of Godin *et al.*, which, besides being a proof of the method's correctness, enables its use in our framework.

---

[11] Here $'$ is computed on the augmented context.

[12] In the sense of the equality between the concept intent Y and the value of $\mathcal{Q}$ for the node.

6.2   Description of the Algorithm

Space limits keep us from providing a completely re-engineered version of Algorithm 1. Thus, we only provide the code of FIND-UPPER-COVERS-BIS which replaces FIND-UPPER-COVERS in the original version. Based on the above structural results, the new subroutine explores a minimal part of $\mathcal{L}^+$. It starts by

```
1: procedure FIND-UPPER-COVERS-BIS(In: new, gen concepts)
2:
3:   Candidates ← ψ(θ^u(Cov^u(gen)));  TrueCovers ← MIN(Candidates)
4:   for all c̄ in TrueCovers do
5:      NEW-LINK(new,c̄)
6:      if c̄ ∈ M(o) then
7:         DROP-LINK(gen, c̄)
```

Algorithm 4: Insertion of a new object into a concept (Galois) lattice.

computing the candidate upper covers, i.e., the images by ψ of all $\theta^u(c)$ where c is an upper cover of the generator. Here neither ψ nor θ^u needs an effective computation. In fact, $\psi(\theta^u(c))$ is the concept of $\mathcal{L}^+$ corresponding to an intent $Intent(c) \cap \{o\}'$. It can be computed as follows: for $c \in M(o)$, it is c itself, for $c \in G(o)$, it is the respective new, and for $c \in U(o)$, the value of $\psi(\theta^u(c))$ is shared with any upper cover of c which is itself in $[c]_Q$. In the last case, the values can be "inherited" if properly stored in a vector during the traversal.

Example 2. Let new be $(134, fgh)$ and gen be $c_{\#11} = (14, efgh)$. The candidate upper covers are the images of $(c_{\#4}, \top_o)$ and $(c_{\#8}, \top_o)$ by ψ, i.e., $Candidates=$ $\{(1345, g), (1346, fh)\}$, and both of them are further kept as actual covers and linked to new. Finally, as $(1345, g)$ is in $M(o)$ (see Example 1), its link to gen is dropped out.

6.3   Complexity Issues

The following table presents basic parameters used in the estimation of the worst-case complexity for both incremental algorithms.

Variable	Stands for	Variable	Stands for
m	$\|A\|$	k	$\|O\|$
l	$\|\mathcal{C}\|$	$d(\mathcal{L})$	$\max_{c \in \mathcal{C}}(\|Cov^u(c)\|)$

The global cost is split into two separate values that are to be summed up. The fist one assesses the traversal of the initial lattice $\mathcal{L}$ (lines $5 - 24$ and 28 of Algorithm 1) and is therefore identical in both cases. The second one focuses on the overhead induced by the processing of new nodes and varies according to the upper cover computation method.

The lattice traversal takes $O(l)$ concept examinations which, when implemented in an optimal way (e.g., using a trie for concept lookup [10]) require only a fixed number of intent manipulations. These are essentially set operations and an appropriate set representation (e.g., as sorted lists) would keep their cost linear in the size of the set ($O(m)$). Hence the first cost value is $O(ml)$.

Actual upper covers of a new concept c are detected in FIND-UPPER-COVERS by examining a candidate set made up of all concepts with intent size less than $\|Intent(c)\|$. Although hard to estimate exactly, the number of such candidates will be in $O(l)$, hence the cost of the upper cover computation for a single concept is in $O(ld(\mathcal{L})m)$. In contrast, FIND-UPPER-COVERS-BIS takes $O(kd(\mathcal{L}))$ time, which is actually the cost of minima selection (see [13] for details). The entire overhead further depends on the actual number of new concepts, say $\Delta(l)$.

Method	Traversal	Single insertion	Entire construction
original method	$O(ml)$	$O(ml\Delta(l)d(\mathcal{L}))$	$O(kml\Delta(l)d(\mathcal{L}))$
improved method	$O(ml)$	$O(ml + \Delta(l)kd(\mathcal{L}))$	$O(k(ml + \Delta(l)kd(\mathcal{L})))$

Thus, for lattice construction, our method is comparable to the (theoretically) most efficient batch algorithm [10] ($O((k + m)lk)$ time).

Concerning practical performances, [8] empirically proved the superiority of the incremental approach over batch algorithms in sparse contexts[13]. However, recent findings [9] indicate that the incremental algorithm may lose its supremacy when the context gets denser. We believe that this shift is due to the inefficient upper cover computation, a flaw that is missing in our own incremental method.

7   Conclusion and Further Research

The present study was motivated by the need for a theoretical and algorithmic contribution to the problem of the incremental update of concept (Galois) lattices when a set of objects need to be added at a time. To achieve the defined goal, we presented an extended framework as a generalization of the work done by Godin *et al.*, and considered two scenarios: (*i*) update the initial lattice by considering new objects one at a time, and (*ii*) first build the partial lattice over the new object set and then merge it with the initial lattice. We provided two algorithms carrying out these two scenarios. The first one is an improvement of a well-known incremental algorithm [8], while the second one is an assembly procedure based on context subposition. The first scenario seems to be the right choice for small sets ΔO, whereas medium-size ΔO might favor the assembly scenario. When a large set of new objects need to be added, constructing the lattice from scratch may be the best solution.

The lattice assembly framework we described in this paper speaks in favor of a theoretical approach towards the algorithmic design of concept lattices. This could be a starting point for the design of an algorithmic meta-strategy which, based on the recognition of particular relation classes, chooses automatically the best algorithm to apply to a concrete context, or even to context fragments before the assembly of partial lattices.

---

[13] i.e., with very few attributes per object.

References

[1] M. Barbut and B. Monjardet. *Ordre et Classification: Algèbre et combinatoire.* Hachette, 1970.

[2] J.-P. Bordat. Calcul pratique du treillis de Galois d'une correspondance. *Mathématiques et Sciences Humaines,* 96:31–47, 1986.

[3] B. A. Davey and H. A. Priestley. *Introduction to lattices and order.* Cambridge University Press, 1992.

[4] B. Ganter and S. Kuznetsov. Stepwise Construction of the Dedekind-McNeille Completion. In *Proceedings of the 6th ICCS,* pages 295–302, Montpellier, 1998.

[5] B. Ganter and R. Wille. *Formal Concept Analysis, Mathematical Foundations.* Springer-Verlag, 1999.

[6] R. Godin and H. Mili. Building and maintaining analysis-level class hierarchies using Galois lattices. In *Proceedings of OOPSLA'93, Washington (DC), USA,* special issue of ACM SIGPLAN Notices, 28(10), pages 394–410, 1993.

[7] R. Godin and R. Missaoui. An Incremental Concept Formation Approach for Learning from Databases. *Theoretical Computer Science,* 133:378–419, 1994.

[8] R. Godin, R. Missaoui, and H. Alaoui. Incremental concept formation algorithms based on galois (concept) lattices. *Computational Intelligence,* 11(2):246–267, 1995.

[9] S. Kuznetsov and S. Objedkov. Algorithms for the Construction of the Set of All Concept and Their Line Diagram. preprint MATH-AL-05-2000, Technische Universität, Dresden, June 2000.

[10] L. Nourine and O. Raynaud. A Fast Algorithm for Building Lattices. *Information Processing Letters,* 71:199–204, 1999.

[11] N. Pasquier, Y. Bastide, T. Taouil, and L. Lakhal. Efficient Mining of Association Rules Using Closed Itemset Lattices. *Information Systems,* 24(1):25–46, 1999.

[12] P. Valtchev. An algorithm for minimal insertion in a type lattice. *Computational Intelligence,* 15(1):63–78, 1999.

[13] P. Valtchev, R. Missaoui, and P. Lebrun. A partition-based approach towards building Galois (concept) lattices. *to appear in Discrete Mathematics,* 2001.

[14] R. Wille. Restructuring the lattice theory: An approach based on hierarchies of concepts. In I. Rival, editor, *Ordered sets,* pages 445–470, Dordrecht-Boston, 1982. Reidel.

A Term-Based Approach to Project Scheduling

Pok-Son Kim and Manfred Schmidt-Schauß

Fachbereich Informatik
Johann Wolfgang Goethe-Universität
Postfach 11 19 32
D-60054 Frankfurt/Main, Germany
Tel.: (+49) 69 798 28597; Fax: (+49) 69 798 28919
{kim, schauss}@cs.uni-frankfurt.de

Abstract. We introduce a new method for representing and solving a general class of non-preemptive resource-constrained project scheduling problems. The new approach is to represent scheduling problems as descriptions (activity terms) in a language called $\mathcal{RSV}$, which allows nested expressions using **pll, seq,** and **xor**. The activity-terms of $\mathcal{RSV}$ are similar to concepts in a description logic. The language $\mathcal{RSV}$ generalizes previous approaches to scheduling with variants insofar as it permits **xor**'s not only of atomic activities but also of arbitrary activity terms. A specific semantics that assigns their set of active schedules to activity terms shows correctness of a calculus normalizing activity terms $\mathcal{RSV}$ similar to propositional DNF-computation.
Based on $\mathcal{RSV}$, this paper describes a diagram-based algorithm for the $\mathcal{RSV}$-problem which uses a scan-line principle. The scan-line principle is used for determining and resolving the occurring resource conflicts and leads to a nonredundant generation of all active schedules and thus to a computation of the optimal schedule.

1 Introduction

Ever since the introduction of the pioneer works of Kelly [13] and Wiest [27] very much has been reported for the *classical* resource-constrained project scheduling ($\mathcal{RCPS}$) but most approaches solving this problem use either heuristic algorithms ([11], [5], [3], [14], [19]) or exact algorithms ([21], [23], [24], [26], [7], [8], [18]), are restricted to the case in which each job could be performed in only one prescribed way. Although Schrage [23] and König et al. [16] suggested the possibility of generalization to "OR" activities, there are only a few publications ([10], [25]) which deal with a class of resource-constrained scheduling problems with *variants*, where "OR" was restricted to atomic activities.

This paper generalizes *resource constrained project scheduling with variants* ($\mathcal{RSV}$) to allow "OR" of compound activities. The use of semantic methods from description logics ([22,9] is the key for understanding the meaning of compound activity terms (for more information see [15]).

Description Logic emerged from KL-ONE-based, terminological knowledge representation systems and is now an active field of research in artificial intelligence

H. Delugach and G. Stumme (Eds.): ICCS 2001, LNAI 2120, pp. 304–318, 2001.
© Springer-Verlag Berlin Heidelberg 2001

with its own conference(s). Characteristics of descriptions logics are a term language for concepts and other notions, a clean denotational semantics, and specific calculi (like subsumption) based on the semantics. There are extension of description logics by temporal operators (see e.g. [1,2]), which allow reasoning about concepts in time.

Scheduling is roughly the problem, given certain tasks, their duration and resource usage to find an optimal conflict free schedule, usually on a discrete time line. It is possible to encode these problems into temporal description logics, though it appears to be inconvenient, since for scheduling problems, time has to be measured and added, and the goal is finding an optimal schedule. Furthermore reasoning is not optimized to deal with the constraint of exclusive resource usage.

The methods proposed so far for solving scheduling problems are mostly based on integer programming. In that approach it is generally difficult to read the flow structure and the content of a scheduling problem (for example, which activity requires what resource).

This motivates to use a term language $\mathcal{RSV}$ for activity terms and to model its semantics in a way best suited to the specific time and resource constraints. Based on this semantics, a calculus is defined which can transform each activity term A into a semantically equivalent, normalized activity term B, which is a nonredundant disjunction of *reduced* activity terms corresponding to A. For every reduced activity term, optimal schedules (with the minimal makespan) can be computed using an algorithm for solving the classical $\mathcal{RCPS}$-problem. For any $\mathcal{RSV}$-activity term the optimal schedules can be computed by this two-step process using a final minimization. It is non-obvious how to compute the active schedules of compound activity terms in another way.

Further we introduce a new diagram-based method, which represents reduced activity terms graphically as $\mathcal{RSV}$-diagrams. Based on this method, an algorithm for generating all *nonredundant* active schedules for a reduced activity term is described, which uses a time-based scan-line principle to determine and resolve the occurring resource conflicts. The principle of using a scan-line is well known in the area of geometrical algorithms ([20], [28]) and may be used for example to solve problems occurring during the design of VLSI circuits.

This paper is structured as follows. First, the scheduling language $\mathcal{RSV}$ is defined in section 2. Then the calculus is described in section 3. The optimal solution algorithm and the correctness proof for it are given in section 4. The last section 5 contains conclusions and future work.

2   The Scheduling Language $\mathcal{RSV}$

A terminological language $\mathcal{RSV}$ that may be used to model a new general class of resource-constrained project scheduling problems with variants is defined as follows:

2.1 The Syntax of the Language $\mathcal{RSV}$

Definition 1. *The vocabulary of $\mathcal{RSV}$ consists of two disjoint sets of symbols:*

- *A set of atomic activities, also called ground activities. Each atomic activity consists of a name and two integer constants (written like arguments of a predicate symbol). The first denotes a resource; the second a duration. For P a name and $r, t \in \mathbb{N}$, $P(r, t)$ is a ground activity. The name of each ground activity is uniquely chosen. For a ground activity A let $r(A)$ and $d(A)$ denote its associated resource and duration, respectively.*
- *Three structural symbols (operators) seq, xor and pll.*

The activity terms of $\mathcal{RSV}$ are inductively defined as follows:

1. *Each ground activity is an activity term.*
2. *If $A_1, A_2, \cdots, A_k$ are activity terms, then*

$$(\text{seq}\, A_1, A_2, \cdots, A_k),$$
$$(\text{xor}\, A_1, A_2, \cdots, A_k),$$
$$(\text{pll}\, A_1, A_2, \cdots, A_k)$$

are also activity terms.

The operators 'seq', 'xor' and 'pll' have the following meaning:

- **'seq'** : This operator specifies the *sequential* processing of an activity term or activity terms (precedence constraints).
- **'xor'** : This operator can be used to select an activity term among several different alternative activity terms. *Exactly one* activity term among the alternatives must be selected and executed.
- **'pll'** : This operator specifies the possibility of parallel processing of activity terms.

2.2 Reduced Activity Terms

An expression of $\mathcal{RSV}$ which is **xor**-free is called *a reduced activity term*. For a project represented by a reduced activity term there is no selection possibility for any part of it.

B is a *reduced activity subterm* of A, if B can be derived from A by repeatedly replacing subterms of the form $(\text{xor}\, C_1, \cdots, C_n)$ by exactly one C_i ($i = 1, \cdots$ or n) so that B is **xor**-free. Associated with any activity term A of $\mathcal{RSV}$, there exist finitely many different reduced activity terms which can be derived from A. These reduced activity terms take partially different paths but complete the same project.

Example 1. We show how to encode the small project of preparing food for two persons and then eating this food. The project is to either putting 2 pizzas in the oven, or putting 4 toasts into the toaster, and then eating them. Let resource

1 be the oven, resource 2 be the toaster, 3 and 4 are the two persons. Then a description could be

$$\text{(seq (xor pizza}(1,15)\ (\text{pll toastfst}(2,2),\ \text{toastsnd}(2,2))))$$
$$(\text{pll arne\_eating}(3,10),\ \text{pokson\_eating}(4,12))$$

Note that we permit several occurrences of the same ground activity in an activity term but this is not permitted for a reduced activity term, i.e. each ground activity can occur once and only once in a reduced activity term.

2.3   Schedules

For a reduced activity term A let $g(A) = \{A_1, \cdots, A_n\}$ be the set consisting of all ground activities occurring in A. The activity term A defines a strict partial order $<_A$ on $\{A_1, \cdots, A_n\}$, using the sequentiality operator. It is generated on the set S of subterms of A as follows:

- $(\text{seq } B_1, \ldots, B_m) \in S \wedge i < j \Rightarrow B_i <_A B_j$.
- $B_1, B_2 \in S \wedge B_1 <_A B_2 \Rightarrow B_1' <_A B_2'$ for every subterm B_i' of $B_i, i = 1, 2$.

Definition 2. *Let A be a reduced activity term and $g(A) = \{A_1, \cdots, A_n\}$. An* active schedule *for A is a set of starting times of ground activities $\{t_{A_i} \in \mathbb{N} \mid A_i \in g(A)\}$ such that:*

- *The precedence constraints are satisfied: $t_{A_h} + d(A_h) \le t_{A_i}$ for each A_i and each immediate predecessor A_h with $A_h <_A A_i$,*
- *The resource constraints are satisfied: $t_{A_m} \ge t_{A_l} + d(A_l)$ or $t_{A_l} \ge t_{A_m} + d(A_m)$ for all $A_l, A_m \in g(A)$ with $r(A_l) = r(A_m)(l \ne m)$ and*
- *No ground activity can be started earlier without changing other start times: There does not exist another set $\{t'_{A_i} | A_i \in g(A)\}$ with a ground activity A_j, which satisfies the precedence and resource constraints, such that $t_{A_i} = t'_{A_i}$ for $i \ne j$ and $t_{A_j} > t'_{A_j}$.*

The makespan *of an active schedule is the duration from the first starting time $\min_i(t_{A_i})$ to the stopping time $\max_i(t_{A_i} + d(A_i))$.*
Let A be an activity term. Then the set of active schedules for A is the union of the set of active schedules for all reduced activity subterms of A.

Since time is discrete it is easy to see that for any activity term A the set of active schedules derived from A is finite. In the following all schedules are assumed to be active.

2.4   The Semantics of the Language $\mathcal{RSV}$

Definition 3. *The model-theoretic semantics of activity terms in $\mathcal{RSV}$ is given by an interpretation $\mathcal{I}$ which consists of the set $\mathcal{D}$ (the domain of $\mathcal{I}$) and a function $\cdot^{\mathcal{I}}$ (the interpretation function of $\mathcal{I}$). The set $\mathcal{D}$ consists of all active schedules derived from activity terms in $\mathcal{RSV}$. The interpretation function $\cdot^{\mathcal{I}}$ assigns to every activity term A the subset of $\mathcal{D}$ that consists of all active schedules derived from A.*

2.5   Scheduling Problem

The objective is minimizing the project makespan, i.e. finding an active schedule with a minimal duration. So, we define the scheduling problem $\mathcal{RSV}$ as follows:

Definition 4 (The scheduling problem $\mathcal{RSV}$).
For a given activity term A of $\mathcal{RSV}$ an (active) schedule corresponding to A which has the minimal project makespan has to be determined.

3   A Calculus for the Scheduling Language $\mathcal{RSV}$

Two activity terms are semantically equivalent, if the interpretations of the two activity terms are identical. For example, the following two activity terms

$$(\text{pll } (\text{seq } P_3(c,15), P_4(c,16)), (\text{xor } P_5(c,3), P_6(d,5))) \tag{1}$$

and

$$\begin{array}{l} (\text{xor } (\text{pll } (\text{seq } P_3(c,15), P_4(c,16)), P_5(c,3)), \\ \quad (\text{pll } (\text{seq } P_3(c,15), P_4(c,16)), P_6(d,5))) \end{array} \tag{2}$$

are semantically equivalent, i.e. the set of all schedules which may be derived from (1) and the set of all schedules which may be derived from (2) are identical. An activity term such as (2) in which the **xor**-operator occurs only once in the leftmost position is called a *normalized* activity term. In this section we will define a calculus that may be used to transform any $\mathcal{RSV}$-expression A into a semantically equivalent *normalized* $\mathcal{RSV}$-expression B. Further we will show the calculus to be correct, i.e. in the calculus only such syntactical derivations which cause no semantical change are permitted. So, if $A \vdash B$, then $A \doteq B$ holds, as illustrated in Figure 1.

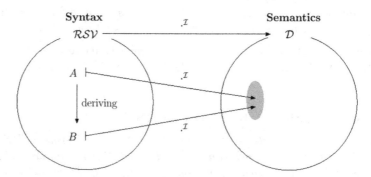

Fig. 1. Graphical representation of a scheduling equation $A \doteq B$

3.1 The $\mathcal{RSV}$-Calculus

Similar to the computation rules for a disjunctive normal form in propositional logic, a $\mathcal{RSV}$- calculus is defined:

Definition 5. *If* $A_1, A_2, \cdots, A_k, B_1, \cdots, B_l, A_{k+2}, \cdots, A_n$ *are activity terms, the calculus has the following 3 associative rules (3,4,5) and 2 distributive rules (6,7). Each associative rule describes that a subexpression combined by 'seq', 'xor' or 'pll' which is an argument of the operator 'seq', 'xor' or 'pll' respectively may be flattened. Each distributive rule describes that if the operator 'xor' occurs as an argument of the operator 'seq' or 'pll', the operator 'xor' may be moved to the leftmost position:*

$$\frac{(\text{seq } A_1, A_2, \cdots, A_k, (\text{seq } B_1, B_2, \cdots, B_l), A_{k+2}, A_{k+3}, \cdots, A_n)}{(\text{seq } A_1, A_2, \cdots, A_k, B_1, B_2, \cdots, B_l, A_{k+2}, A_{k+3}, \cdots, A_n)} \tag{3}$$

$$\frac{(\text{xor } A_1, A_2, \cdots, A_k, (\text{xor } B_1, B_2, \cdots, B_l), A_{k+2}, A_{k+3}, \cdots, A_n)}{(\text{xor } A_1, A_2, \cdots, A_k, B_1, B_2, \cdots, B_l, A_{k+2}, A_{k+3}, \cdots, A_n)} \tag{4}$$

$$\frac{(\text{pll } A_1, A_2, \cdots, A_k, (\text{pll } B_1, B_2, \cdots, B_l), A_{k+2}, A_{k+3}, \cdots, A_n)}{(\text{pll } A_1, A_2, \cdots, A_k, B_1, B_2, \cdots, B_l, A_{k+2}, A_{k+3}, \cdots, A_n)} \tag{5}$$

$$\frac{(\text{seq } A_1, A_2, \cdots, A_k, (\text{xor } B_1, B_2, \cdots, B_l), A_{k+2}, A_{k+3}, \cdots, A_n)}{\begin{array}{l}(\text{xor } (\text{seq } A_1, A_2, \cdots, A_k, B_1, A_{k+2}, A_{k+3}, \cdots, A_n), \\ \quad (\text{seq } A_1, A_2, \cdots, A_k, B_2, A_{k+2}, A_{k+3}, \cdots, A_n), \\ \qquad \vdots \\ \quad (\text{seq } A_1, A_2, \cdots, A_k, B_l, A_{k+2}, A_{k+3}, \cdots, A_n))\end{array}} \tag{6}$$

$$\frac{(\text{pll } A_1, A_2, \cdots, A_k, (\text{xor } B_1, B_2, \cdots, B_l), A_{k+2}, A_{k+3}, \cdots, A_n)}{\begin{array}{l}\text{xor } ((\text{pll } A_1, A_2, \cdots, A_k, B_1, A_{k+2}, A_{k+3}, \cdots, A_n), \\ \quad (\text{pll } A_1, A_2, \cdots, A_k, B_2, A_{k+2}, A_{k+3}, \cdots, A_n), \\ \qquad \vdots \\ \quad (\text{pll } A_1, A_2, \cdots, A_k, B_l, A_{k+2}, A_{k+3}, \cdots, A_n))\end{array}} \tag{7}$$

Lemma 1. *The $\mathcal{RSV}$-calculus is a correct calculus.*

Proof. A rule in the form

$$\frac{A}{B}$$

is "correct" iff the interpretation of the upper expression A and the lower expression B is identical ($A^{\mathcal{I}} = B^{\mathcal{I}}$). In the following we show this for each of the 5 rules.

Rule (3): The set of all active schedules derived from the upper and lower expression is obviously identical, because both expressions describe the same ordering of the activity terms $A_1, \cdots, A_k, B_1, \cdots, B_l, A_{k+2}, \cdots, A_{n-1}$ and A_n. Otherwise the expression transformation doesn't make changes. Therefore, the following equation holds:

$$(\text{seq } A_1, A_2, \cdots, A_k, (\text{seq } B_1, B_2, \cdots, B_l), A_{k+2}, A_{k+3}, \cdots, A_n)^{\mathcal{I}}$$
$$= (\text{seq } A_1, A_2, \cdots, A_k, B_1, B_2, \cdots, B_l, A_{k+2}, A_{k+3}, \cdots, A_n)^{\mathcal{I}}$$

Rule (4) and *Rule* (5) may be proved similarly to *rule* (3).

Rule (6): For the $k+1$-st argument of the operator 'seq' a choice possibility exists. One of the l activity terms $B_1, \cdots B_{l-1}$ and B_l must be selected. For each choice of B_i $(i = 1, \cdots, l)$ a set of all schedules derived from the expression

$$(\text{seq } A_1, A_2, \cdots, A_k, B_i, A_{k+2}, A_{k+3}, \cdots, A_n)$$

denoted by M_{B_i} is determined. Then the set of all schedules which may be derived from the upper expression

$$(\text{seq } A_1, A_2, \cdots, A_k, (\text{xor } B_1, B_2, \cdots, B_l), A_{k+2}, A_{k+3}, \cdots, A_n)$$

corresponds to the union of the sets $M_{B_1} \cdots M_{B_{l-1}}$ and M_{B_l}. Further this union corresponds to the set of all schedules which may be derived from the lower expression. So, the following equation holds:

$$\begin{aligned}
&(\text{seq } A_1, A_2, \cdots, A_k, (\text{xor } B_1, B_2, \cdots, B_l), A_{k+2}, A_{k+3}, \cdots, A_n)^{\mathcal{I}} \\
&= (\text{xor } (\text{seq } A_1, A_2, \cdots, A_k, B_1, A_{k+2}, A_{k+3}, \cdots, A_n), \\
&\qquad (\text{seq } A_1, A_2, \cdots, A_k, B_2, A_{k+2}, A_{k+3}, \cdots, A_n), \\
&\qquad \vdots \\
&\qquad (\text{seq } A_1, A_2, \cdots, A_k, B_l, A_{k+2}, A_{k+3}, \cdots, A_n))^{\mathcal{I}}
\end{aligned}$$

Rule (7): This may be proved similar to *rule* (6).

The correctness of the $\mathcal{RSV}$-calculus permits to formalize the following theorem:

Theorem 1. *For any activity description A of $\mathcal{RSV}$ all operators 'xor' in the interior of A always can be moved to the leftmost position such that A is transformed to a semantically equivalent, normalized activity term A' in which the operator 'xor' can occur uniquely once in the leftmost position combining all reduced activity terms derived from A.*

Proof. It is sufficient to show that for any expression A of $\mathcal{RSV}$ all derivations terminate in a normalized expression. This is the case when no further $\mathcal{RSV}$-rules can be applied.

Using an innermost strategy and induction on the number of occurrences of **xor**'s it is easy to see that in every expression containing the 'xor'-operator, it can be shifted to the topmost position.

Thus all 'xor'-operators in the interior of the activity term A may be moved stepwise to the left until at most one topmost **xor** remains. □

Theorem 1 shows that the $\mathcal{RSV}$-problem can be solved by first transforming each activity term A into a semantically equivalent, normalized activity term B and then computing the schedules with the minimal project makespan for every reduced activity term of B separately. A final minimizing step computes the minimum makespan for A.

In the following section we describe a solution algorithm that nonredundantly computes all active schedules for each reduced activity term of $\mathcal{RSV}$.

4   Solving the $\mathcal{RSV}$-Problem Using Diagram-Based Calculation

Many varieties of branch-and-bound-based implicit enumeration methods ([23], [24], [26], [4], [7], [8], [18], [6]) for solving the $\mathcal{RCPS}$-problem which may be also used for determining the optimal schedules for reduced activity terms of $\mathcal{RSV}$ have been reported. In this section we introduce a new diagram-based method for representing reduced activity terms graphically. The resulting diagrams are called $\mathcal{RSV}$-diagrams. Further we show that based on the representation method, a solution algorithm is described for explicit generation of all nonredundant active schedules. This is illustrated graphically using $\mathcal{RSV}$-diagrams.

4.1   The $\mathcal{RSV}$-Diagram

A $\mathcal{RSV}$-diagram has a time axis and a scan-line. The operator 'seq' is specified using a continuous line while the operator 'pll' is specified using a broken line. In the following two reduced activity terms are for example represented by a $\mathcal{RSV}$-diagram.

Example 2. The following two reduced activity terms

$$\begin{array}{ll}
\begin{array}{l}
\text{pll (seq } P_1(b,3), P_2(c,4)), \\
\quad (\text{seq } P_3(a,2), P_4(b,3)), \\
\quad (\text{seq } P_5(a,4), P_6(b,2))
\end{array}
& \quad\text{and}\quad
\begin{array}{l}
\text{pll (seq (pll } P_1(a,1), P_2(b,2)), P_3(c,2)), \\
\quad (\text{seq } P_4(b,1), P_5(a,1)), \\
\quad P_6(a,2), \\
\quad (\text{seq } P_7(d,1), P_8(b,2))
\end{array}
\end{array}$$

may be represented by the $\mathcal{RSV}$-diagram 1 and 2 of Figure 2 respectively.

4.2   Solution Algorithm $\mathcal{A_{RSV}}$ Based on a Scan-Line principle

In a $\mathcal{RSV}$-diagram each ground activity has a left and a right end point (a start and end time). The left and right end point of any ground activity $P(r,t)$ denoted by $LE(P(r,t))$ and $RE(P(r,t))$ are referred to as *stopping times* of the scan-line. (D,t) with $t \geq 0$ denotes that the scan-line is found at the stopping time $t_{SL} = t$ in the $\mathcal{RSV}$-diagram D. The scan-line is used for determining and resolving resource conflicts. Instead of continuously moving, the scan-line jumps from one stopping time to the next right stopping time while determining and then resolving resource conflicts.

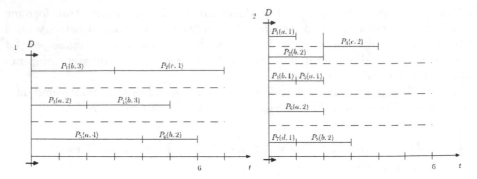

Fig. 2. $\mathcal{RSV}$-diagrams with the scan-line on the stopping time $t_{SL} = 0$.

In the beginning the scan-line is found at the time $t_{SL} = 0$ and the diagram is empty. The diagram is accompanied by a given input activity term A and an actual activity term T, where in the beginning $A = T$ holds.

Step 1: **Attaching start ground activities to the scan-line:** First all start ground activities of A (or T) which have no predecessors in A are attached to the scan-line. "Attaching a ground activity P to the scan-line" means that P is placed in the diagram so that the time at which the scan-line is found is assigned to P as its start time.

Step 2: **Moving the scan-line:** The scan-line jumps to the next stopping time.

Definition 6. *In a $\mathcal{RSV}$-diagram (D, t_{SL}) with $t_{SL} > 0$ a ground activity $P(r,t)$ is called a (t_{SL}-time) scan-line activity iff $LE(P(r,t)) < t_{SL}$ and $RE(P(r,t)) \geq t_{SL}$ holds. A scan-line activity $P(r,t)$ with $RE(P(r,t)) = t_{SL}$ is called a (t_{SL}-time) direct scan-line activity. A direct scan-line activity $P(r,t)$ is called a (t_{SL}-time) conflict-free activity iff there is no other scan-line activity that requires the resource r, i.e. $P(r,t)$ is a unique scan-line activity requiring the resource r. The resource r which is required by a direct scan-line activity is called a (t_{SL}-time) direct scan-line resource.*

Definition 7. *If in a $\mathcal{RSV}$-diagram (D, t_{SL}) with $t_{SL} > 0$, ground activities $P_1(r,t_1), P_2(r,t_2), \cdots$ and $P_n(r,t_n)(n \geq 2)$ and are all (t_{SL}-time) scan-line activities requiring the same resource r, $P_1(r,t_1), P_2(r,t_2), \cdots$ and $P_n(r,t_n)$ $(n \geq 2)$ are called to be involved in a (t_{SL}, r)-resource conflict or t_{SL}-time resource conflict iff r is a direct scan-line resource . This resource r is called a t_{SL}-time conflict resource. Further tthese activities $P_1(r,t_1), P_2(r,t_2), \cdots$ and $P_n(r,t_n)$ $(n \geq 2)$ are called (t_{SL}, r)-conflict activities or t_{SL}-time conflict activities.*

Step 3: **Determining and resolving resource conflicts; Freezing all definitely placed ground activities:** First, because the begin and end times of all t_{SL}-time conflict-free activities have been definitely determined, all t_{SL}-time conflict-free activities are frozen. If several scan-line activities require a conflict resource r simultaneously, a resource conflict occurs. A resource conflict is resolved by selecting an activity and shifting all the other activities behind the selected activity. In this case the begin and end time of this selected activity are definitely determined. In order to mark that a selected activity must no longer be moved, it is frozen.

At any stopping time t_{SL}, several different t_{SL}-time resource conflicts can simultaneously occur. In this case exactly one (t_{SL}, r)-conflict activity for *each* t_{SL}-time conflict resource r is selected in order to freeze it. There exist several different combinational possibilities for selecting activities. Such a combination is called *a conflict combination* and is formally defined as follows:

Definition 8. *Let* $r_1, r_2, \cdots, r_n$ *be* t_{SL}-*time conflict resources in a* $\mathcal{RSV}$-*diagram* (D, t_{SL}). *Let* $P_{r_1,1}, \cdots, P_{r_1,m_1}$ *be all* (t_{SL}, r_1)-*conflict activities,* $P_{r_2,1}, \cdots, P_{r_2,m_2}$ *be all* (t_{SL}, r_2)-*conflict activities,* $\cdots$, *and* $P_{r_n,1}, \cdots, P_{r_n,m_n}$ *be all* (t_{SL}, r_n)-*conflict activities. An element of the following set*

$$\{[P_{r_1,i_1}, P_{r_2,i_2}, \cdots, P_{r_n,i_n}] | i_1 = 1, \cdots, m_1, i_2 = 1, \cdots, m_2, \cdots, i_n = 1, \cdots, m_n\}$$

is called a t_{SL}-*time conflict combination.*

In the case of the definition 8 there exist altogether $m_1 \times m_2 \times \cdots \times m_n$ t_{SL}-time conflict combinations. In order to pursue all possible precedence orderings, the actual $\mathcal{RSV}$-diagram (D, t_{SL}) is *multiplied* by the number of the existing conflict combinations. Every conflict combination is assigned to one of the multiplied diagrams respectively. In every diagram, the assigned conflict activities are frozen and all the other t_{SL}-time conflict activities are moved behind each corresponding frozen activity respectively. We proceed with the step 4 for *every* diagram accompanied by A and T.

Step 4: **Deleting all t_{SL}-time direct scan-line activities from the actual activity term T:** If in the diagram (D, t_{SL}) t_{SL}-time direct scan-line activities exist, they surely have been frozen in the last step 3. Now all t_{SL}-time direct scan-line activities in (D, t_{SL}) are deleted from T of (D, t_{SL}). So T may become smaller.

Step 5: **Attaching further ground activities to the scan-line:** Further ground activities from the actual activity term T which can be attached to the scan-line are determined in order to place them. If in an actual diagram (D, t_{SL}) a scan-line activity $P(r, t)$ with $RE(P(r, t) > t_{SL}$ has been frozen, the resource r is being blocked until the time $RE(P(r, t))$. So, all further ground activities requiring *the t_{SL}-time blocked resource* r which have not yet been placed in the diagram and have no predecessor in T must wait until the scan-line has jumped to the time $RE(P(r, t))$. For $(D, t_{SL})(t_{SL} > 0)$ with an input activity term A and an actual activity term T, a ground activity $P(r, t)$ of T can be attached to the scan-line iff

1. P isn't from the diagram (D, t_{SL}),
2. in (D, t_{SL}) there exists no frozen activity $Q(r, l)$ for which $LE(Q) < t_{SL}$ and $RE(Q) > t_{SL}$ hold.
3. in T P has no predecessor.

Furthermore the steps 2, 3, 4 and 5 are recursively applied until all ground activities have been placed in the diagram and all activities in the diagram have been frozen so that T is empty and an active schedule is completed. Among all computed schedules, those that have the minimal project makespan are delivered as the optimal schedules for A.

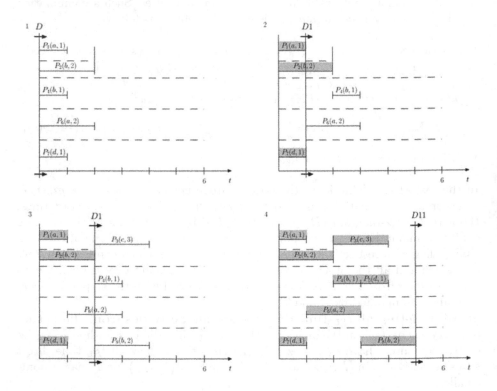

Fig. 3. $\mathcal{RSV}$-diagram-based calculation of active schedules for the reduced activity term of example 2

Example 3. In order to demonstrate that computing with the aid of $\mathcal{RSV}$-diagrams makes the algorithm easily understandable, we consider the second reduced activity term of example 2. In the beginning, the input activity term and the actual activity term are identical. There are start activities P_1, P_2, P_4, P_6 and P_7. The diagram 1 of figure 3 shows the resulting diagram after applying step 1, in which the scan-line time 0 has been assigned to these start activities P_1, P_2, P_4, P_6 and P_7 as their start time. In step 2, the scan-line jumps into the next stopping time $t_{SL} = 1$ and we have the resulting diagram $(D, 1)$, in which there is one 1-time conflict free activity P_7 and two 1-time conflict resources a and b. So there exist four 1-time conflict combinations $[P_1, P_2], [P_1, P_4], [P_2, P_6]$ and $[P_4, P_6]$ altogether. The diagram $(D, 1)$ is multiplied 4 times, let these be

$D1, \cdots, D4$ and $[P_1, P_2], [P_1, P_4], [P_2, P_6], [P_4, P_6]$ are assigned to $D1, \cdots, D4$ respectively. In every diagram, the 1-time conflict free activity P_7 and the assigned 1-time conflict activities are frozen and the other 1-time conflict activities are moved behind each corresponding frozen activity. Subsequently we proceed with the next step 4 in every diagram accompanied by the input activity term and the actual activity term.

If we pursue $(D1, 1)$ to which the combination $[P_1, P_2]$ is assigned, we have diagram 2 of figure 3 where P_1, P_2 and P_7 have been frozen and P_4 and P_6 have been moved behind P_2 and P_1 respectively. Now the two 1-time direct scan-line activities P_1 and P_7 in $(D1, 1)$ (diagram 2) have to be deleted from the actual activity term. After deleting both activities we have the following *new* actual activity term for $(D1, 1)$:

$$\text{pll (seq (pll } P_2(b, 2)), P_3(c, 2)),$$
$$(\text{seq } P_4(b, 1), P_5(a, 1)),$$
$$P_6(a, 2),$$
$$(\text{seq } P_8(b, 2))$$

P_8 is here the unique activity which has no predecessor and isn't yet included in the diagram, but it requires the 1-time blocked resource b. So in the next step 5 there is no activity to be attached to the scan-line. After applying the further steps 2, 3, 4 and 5 to the diagram 2, we have the diagram $(D1, 2)$ the diagram 3 of figure 3 shows, where the two further activities P_3 and P_8 newly have been attached to the scan-line. The information that P_7 corresponds to the immediate predecessor of P_8 could be read from the input activity term accompanied. So, P_8, for example, has been placed behind P_7 in the diagram. After applying the next step 2 to the diagram 3 we get $(D1, 3)$, in which there is a 3-time conflict resource b. $(D1, 3)$ is duplicated 2 times, let these be $(D11, 3)$ and $(D12, 3)$ where 3-time conflict activities P_4 and P_8 are assigned to $(D11, 3)$ and $(D12, 3)$ respectively. In $(D11, 3)$ the 3-time conflict free activity P_6 and the assigned activity P_4 are frozen etc. Finally, from the diagram $(D11, 3)$, one active schedule requiring the project makespan 5 is generated which the diagram 4 of figure 3 shows while from the diagram $(D12, 3)$, one active schedule requiring project makespan 6 is generated. For this example, there are 4 different optimal active schedules with project makespan 5 altogether.

4.3   Proving Correctness of $\mathcal{A}_{\mathcal{RSV}}$

A correctness proof for $\mathcal{A}_{\mathcal{RSV}}$ is given as follows:

Theorem 2. *For any given reduced $\mathcal{RSV}$-activity term A, $\mathcal{A}_{\mathcal{RSV}}$ generates nonredundantly all active schedules which may be derived from A.*

Proof. We show the theorem through structural induction on the term construction of A.

Induction base: If A is a ground activity, the proof is trivial.

Induction step: In the beginning the scan-line is found at the time $t_{SL} = 0$. After applying the first step all start activities of A are attached to the scan-line. In the following the scan-line jumps to the next stopping time l. Let $G_1, \cdots, G_n$ be all l-time direct scan-line activities, i.e. it holds that $RE(G_1) = RE(G_2) = \cdots RE(G_n) = l$. Now, the following 2 different cases have to be distinguished:

Case 1: There is at least one activity G_i which corresponds to a l-time conflict-free activity. First, all l-time conflict-free activities are frozen and then the occurring resource conflicts are resolved. Here, let the diagram be multiplied to k diagrams $D_1, \cdots, D_k$ so that each l-time conflict combination is assigned to a diagram respectively. After resolving the resource conflicts, for every diagram all l-time direct scan-line activities are deleted from the actual activity term T respectively. So, in every diagram the corresponding actual activity term T becomes smaller since at least one l-time conflict-free activity G_i is deleted from T. Furthermore, $\mathcal{A}_{\mathcal{RSV}}$ is applied recursively to every diagram accompanied by the corresponding actual activity term T. By induction hypothesis, $\mathcal{A}_{\mathcal{RSV}}$ generates nonredundantly all active schedules for every diagram since every corresponding actual activity term T is smaller than A. Moreover, $\mathcal{A}_{\mathcal{RSV}}$ generates nonredundantly all active schedules for A since $D_1, \cdots, D_k$ are pairwise different. It is obviously true for the case $k = 0$, i.e. all $G_1, \cdots, G_n$ are l-time conflict-free too.

Case 2: Each activity G_i is involved in a l-time resource conflict, i.e. there is no l-time conflict-free activity. First the occurring resource conflicts are resolved. Let the diagram be multiplied to k diagrams $D_1, \cdots, D_k$ so that each l-time conflict combination is assigned to a diagram respectively. For any D_i the following two subcases have to be distinguished:

Case 2.1: There is at least one activity G_i which is frozen. This case is very similar to the case 1.

Case 2.2: None of the l-time direct scan-line activities $G_1, \cdots, G_n$ is frozen. Then, there are further l-time conflict activities which are frozen. Let these be $H_1, \cdots, H_m$ where $RE(H_j) > l$ for each j must hold. Eventually $H_1, \cdots, H_m$ will be deleted from the actual activity term T and in the result T will become structurally smaller. So, by induction hypothesis, $\mathcal{A}_{\mathcal{RSV}}$ generates nonredundantly all active schedules.

Consequently, for the case 2, $\mathcal{A}_{\mathcal{RSV}}$ generates nonredundantly all active schedules for A since $D_1, \cdots, D_k$ are pairwise different.

$\square$

5   Summary and Future Work

The methods of description logics have been applied in order to formulate and solve a new general class of resource-constrained scheduling problems. Scheduling problems with variants are defined as activity terms of a concept language $\mathcal{RSV}$. The logic of $\mathcal{RSV}$ offered an effective approach for solving the $\mathcal{NP}$-complete $\mathcal{RSV}$-problem. Furthermore, based on the language $\mathcal{RSV}$ a new diagram-based method for representing reduced activity terms of $\mathcal{RSV}$ has been introduced. The nonredundant generation of all active schedules for any reduced activity term could be described graphically using $\mathcal{RSV}$-diagrams, in whose center a scan-line principle stands.

The resource availability we discussed in this paper falls into category of type $n/1/1$, according to Holloway et al.'s [12] notation, where the most general category of type $n/n/n$ stands for multiple resource types, multiple units of resources and multiple number of resource types required by a ground activity. Until now, many models which deal with the classical $\mathcal{RCPS}$-problem and fall into the category of type $n/n/n$ (e. g. [17], [7], [8], [18]) have been introduced. In these models, a constant amount of each resource is assumed to be available throughout the duration of the project and to be also demanded by a ground activity throughout the duration of the ground activity.

Future work may investigate a generalization of resource availability of type $n/n/n$ for the language $\mathcal{RSV}$. Such general problems may be easily formulated by generalizing the syntax $P(r,t)$ to $P((r_1, r_2, \cdots, r_n), t)$, where n corresponds to the number of resource types and r_i $(i = 1, \cdots, n)$ and $0 \le r_i \le b_i$ describes required units of resource type i by P. Here, each resource type i $(i = 1, \cdots, n)$ is assumed to be available in a constant amount b_i throughout the duration of the project. Otherwise the three structural symbols (operators) 'seq', 'xor' and 'pll' and the inductive rules for constructing activity terms may be applied unchanged.

References

1. A. Artale and Franconi E. A temporal description logic for reasoning about actions and plans. *J. Artificial Intelligence Research*, 9:463–506, 1998.
2. A. Artale and Franconi E. Temporal description logics. In *Handbook of Time and Reasoning in Artificial Intelligence*. MIT Press, 2000. forthcoming.
3. C. E. Bell and J. Han. A New Heuristic Solution Method in Resource-Constrained Project Scheduling. *Naval Research Logistics*, 38:315–331, 1991.
4. C. E. Bell and K. Park. Solving Resource-Constrained Project Scheduling Problems by A^* Search. *Naval Research Logistics Quarterly*, 37(1):61–84, 1990.
5. Fayez F. Boctor. Some efficient multi-heuristic procedures for resource-constrained project scheduling. *European Journal of Operational Research*, 49:3–13, 1990.
6. P. Brucker, S. Knust, and O. Schoo, A. Thiele. A Branch and Bound Algorithm for the Resource-constrained Project Scheduling Problem. *European Journal of Operational Research*, 107:272–288, 1998.
7. E. Demeulemeester and W. Herroelen. A Branch-and-Bound Procedure for the Multiple Resource-Constrained Project Scheduling Problem. *Management Science*, 38(12):1803–1818, 1992.
8. E. Demeulemeester and W. Herroelen. New Benchmark Results for the Resource-constrained Project Scheduling Problem. *Management Science*, 43(11):1485–1492, 1997.
9. Francesco M. Donini, Maurizio Lenzerini, Daniele Nardi, and Werner Nutt:. The complexity of concept languages. *Information and Computation*, 134(1):1–58, 1997.
10. S. E. Elmaghraby. *Activity Networks: Project Planning and Control by Network Models*. Wiley, New York, 1977.
11. E. A. Elsayed and N. Z. Nasr. Heuristics for Resource Constrained Scheduling. *International Journal of Production Research*, 24(2):299–310, 1986.

12. C. A. Holloway, R. T. Nelson, and V. Suraphongschai. Comparison of a Multi-Pass Heuristic Decomposition Procedure with other Resource-Constrained Project Scheduling Procedures. *Management Science*, 25(9):862–872, 1979.
13. J. E. Jr. Kelly. *The Critical Path Method: Resource Planning and Scheduling, Ch 21 in Industrial Scheduling, Muth, J. F. AND Thompson, G. L. (eds.)*. Prentice Hall, Englewood Cliffs, NJ, 1963.
14. M. M. Khattab and F. Choobineh. A New Approach for Project Scheduling with a Limited Resource. *International Journal of Production Research*, 29(1):185–198, 1991.
15. Pok-Son Kim. *Terminologische Sprachen zur Repräsentation und Lösung von ressourcenbeschränkten Ablaufplanungsproblemen mit Prozeßvarianten*. PhD thesis, Fachbereich Wirtschaftswissenschaften, Universität Frankfurt, 2001.
16. W. König, O. Wendt, and P. Rittgen. Das Wirtschaftsinformatik-Schwerpunktprogramm "verteilte DV-Systeme in der Betriebswirtschaft" der Deutschen Forschungsgemeinschaft -Frankfurt am Main, 1993-21 Bl. Technical Report 13, Institut für Wirtschaftsinformatik, 1993.
17. I. Kurtulus and E. W. Davis. Multi-Project Scheduling: Categorization of Heuristic Rules Performance. *Management Science*, 28(2):161–172, 1982.
18. A. Mingozzi, V. Maniezzo, and L. Ricciardelli, S. Bianco. An exact Algorithm for Project Scheduling with Resource Constraints based on a New Mathematical Formulation. *Management Science*, 44(5):714–729, 1998.
19. O. Oguz and H. Bala. A Comparative Study of Computational Procedures for the Resource Constrained Project Scheduling Problem. *European Journal of Operational Research*, 72:406–416, 1994.
20. T. Ottmann and P. Widmayer. *Algorithmen und Datenstrukturen*. Wissenschaftsverlag, Mannheim/Wien/Zürich, 1990.
21. A. B. Pritsker, L. J. Watters, and P. M. Wolfe. Multiproject scheduling with limited resources: A zero-one programming approach. *Management Science*, 16(1):93–108, September, 1969.
22. M. Schmidt-Schauß and G. Smolka. Attributive Concept Descriptions with Unions and Complements. Technical Report SEKI Report SR-88-21, FB Informatik, Universität Kaiserslautern, D-6750, Germany, 1988.
23. L. Schrage. Solving Resource-Constrained Network Problems by Implicit Enumeration-Nonpreemptive Case. *Operations Research*, 10:263–278, 1970.
24. J. P. Stinson, E. W. Davis, and B. M. Khumawala. Multiple Resource-Constrained Scheduling Using Branch and Bound. *AIIE Transactions*, 10(3):252–259, 1978.
25. F. B. Talbot. Resource-Constrained Project Scheduling with Time-Resource Trade-offs: The Nonpreemptive Case. *Management Science*, 28(10):1197–1210, 1982.
26. F. B. Talbot and J. H. Patterson. An Efficient Integer Programming Algorithm with Network Cuts for Solving Resource-Constrained Scheduling Problems. *Management Science*, 24(11):1163–1174, 1978.
27. J. D. Wiest. *The Scheduling of Large Projects with Limited Resources*. PhD thesis, Carnegie Institute of Technology, 1963.
28. D. Wood. An Isothetic View of Computational Geometry. Technical Report CS-84-01, Department of Computer Science, University of Waterloo, Jan. 1984.

Browsing Semi-structured Web texts using Formal Concept Analysis

Richard Cole and Peter Eklund

Distributed Systems Technology Centre (DSTC)
Knowledge, Visualization and Ordering Laboratory (KVO)
GRIFFITH UNIVERSITY
PMB 50 Gold Coast 9726, Australia
r.cole@gu.edu.au, peklund@dstc.edu.au

Abstract. Query-directed browsing of unstructured Web-texts using Formal Concept Analysis (FCA) confronts two problems. Firstly on-line Web-data is sometimes unstructured and any FCA-system must include additional mechanisms to structure input sources. Secondly many on-line collections are large and dynamic so a Web-robot must be used to automatically extract data. These issues are addressed in this paper. We report on the construction of a Web-based FCA system for browsing classified advertisements for real-estate properties[1]. Real-estate advertisements were chosen because they are typical of semi-structured textual information sources accessible on the Web. Furthermore, the analysis of real-estate data using FCA is a classic example used in introductory courses on FCA. However, unlike the classic FCA real-estate example, whose input is a structure relational database, we automatically mine Web-based texts for their structure.

1 Introduction

Mixed initiative[6] is a process from human-computer interaction involving humans and machines sharing tasks best suited to their individual abilities. The computer performs computationally intensive tasks and prompts human-clients to intervene when either the machine is unsuited to make a decision or resource limitations demand human intervention. Mixed initiative requires that the client be able to determine trade-offs between different attributes and alter search constraints to locate objects that satisfy an information requirement. This process is well suited to FCA and demonstrated in our previous work[1–3]. This paper reinforces these ideas by re-using the classic real-estate browsing application domain. The main difference with our previous work is that the browsing program for real-estate advertisements is more primitive than the CEM[1–3], which uses concept lattices to browse Email, and is also Web-based. However, when analysis becomes more demanding, we demonstrate how CEM can be re-used to produce

---

[1] In FCA the word *property* has a meaning similar to *attribute*. In this paper *property* is only be used with the meaning of *real-estate property*, e.g. a house or apartment.

H. Delugach and G. Stumme (Eds.): ICCS 2001, LNAI 2120, pp. 319–332, 2001.
© Springer-Verlag Berlin Heidelberg 2001

Fig. 1. The Homes On-line home-page. The site acts as the source of unstructured texts for our experiment.

nested-line diagrams for real-estate data imported from the Web. Other related work demonstrates mixed initiative extensions by using concept lattice animation, notably the algorithms used in CERNATO[2] and joint work in the GODA project[3] is underway to engineer tools for Web-based FCA solutions that feature nested line-diagrams as well as lattice animation.

This paper is structured as follows. Section 2 describes practical FCA systems and their coupling to relational database management systems (RDBMS). This highlights the necessity of structured input when using FCA. Section 3 describes the Web-robot used to mine structure from real-estate advertisements. This describes the methods required to extract structured data from unstructured Web-collections and measures their success. Section 4 shows the Web-based interface for browsing structured real-estate advertisements. Section 5 demonstrates how real-estate data can be exported and the CEM program re-used to deploy nesting and zooming[7].

2   Formal Concept Analysis and RDBMSs

FCA[5] has a long history as a technique for data analysis. Two software tools, TOSCANA[8] and ANACONDA embody a standard methodology for data-analysis based on FCA. Following this methodology, data is organized as a table in a

---

[2] See the paper "Multi-dimensional Representations of Conceptual Hierarchies" by Peter Becker in these proceedings where the nested-line diagram is animated to reveal structure.

[3] GODA is a Griffith/Darmstadt collaboration funded by the ARC and DFG.

RDBMS (see Fig. 2) and is modeled mathematically as a multi-valued context, (G, M, W, I) where G is a set of objects, M is a set of attributes, W is a set of attribute values and I is a relation between G, M, and W such that if (g, m, w_1) and (g, m, w_2) then $w_1 = w_2$. In the RDBMS table there is one row for each object, one column for each attribute, and each cell can contain an attribute value.

Organization over the data is achieved via conceptual scales that map attribute values to new attributes and are represented by a mathematical entity called a formal context. A conceptual scale is defined for a particular attribute of the multi-valued context: if $\mathbb{S}_m = (G_m, M_m, I_m)$ is a conceptual scale of $m \in M$ then we require $W_m \subseteq G_m$. The conceptual scale can be used to produce a summary of data in the multi-valued context as a derived context. The context derived by $\mathbb{S}_m = (G_m, M_m, I_m)$ w.r.t. to plain scaling from data stored in the multi-valued context (G, M, W, I) is the context (G, M_m, J_m) where for $g \in G$ and $n \in M_m$

$$g J_m n \iff: \exists w \in W : (g, m, w) \in I \quad \text{and} \quad (w, n) \in I_m$$

Scales for two or more attributes can be combined together in a derived context. Consider a set of scales, S_m, where each $m \in M$ gives rise to a different scale. The new attributes supplied by each scale can be combined together using a special type of union:

$$N := \bigcup_{m \in M} \{m\} \times M_m$$

Then the formal context derived from combining all these scales together is (G, N, J) with

$$g J(m, n) \iff: \exists w \in W : (g, m, w) \in I \quad \text{and} \quad (w, n) \in I_m$$

The derived context is then displayed to the user as a lattice of concepts.

A formal context is a triple (G, M, I) where G is a set of objects, M is a set of attributes, and I is a relation between the objects and the attributes, i.e. $I \subseteq G \times M$. A concept of a formal context (G, M, I) is a pair (A, B) where $A \subseteq G$, $B \subseteq M$, $A = \{g \in G \mid \forall m \in B : (g, m) \in I\}$ and $B = \{m \in M \mid \forall g \in A : (g, m) \in I\}$. For a concept (A, B), A is called the extent and is the set of all objects that have all of the attributes in B. Similarly, B is called the intent and is the set of all attributes possessed in common by all the objects in A. As the number of attributes in B increases, the concept becomes more specific, i.e. a specialization ordering is defined over the concepts of a formal context by:

$$(A_1, B_1) \leq (A_2, B_2) :\iff B_2 \subseteq B_1$$

More specific concepts have larger intents and are considered "less than" $(<)$ concepts with smaller intents. The same partial ordering is achieved by considering extents, in which case more specific concepts have smaller extents. The partial ordering over concepts is always a lattice and commonly drawn using

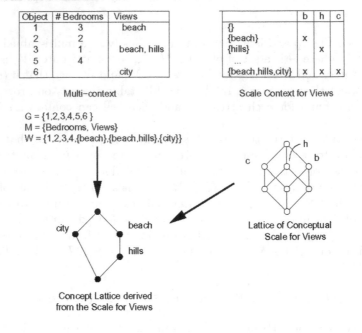

Object	# Bedrooms	Views
1	3	beach
2	2	
3	1	beach, hills
5	4	
6		city

Multi–context

	b	h	c
{}			
{beach}	x		
{hills}		x	
...			
{beach,hills,city}	x	x	x

Scale Context for Views

G = {1,2,3,4,5,6 }
M = {Bedrooms, Views}
W = {1,2,3,4,{beach},{beach,hills},{city}}

Lattice of Conceptual
Scale for Views

Concept Lattice derived
from the Scale for Views

Fig. 2. Example showing the process of generating a derived concept lattice from a multi-context and a conceptual scale for the attribute *Views*.

a Hasse diagram. Attribute and object labels are disambiguated by attaching object labels from below and attribute labels from above.

Consider Fig. 2. A RDMS table contains a list of real-estate properties (objects 1–6), the number of bedrooms and the type of views the properties afford. The multi-valued context has two attributes: *#Bedrooms* and *Views*. *Views* is organized by the scale context shown on the top-right of Fig. 2. The scale context has all possible combinations of *beach, hills,* and *city* views as objects and introduces three new attributes: *b, h* and *c*. The set of scale objects must contain all the attribute values taken on by objects for the attribute being scaled. The scale is applied to the multi-valued context to produce a derived context giving rise to the derived concept lattice shown in Fig 2 (lower-left). This lattice reveals that there are no objects having both views of the *hills* and *city* since the most specific concept (the concept at the bottom of the lattice) has an empty extent. Furthermore any object in the data set that has a view of the *hills* (there is only one, object 3) will also have a view of the *beach*. With large data sets (small numbers of attributes of interest, large number of objects) concept lattices are vastly superior to tables in their ability to communicate such information.

In practice it is easier to define a scale context by attaching expressions to objects rather than attribute values. The expressions denote a range of attribute values all having the same scale attributes. To represent these expressions in the mathematical description of conceptual scaling we introduce a function called the composition operator for attribute m, $\alpha_m : W_m \to G_m$ where $W_m = \{w \in$

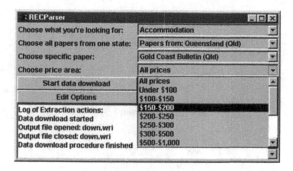

Fig. 3. Interface to WebRobot that "extracts" the advertisements from the Newclassifieds Web-site.

$W \mid \exists g \in G : (g, m, w) \in I\}$. This maps attribute values to scale objects. The derived scale then becomes (G, N, J) with:

$$gJ(m, n) \Leftrightarrow: \exists w \in W : (g, m, w) \in I \quad \text{and} \quad (\alpha_m(w), n) \in I_m$$

The main purpose of this summary of FCA is to reinforce that in practice FCA works with structured object-attribute data in RDBMS form, in conjunction with a collection of conceptual scales.

3 Web-robot for extracting structured data from unstructured sources

We were unable to negotiate cooperation with Newslimited, the owner of copyright on the real-esttate advertisements. Our initial intention was to obtain access to their structured classifieds database for student experiments introducing FCA. Undeterred, a purpose built script and interface to determine the query parameters from the Newslimited Web-site is shown in Fig. 3.

We therefore began with a sequence of real-estate advertisements in an HTML file rather than with the ideal format, an RDBMS export format. The first task is to separate the advertisements from the surrounding HTML mark-up and segment the advertisements into self-contained objects, one for each property. This was done using a string processing algorithm. An example advertisement is shown in Fig. 5. The text refers to six different properties, three with a rental price $250 per week and three with a price of $300. All properties are located in the suburb of *Arundel*. The format of the advertisements presents three main challenges: (i) the information about properties overlap, i.e. the single instance of the word *Arundel* indicates that all six properties are in Arundel, (ii) there are many aliases for the same basic attribute, e.g. double garage and dble garage, (iii) some information is very specific, e.g. 1.up garage or near golf course.

An LL(1) parser was constructed using the Metamata Java Compiler Compiler (JavaCC[4]) to parse advertisements of this type. The parser is able to handle the first two of these challenges with reasonable success. The parser recognizes pre-defined attributes and discards all unrecognized information.

---

[4] see http://www.metamata.com/JavaCC/

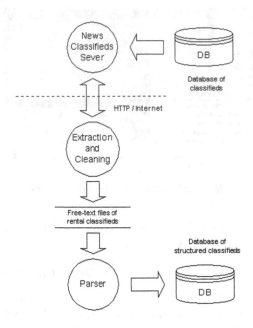

Fig. 4. Extraction DataFlow System Diagram.

The initial segmentation of the advertisements was able to extract 89% of advertisements. The remaining 11% were of low quality and omitted, they did not include a rental price and were therefore not meaningful. The parser recognized 64 attributes of which 53 were single valued, i.e. true or false. The remaining 11 attributes, including rental price, number of bedrooms and car park type, were multi-valued. To assess the accuracy of the parser, precision and recall were measured for each attribute and then aggregated. A summary of the most common and most important attributes for 53 rental properties is given in Table 1.

	Location	Price	Bedroom	Furnished	Car Park	Other
Frequency	100%	100%	100%	26.4%	50.9%	88.7%
Precision	100%	100%	100%	100%	100%	100%
Recall	94.3%	100%	98.1%	71.4%	96.3%	68.1%

Table 1. Recall and Precision for 53 unseen real-estate adverts.

N_A is the number of identified words and N_B the number of correct words. The *precision* of multi-valued attributes is calculated as the number of correctly identified attribute values ($|N_A \cap N_B|$) as a proportion of the number of identified attributes values ($|N_A|$). The *recall* is the number of correctly identified attribute values ($|N_A \cap N_B|$) as a proportion of the number of correct attribute values ($|N_B|$).

```
1 FOR RENT - ARUNDEL - Phone 55948184
2 $300
3 4 Bedrm, in-grnd pool, dble garage, near shops and school
4 3 bedrm, tripple garage, immac. presented, close to transport
5 Exec. 3 Bedrm + study, pool, dble garage, all ammen. close to school
6 $250
7 Leafy 3 bedrm, double garage, avail. Aug.
8 3 bedrm townhouse, resort fac. l.up garage, 2 bathroom and on-suite.
9 Townhouse, 2 bedroom, resort fac. garage, near golf course and transport.
```

Fig. 5. A rental classified advertisement illustrating multiple aliases for attributes (as in abbreviations such as Bedrm=bedroom), multiple objects (as rental properties described on lines 3, 4, 5, 7, 8 and 9) in a single advert (all lines) clustered on an primary key attribute: in this case the two prices $300 and $250.

Averaging the most important attributes — *Location, Price, Bedroom, Furnished, and Car parking* – weighted by their frequency yields a precision of 100% and a recall of 95% while the inclusion of the *Other* attribute reduces the recall to <70%. All real-estate advertisements leave out some information about the property they advertise, presumably because of the per word cost of advertising space. As a result we would expect the recall of actual information about the property being advertised to be much less w.r.t. the actual property.

One of the strengths of FCA is that it allows the user to compose views of the data that separate objects at different levels of detail. For example the user may have a coarse distinction based on price, but a fine-grain distinction based on proximity to facilities. Table 1 shows poor recall for attributes in the group *Other*. When combined with the knowledge that the adverts contain only partial descriptions of the data this places a practical limit on the level of detail that can be usefully explored. This limit could be extended if the initial data source was a database or XML file containing more extensive information about the features of properties for rent.

The LL(2) parser was very fast, building the relational database and storing the multi-valued context in under 8 seconds on a Pentium-III 300 MHz for an entire week's worth of adverts, approximately 3,400 properties listed in the local newspaper.

4   RFCA - the Web-based FCA Interface

The Web-based user interface presents a Web page with a scale selector as shown in Fig. 6. The client selects a suitable scale to browse through the newsclassified advertisements. The scales are pre-defined. The newsclassifieds are now in a structured database form after the parsing described in the previous section. When the user selects a scale, a new Web page is loaded containing the scale image. This image now contains all the resulting extent numbers from the scale's interrogation of the database. The number of objects in the extents are displayed

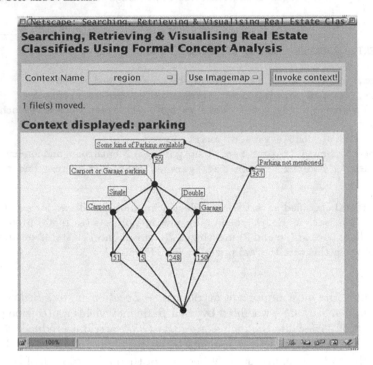

Fig. 6. The RENTAL-FCA prototype: Scales are pre-defined and selected off the "Context Name" menu (top-left). This figure shows the conceptual scale used for car parking and the extent numbers for each formal concept.

over each vertex in the usual way. The same scale selector is also available on the Web page displaying the scale image. This allows the user to select a new scale without having to go back to the previous page. In other words the same scale selection should be present on each of the pages displaying a selected scale.

A process that reproduces the Web page dynamically with different scales and extent numbers was implemented. This program creates the scale images after each selection by the user. A database connection and support for reading the scale files from the server are supported.

The Web pages with extent numbers do not exist as files but are generated on demand. When a scale is selected, the script calls the graph drawing program as a system command with the new scale name as parameter. This drawing program draws a concept lattice corresponding to the context. The result is stored as a PNG file representing the scale image and an image map representing the coordinates for the vertices in the graph. The image map also contains SQL queries extracted from the current context file. Queries in the image map corresponding to vertices in the PNG file are used to interrogate the database. After executing the graph drawing program the script starts to build the client-side image map. All the vertex coordinates are read in sequence from the image map and transformed to "hot" regions in the click-able image. Each hot region is linked to a CGI script with the SQL queries also read from the image map. When the user clicks a vertex in the scale image the browser loads another database extraction

Browser   Server

Page Request

HTTP   Query

Page & Image   CGI   ODBC
                                      program

                                                                        RDBMS

Fig. 7. System diagram of the Web-based FCA browser for real-estate classifieds.

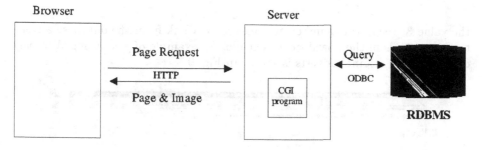

Fig. 8. From the scales view shown in Fig. 6 the user can navigate to the objects which are displayed in the structured extracted form as a database table.

script which produces a new Web page displaying the selected data. Such a Web page with a scale presenting classified data is displayed in Fig. 6.

The systems diagram for the Web-based front-end is shown in Fig. 7. Results must be displayed in the form of a table with the data extracted from the structured database. The Win32::ODBC module provided a secure way to establish a connection between the data extraction script and ODBC under Windows NT. A HTML table is built using the adverts received as rows from the database. A Web-page with the resulting HTML table of adverts is showed in Fig.8. All attributes are listed for each advert, boolean attributes replaced with an image-hook and abbreviated attributes replaced with full descriptions. Background colors for each advert row are alternated so the user can follow an advert when scrolling sideways.

Sometimes the original advert contains attributes not included in the database that can be of additional interest. The first column in the table is a running number that uniquely identifies the advert. This number is inserted when parsing the adverts. The table contains a column named *Id*. *Id* contains the number of the section from where the advert was parsed. So, if the advert was originally located in the third section in the free-text of the rental classifieds file, the column has the value **3**. Using this number, we can create a link from the database adverts to the originals in the downloaded text file. An example of a resulting Web page displaying the original adverts is shown in Fig. 9.

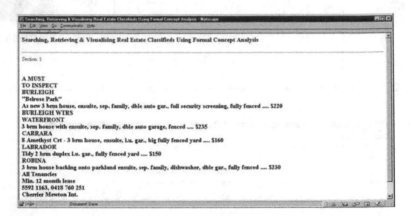

Fig. 9. By navigating by the *Id* in Fig. 8 the user can recover the text of the original unstructured text. This text can be dynamically generated by a query against the Newclassified Web-site if copyright is a concern.

5   Reusing CEM for Nesting and Zooming

Nesting and zooming[8, 7] are two well established techniques used in FCA. Together these techniques allow a user to wander around in a conceptual landscape[9] attempting to find concepts that satisfy their constraints. When searching for a real-estate property, there will obviously be compromises between location, price and other factors. By using concept lattices to show how constraints can be satisfied users are able to adapt their search to areas more likely to bear fruit. We re-used the CEM program to reinforce these ideas although the same approach could be implemented with some effort in the RENTAL-FCA prototype.

This contrasts with current on-line real-estate systems which ask the user to provide a specification for the type of property they are interested in and then (in most cases) provide either an empty list or a very long list of candidate properties. Using nesting and zooming in FCA allows questions like, "What are the possibilities for a mid-range house close to the city with a view, maybe close to park, shops or transport" as opposed to a question like: "List all mid-range houses that are close to the city, have a view, are close to a park, close to shops, and close to transport."

Consider a person who is new to a city and looking for accommodation. A good place to start is a decision about price. Fig. 10 shows a conceptual scale defined for price. The scale shows that most properties are either mid-range or expensive and that roughly 3/5's of each of the mid-range and cheap houses are in the intersection of mid-range and cheap. Consider that without more information the user is uncertain of what price range they are interested in.

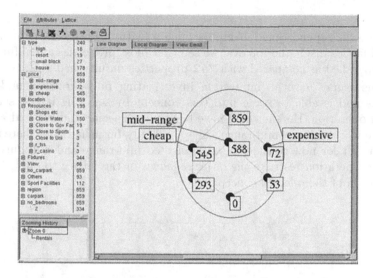

Fig. 10. Derived concept lattice showing how the properties are distributed with respect to three linguistic variables (scale attributes): *cheap*, *mid-range*, and *expensive*.

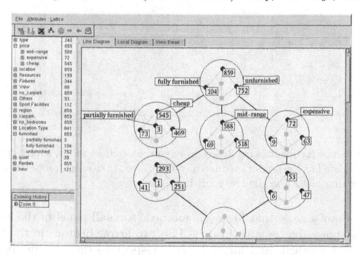

Fig. 11. A combination of the scales for *price* and *furniture* using a nested line-diagram.

They decide to add more information to the lattice by combining it with a scale specifying whether or not a property is furnished.

Fig. 11 combines the *price* scale with a scale for *furnished*, using a nested line-diagram. The rules for reading a nested line diagram are similar to reading a normal lattice. Thick lines connect ovals containing small lattices. The small grey circles show a location for a potential concept which is not instantiated by the data. The first thing to notice about the diagram is the large number of times that the middle concept of the inner diagram (the small lattices) is grey. The grey vertex indicates no mid-range or expensive partially furnished properties so the user needn't spend time looking for such a feature. Furthermore looking at the small lattice inside the top oval we see that most properties are unfurnished — **104** furnished as compared with **752** properties unfurnished.

The user may have an interest in investigating fully-furnished mid-range properties and is able to zoom into this concept by selecting it and selecting the zoom operation. He/she could have been more specific and selected a property in the intersection of mid-range, cheap and fully-furnished since there is such a concept but for now consider he/she selected mid-range and fully-furnished. The zoom operation restricts the objects shown in the lattice to only those in the extention of the selected concept.

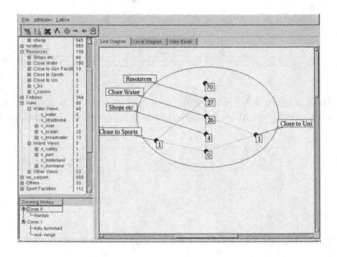

Fig. 12. A concept lattice showing access to resources such as water, shops, sports etc. The set of objects has been restricted to fully-furnished, mid-range properties, evident from the zooming history in the lower left-hand corner.

Fig. 12 shows a scale that has been zoomed. A small panel in the lower left hand corner shows the zooming history. The two arrow buttons in the tool-bar allow moving backwards and forwards with respect to the zooming operation in a manner similar to forward and backwards in Web browsers. The concept lattice now contains only **69** real estate properties since the zooming operation has restricted the object set to the extent of the concept for fully-furnished and mid-range. The conceptual structure in this lattice is different from the general picture without zooming. In the **69** properties shown, proximity to shops implies proximity to water (*Close Water*), and it is impossible to satisfy a desire to be close to University and close to shops in this restricted set of properties.

The user is now able to make a decision between different criteria, perhaps zooming further into the concept labeled *Close Water* or alternatively retrieving all four properties that are close to shops. Similarly the user is free to go back and make different zooming choices or include another scale with still more criteria to the current scale.

6 Conclusion

The paper demonstrates how FCA can be of use to search for rental properties on the Web even when the structure of the source data is unknown or unavailable. We believe the same technique is of use in browsing other unstructured legacy data on the Web.

A number of problems remain to be solved. The current browsing system is implemented as a stand-alone application and can only browse real-estate adverts with pre-defined scales. In order to be widely available it would have to extend to a distributed framework. A good candidate would be a Java Applet implementation of the graphical user interface communicating with a server. A Web-based FCA implementation of this sort is presently being engineered as part of the GoDa collaboration.

Another difficulty is that many users are unfamiliar or uncomfortable with concept lattice diagrams and require a form-based interface. In this way, the process and interpretation of the diagrams can be taught to the user while using the tool for the first time. The advantage of FCA, even without the concept lattice, is that feedback can be given on the volume of data satisfying search constraints.

The system we implemented obtained its data by parsing small textual descriptions of objects. The increasing use of the Internet is encouraging the storage of more structured information and thus in the future we expect the difficult task of constructing one-off parsers to suite specific textual descriptions will disappear as data is directly entered with structure. In other words, browsing XML data using FCA on the Web is vastly more simple than what has been described here although the techniques for mining structure from unstructured textual sources will be of value in various intelligence applications.

Acknowledgement The GoDa project is supported by the Australian Research Council and the DFG. This research also benefits from the support of the Distributed Systems Technology Research Centre (DSTC Pty Ltd) which operates as part of the Australian Government's CRC program. The authors also gratefully acknowledge the input of Åge Strand, Bernd Groh and Peter Becker.

References

1. Cole, R and P. Eklund, Analyzing an Email Collection using Formal Concept Analysis, *European Conf. on Knowledge and Data Discovery, PKDD'99*. pp. 309-315, LNAI 1704, 1999. Springer Verlag, 1999.
2. Cole, R. and G. Stumme: CEM: A Conceptual Email Manager *Proceeding of the 8th International Conf. on Conceptual Structures, ICCS'00*, LNAI 1867, pp. 438-452, Springer Verlag, 2000.
3. Cole, R, P. Eklund, and G. Stumme, CEM — A Program for Visualization and Discovery in Email, In D.A. Zighed, J. Komorowski, J. Zytkow (Eds), *Proc. of PKDD'00*, LNAI 1910, pp. 367-374, Springer-Verlag, 2000
4. Cole, R., P. Eklund and Å. Strand, Recovering Structure from Unstructured Web-accessible Rental Classifieds, Proceedings of the 4th Australian Document Computing Symposium (ADCS'00), DSTC Pty Ltd, pp. 131-140, 2000.

5. Ganter, B and R. Wille: Formal Concept Analysis: Mathematical Foundations Springer Verlag, 1999.
6. E. Horvitz, Uncertainty, Action and Interaction: In pursuit of Mixed-initiative Computing *Intelligent Systems* IEEE, pp. 17-20, September, 1999 http://research.microsoft.com/~horvitz/mixedinit.HTM
7. Vogt, F. C. Wachter and R. Wille, Data Analysis based on a Conceptual File, *Classification, Data Analysis and Knowledge Organization*, Hans-Hermann Bock, W. Lenski and P. Ihm (Eds), pp. 131-140, Springer Verlag, Berlin, 1991.
8. Vogt, F. and R. Wille, TOSCANA: A Graphical Tool for Analyzing and Exploring Data In: R. Tamassia, I.G. Tollis (Eds) *Graph Drawing '94*, LNCS 894 pp. 226-233, 1995.
9. Wille, R. Conceptual Landscapes of Knowledge: A Pragmatic Paradigm for Knowledge Processing In: W. Gaul, H. Locarek-Junge (Eds) *Classification in the Information Age*, Springer, Heidelberg, 1999.

Ossa - A Conceptual Modelling System for Virtual Realities

Finnegan Southey and James G. Linders

[1] Dept. of Computer Science, University of Waterloo Waterloo, Ontario, Canada, N2L 3G1 fdjsouthey@uwaterloo.ca
[2] Dept. of Computing and Information Science University of Guelph, Guelph, Ontario, Canada, N1G 2W1 jgl@snowhite.cis.uoguelph.ca

Abstract. As virtual reality systems achieve new heights of visual and auditory realism, the need for improving the underlying conceptual modelling facilities becomes increasingly apparent. The Ossa system provides a media-independent modelling environment based on a production system model that uses conceptual graphs to represent both the facts and the rules. Using conceptual graphs allows for interaction with the virtual world using multiple modalities (e.g. graphics and natural language). Conceptual graphs also allow for highly expressive facts and rules, and a diagrammatic programming technique. The motivation, design, and implementation of the Ossa system are discussed.

1 Introduction

In recent years virtual reality (VR) research has produced incredible sensory realism for large and complex worlds. In scientific visualization, engineering, education, and entertainment it is making rapid advances. The construction of virtual worlds may be divided into two broad areas: presentation and modelling. The presentation domain is primarily concerned with the rendering and interface presented to the user. The modelling domain is concerned with the simulation of behaviours and relationships amongst objects. It focusses on representing and manipulating knowledge about virtual worlds, often involving a variety of broad domains.

On the presentation side, significant progress has been made in producing realistic graphical and auditory environments. User interfaces have also improved, although not to the same degree. However, on the modelling side there has been less progress, except within specific sub-areas. For example, numerical modelling of physical laws has seen substantial advances, but the conceptual modelling of arbitrary relationships between objects has been neglected. This leads to virtual worlds with sensory verisimilitude and convincing physical behaviour but with "shallow" causal relationships and limited, often "scripted", interaction between virtual agents and objects.

To address these shortcomings, we present the design and implementation of Ossa [1], a conceptual modelling system for the development of virtual worlds which employs conceptual graphs (CG's) as its key method for representing knowledge and employs a production system to handle the world dynamics.

H. Delugach and G. Stumme (Eds.): ICCS 2001, LNAI 2120, pp. 333–345, 2001.

2   Background

2.1   Conceptual Modelling in Contemporary VR

Among the earliest examples of conceptual modelling that may be related to modern virtual reality are those found in traditional planning and robotics research. These problem domains involve reasoning about a model of the real world (e.g. STRIPS [2]). In such models, an agent interacts with the world producing a sequence of changes in that world, often with some sort of feedback to the agent. In selecting this comparison, we seek to introduce a broader conception of the term "virtual reality" than is popularly accepted. We will here state a simple definition to clarify our position:

> *virtual reality*: a computer-based model of some aspects of reality including the virtual presence of an agent that can perceive and interact with the model

We do not assert that this definition is more or less appropriate than any other definition. It is only provided to delimit the scope of our discussion. Note that we place no restrictions on how the virtual world may be presented or manipulated. We also state only that "some agent" interacts with the world. While the agent may often be human, it could be non-human (e.g. an animal employed in a pyschological study or a robot being trained or tested). We specify an interactive world to distinguish virtual realities from the broader class of "simulations".

With this definition in hand, we can again consider the kind of virtual worlds that bear a resemblance to early planning research. While not directly inspired by that research, a significant body of text-based virtual realities exist that offer similar conceptual models of reality, focussing on relationships between objects rather than on numerical simulation.

Originating in simple, single-user games and amusements, such systems underwent a rapid advancement in the early 90's forming a set of text-based, multi-user, internet-accessible virtual worlds commonly known as MUD's (from the name "multi-user dungeon") [3]. These systems originally described the world using only text and accepted interactions in the form of highly simplified natural language commands (e.g. "get the ball", "look", "give the apple to the moose"). More recently, people have experimented with adding two- and three-dimensional graphical interfaces [4][5], and are rapidly advancing towards the more popular conception of VR.

The conceptual modelling facilities offered in modern MUD's are, in most aspects, considerably more advanced than those offered by multimedia systems. With few exceptions (the most notable of which is LogiMOO [6]), the most powerful MUD's use an object-oriented (OO) language to build their conceptual models. The languages offered in LambdaMOO [7] and its like are very full-featured, offering a variety of features such as multiple-inheritance, polymorphism, garbage-collection, dynamic class-loading, and more. They have allowed the creation of highly sophisticated virtual worlds that deal in both concrete (e.g. walls, doors) and abstract (e.g. hostility, emotional state, desire) concepts.

2.2 Limitations of the Object-Oriented Approach

When exploring the design and construction of existing environments, we have found that the object-oriented model can serve as a hindrance to the ongoing expansion and refinement of a virtual world. Rome was not built in a day, and neither are complex virtual worlds. Their design is frequently revised over time to include new domains and new levels of detail. We found object-oriented models difficult to maintain in these circumstances, owing chiefly to the necessity of rewriting class interfaces or rearranging the inheritance hierarchy. The problems with object-oriented design in conceptual modelling and software design have been noted elsewhere [8][9], but we will briefly describe the chief problems for VR design here.

In many cases, relationships amongst entities are more plentiful and more complex than the entities themselves in a virtual world. The object-oriented model can complicate the implementation of relationships amongst entities, especially relationships involving many objects. Encapsulation can introduce unnatural and conceptually unnecessary boundaries into the description of a relationship. Such boundaries arise when the designer must decide how to divide responsibility for the relationship between classes (e.g. which classes should store the different pieces of information about the relationship). Such decisions can be complex and time-consuming, and subsequent revisions even more so.

Finally, object-oriented systems require an event model to detect changes in the world and respond to them. Events are typically modelled as messages originating from the object(s) involved in the event. It is difficult to foresee all the events that may be of interest in the future and the facilities for generating a given event must be explicitly provided. Adding support for events that may never be needed is wasteful but adding them at a later date may require extensive redesign.

In dealing with these issues we concluded that the best idea was to "promote" the role of relationships within the system to have a status equal to that of other entities and, at the same time, shed the encapsulation of the object-oriented approach while preserving inheritance capabilities. Additionally, we decided to replace the message-based event model with an execution model that can readily and flexibly detect and respond to any changing patterns (events) within the world's knowledge base. These two goals led us to the choose a production system for the execution model and conceptual graphs to fill the role of the basic unit of knowledge representation.

3 Alternate Approaches

3.1 Conceptual Graphs

Conceptual graphs offer an attractive set of features for knowledge representation. First and foremost, relations and concepts have an essentially equal footing in CG's. Relations have no constraints on their valence other than those imposed by the designer, and have a type hierarchy of their own allowing for specialization/generalization, a key feature offered by the object-oriented approach. Furthermore, the use of nested contexts allows for a great deal of expressive power while retaining some of the benefits found in an OO language's encapsulation.

In terms of interface modalities, CG's are attractive because of their close relationship with natural language processing. This means that natural language interfaces for designing or interacting with a virtual world are straight-forward compared to the object-oriented approach, and by separating the semantic information in the CG's from the actual words used, descriptions of the world may be rendered into several human languages. CG's are also mathematically formalized, which allows their expression using formal languages and mathematical formulae.

There is a diagrammatic representation of CG's which has reasonably wide acceptance. This allows for diagrammatic programming of the virtual world. While diagrammatic programming can easily become cumbersome in large systems, it may provide a useful way to browse the VR system. Diagrams may also offer a gentler learning curve to new programmers of the system (in the authors' experience, it is not uncommon for members of a VR user community to participate in development efforts and occasionally progress from complete novices to competent programmers).

Finally, work has been done on constructing three-dimensional graphical models[10] from CG's. While still in its infancy, this work is very important for using the world model in multimedia environments.

We envision that a CG-based virtual world would offer multi-modal interaction and rendering. An architectural project could begin with two-dimensional schematics of a building constrained by physical laws represented through mathematical formulae. The architect could then navigate through a three-dimensional rendering of the building and use simple verbal and gestural instructions to make minor alterations. Even details such as decoration and furniture layout could be included in the virtual specification. Naturally, this is far beyond our current reach, but we regard this as the goal towards which we should strive and which is best served by using an interface-independent knowledge representation like conceptual graphs.

3.2   Production Systems

In object-oriented VR work, there is a great tendency to "script" sequences of actions in a deterministic and fairly rigid manner. This is a natural side-effect of using the procedural programming approach. The result is a world in which only prescribed chains of events occur. This situation can be improved but it requires the creation of a sophisticated and time-consuming messaging event model that offers sufficient detail in its events to allow for a wide variety of "cause-and-effect" relationships.

Another popular means for describing dynamics is a *Petri net* [11]. While these provide a clean representation for a fully described set of processes, and moreover have a mapping to CG's, it is not immediately obvious how they can be applied without giving rise to the same "scripting" effect described above. Directly mapping an entire world state to a subsequent state is precisely what we strive to avoid. Rather, we wish to allow several distinct aspects of the world state give rise to several distinct consequences in an approximately parallel fashion, as occurs in the real world. Only a small portion of the world's state would provide the basis for the kind of simple cause-and-effect events we are attempting to capture, and so the Petri net's enumeration of the relevant states would be complicated and contrary to our goals.

The choice of a production system to provide the execution model stems directly from this simple view of "cause-and-effect" dynamics. Some subset of the world's state constitutes a cause which automatically gives rise to a change in the world state (an effect). Rather than explicitly scripting chains of events, the dynamics may be factored into simple pre-/post-condition pairs.

We believe that the production system approach offers this complexity of cause-and-effect in a relatively straight-forward fashion. Effects and their causes are expressed using production system rules. Adding a new cause-and-effect relationship is simply a matter of adding a new rule. Since a production system typically has unlimited access to its knowledge base, these rules can be based upon any set of facts the programmer chooses. This allows for virtually unlimited development of the system, unfettered by the need to elaborate an event model. Also, since production rules are independent of the facts they deal with, changing behaviours requires only the modification of a small number of rules, unlike the object-oriented approach where methods are bound to data structures and interfaces in such a manner that changes are often difficult or costly to implement.

4 The Ossa System

4.1 The Architecture of Ossa

Having identified CG's as our knowledge representation and production systems as our execution model, it was decided to fuse the two by using CG's to represent the facts and rules of a production system. Similar work has been done in the past within the CG community [12][13][14]. In our case, we decided to maximize the expressive power of our facts and rules by allowing the use of compound CG's. This can improve efficiency in the system by hiding details within nested graphs and by making rules that deal with these more focussed contexts. The result is the Ossa system. The name is taken from the Latin word for "skeleton", representing our belief that VR systems should have a strong conceptual model at the core of their construction.

Ossa is built using the three-layer architecture shown in Figure 1. The horizontal layers represent the three software layers which comprise the system. The two columns on the right represent the conceptual division of each layer into the two components of a production system, the working memory and the knowledge base (or rule set). The downward arrow represents the downward dependence of the layers, with each layer depending only on the layer below it and knowing nothing of the layer above it.

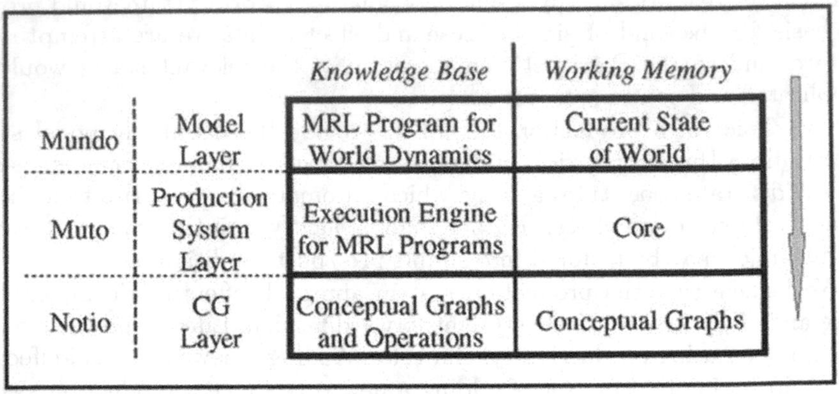

		Knowledge Base	*Working Memory*
Mundo	Model Layer	MRL Program for World Dynamics	Current State of World
Muto	Production System Layer	Execution Engine for MRL Programs	Core
Notio	CG Layer	Conceptual Graphs and Operations	Conceptual Graphs

Fig. 1. Architecture of Ossa

The bottom layer, which is called the Notio layer, provides the raw CG representation and handling ability. It is the reference implementation of our Notio API [15]. It provides facilities for constructing, manipulating, and matching CG's. It also provides a Conceptual Graph Interchange Format (CGIF) parser and generator for input and output. It is a general-purpose API which is flexible enough to be used for Ossa. From the production system perspective, CG's in Ossa form the facts of the working memory and are used in the construction and evaluation of rules for the knowledge base.

The middle layer, the Muto layer, provides the basic production system capabilities of the Ossa system. It is responsible for managing the CG's which make up the working memory, and for parsing and evaluating the rules that make up the knowledge base. It uses the classes and methods of the Notio layer to accomplish these tasks. It is divided into two basic parts. The *core* is essentially the working memory. It stores the CG's that are the facts describing the current state of the virtual world. The other part is the execution engine which evaluates and executes the production rules. It is called the *production agent* and operates by parsing and executing a set of rules written in the Muto Rule Language (MRL) . The MRL is described in greater detail below. The overall architecture of the Muto layer is shown in Figure 2.

The topmost layer is the model layer, which we have called the Mundo layer. This is the layer which is used to describe the conceptual model of the virtual world. It is created by writing a set of MRL rules that describe the dynamics of the world. There is typically a "bootstrap" rule as well which is automatically executed when the system is started. It is used to initialize the state of the

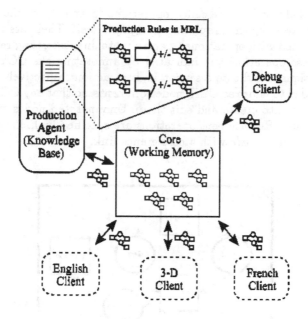

Fig. 2. Architecture of the Production System Layer

world by asserting a set of initial facts. The Mundo layer is unique for each world, although they may well share rules and facts in common.

Interaction with the virtual world is accomplished through clients that directly access the core of the Muto layer and assert facts about user actions. Any change in the Muto core is broadcast to the production agent, so it can respond, and to the clients so the world view can be updated for users.

4.2   The Muto Layer

MRL Rules. A very important feature of the Ossa system is the Muto Rule Language which is used to describe a virtual world's model. The MRL uses a small set of keywords in combination with CG's expressed using CGIF to describe the production system rules. Typical production system rules are composed of a precedent and an antecedent. The precedent is a pattern which is matched against the working memory to determine if the rule fires. When a rule fires, the antecedent determines the changes made to the working memory. Any relationships between the facts matching the precedent and the changes made by the antecedent are usually described using labeled variables.

The use of variables adds considerable complexity when CG's are used to describe the rules and facts. It requires variables linking multiple graphs which lead to confusing rules. We solved this problem by combining both precedent and antecedent in a single graph. The graph describes the elements that must be matched, and what changes are made once the match is found. Thus, a single graph contains both the precedent and antecedent for a single rule.

The graphs themselves are called operational graphs since they incorporate all operations involving the graph into the graph itself. There are four basic operations associated with operational graphs: matching, asserting, retracting, and replacing. These operations are indicated by annotating the CG's with "plus" and "minus" signs to indicate assertion and retraction respectively. These annotations are added to nodes as CGIF node comments, so that any CGIF-compliant editor can be used to create and edit them. Unannotated nodes are considered to be *match nodes*. Replacement operations are specified by linking an *asserting node* and a *retracting node* with a coreference link.

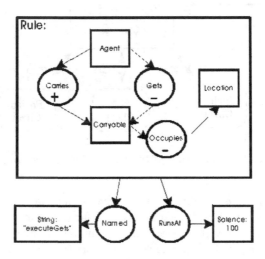

Fig. 3. A graphical view on an MRL rule

The diagram in Figure 3 shows a graphical view of an MRL rule which represents the action of picking up an object. The Rule concept is given both a name and salience (see below) using relations. The graph nested within the Rule concept is an operational graph with matching, asserting, and retracting nodes. The rule in Figure 3 and the rest of an associated MRL world file are shown in linear MRL notation in Algorithm 1[1].

Operational Graphs. The basic way in which an operational graph is evaluated is as follows:

1. find all projections of the operational graph, O, in the working memory graph, M, ignoring any asserting nodes in O
2. if one or more projections exist, the operation graph is said to have fired
3. for each projection found for a fired graph

---

[1] Note that the CGIF syntax used in MRL is based on an early draft of the CGIF standard and does not conform completely to more recent versions of CGIF.

Algorithm 1 A sample MRL world file

```
world SimpleSample;
// A rule to initialize the world.
// It detects the presence of a [_Start] concept,
// retracts it, and creates some
// concept and relation types.
rule initializeWorld(salience 1000) {
  /* Match, and if found, retract a _Start concept. */
  if {{ [_Start;-] }} then {
    concepttype Agent;
    concepttype Gender;
    concepttype Location;
    concepttype Carryable;
    concepttype Male < Gender;
    concepttype Female < Gender;
    concepttype Brick < Carryable;
    concepttype Person < Agent;
    concepttype Man < Person, Male;
    relationtype Acts;
    relationtype Gets < Acts;
    relationtype Carries;
    relationtype Occupies;
  }
}
// Detects a "Gets" act, retracts the [Gets] concept,
// retracts the (Occupies)relation of the Carryable object
// and asserts the new (Carries) relationship
// between the Agent and the Carryable.
rule executeGets(salience 100) {
  execute {{ [Agent*x][Carryable*y] (Gets?x?y;-)
             [Location*z](Occupies?x?z) (Occupies?y?z;-)
             (Carries?x?y;+)
  }}
}
```

---

a) any nodes in M that correspond to retracting nodes in O are removed from M

b) any asserting nodes in O are copied and joined to M using matching nodes as join points

If multiple rules are matched by the current state of the working memory, conflict resolution is applied. There are many kinds of conflict resolution [16], but we have implemented an extremely simple form which establishes a precedence amongst rules by assigning them a *salience* value. Rules of higher salience are executed first. If two or more rules of the same salience are matched, they are all fired. For example, the rule shown in Figure 3 has been assigned a salience of 100.

MRL Statements. The MRL contains four statements that control the effect of a rule.

An *execute* statement takes an operational graph as an argument. It simply evaluates the graph and fires it if appropriate. An *if-then-else* statement uses an operational graph as a condition. If the operational graph fires, the statement block following the *then* keyword is executed. If it does not fire (no projection is found), the *else* block is executed.

This allows for a rule to contain several, nested sections with different operational graphs. Graphs in different statement blocks can be linked via coreference labels. This forms an extremely powerful means for producing multi-function rules.

The last two statements, *concepttype* and *relationtype*, are used for declaring new types and can optionally specify sub- and supertypes for the new type.

Muto Clients. The Muto system consists of the *core*, which holds the current state of the world, and a set of *clients* (or agents) that interact with the core to produce changes. The clients typically represent user interfaces which present some information about the world state to the user by querying the core and rendering the result. The rendering depends on how the client is designed to present the world. For a text-based client, the world may be presented in English text, or perhaps in some other natural language. Alternately, the client could provide a 3-dimensional rendering[2]. The clients also take user commands and translate them into operational graphs which are submitted to the core for possible execution. This clearly requires another level of translation depending on the type of client. This issue of client translation is clearly important, but is not addressed in this paper which concerns only the internal representation.

One special client, called the *production agent*, is responsible for executing the MRL rules that drive the world and respond to changes introduced by the user. This functionality is separated from the core to allow for several separate agents that would govern world control. This offers the possibility of future modularity and allows for control schemes other than MRL. At present only a single production agent is used in practice.

The Muto Core. All of the facts about the world are contained within the Muto *core*. In the current implementation, this is simply one, large (usually disconnected) CG represented using instances of the Notio API classes. The core is manipulated by clients issuing *core requests*. These are instructions for the core, but are called "requests" because the core may opt to ignore or only partially fulfill a request. This allows the core to prevent clients from causing problems or exceeding some fixed set of capabilities. There are four kinds of request:

---

[2] In Figure 1, the clients shown in dotted have not been implemented. Only the debug client, which uses annotated CGIF has been implemented at the time of writing.

1. *Query* Requests: The requests consist of a query graph. The results of matching this graph are returned to the issuing client only. Such graphs might be used to obtain information required for rendering a user's view.
2. *Change* Requests: These requests usually consist of an operational graph which is executed against the graph stored in the core. If the changes are accepted by the core, they are not immediately made, but are instead added to a *schedule* of changes. This prevents the results of a series of changes from interfering with each other. Commands from users and triggered events in the world typically result in change requests.
3. *Commit* Request: These requests simply ask the core to commit the changes currently pending in the schedule. For example, a client may issues several change requests that establish some effect they wish to have on the world. Once all have been submitted and accepted, the client would issue a commit request to activate the changes.
4. *Rollback* Request: These requests ask the core to clear all pending changes from the schedule. This can be used by a client to cancel a partially issued series of changes as the result of user input or the failure of a prior request.

Query and change request graphs are issued to the core using annotated CGIF. When changes are commited, the effects of the changes (i.e. the projections and asserted/retracted nodes) are broadcast to all clients via a Muto *event*. Clients can use these events to update their rendering of the world or as a basis for requesting new information. The production agent uses the events to detect changes in the world and test the appropriate rules in its world file to see if some response is required. Any response by the production agent is submitted in turn as a change request.

5   Conclusions

Ossa, as specified above, has been fully implemented and used to create a simple virtual world. This world offers features commonly found in MUD systems, such as taking and carrying objects and moving between different locations (a model of several locations at the University of Guelph campus). It also features some more advanced abilities which are non-trivial, even in advanced MUD's, such as preventing entrance to a location. The world proved very easy to develop and involved considerably less code than equivalent object-oriented approaches (tens of lines vs. hundreds of lines). Making substantial changes to the behaviour of the world was easy as well, since they generally required the modification or addition of a very few rules, or simple changes to the type hierarchies. While not nearly as extensive as typical MUD's, this simple world has served as a successful proof-of-concept for the modelling approach and revealed some advantages and disadvantages to the approach.

Some care has to be taken while building in order to understand the interaction of the various rules. Unforeseen interactions can be a major problem in any production system. However, we think that the complexity gained through the interactions of simple rules can greatly enhance the resulting world. We also believe that tools can be developed to help designers detect and understand the interactions within a rule set so that they can be exploited rather than allowed to cause problems.

The current implementation offers only a single client for interacting with the system, called the debug client. The interaction is entirely via annotated CGIF statements that are used to issue commands and see the results. Natural language and graphical interfaces are clearly of great interest but a richer conceptual model is required before they can be attempted. In particular, spatial and temporal ontologies need to be explored that can be effectively rendered into different forms.

Some possible enhancements to the MRL and the Muto core include more sophisticated conflict resolution, more control over criteria used to find projections into the working memory, new operations that allow direct changes to the type or referent of a concept (this can already be achieved in a somewhat crude fashion using replacement), and improving the efficiency of the rule matching.

Conceptual modelling using CG's is an active area of research. Probably the greatest problem we encountered while working with Ossa was the temporal arrangement of events. In abandoning the procedural approach and adopting a production system model, we have to contend with the complex execution patterns that can occur. Guaranteeing that events occur in a sensible fashion is a task which we believe depends greatly on the modelling technique and ontology. We hope to exploit existing research on temporal modelling [17][18] using CG's in future efforts.

Finally, the Ossa system has served to demonstrate that VR may be built on the principled foundations of knowledge representation and logic. It is our hope that other researchers will apply advanced conceptual modelling technologies to VR systems and help to breach one of the most significant barriers to a complex virtual world.

Acknowledgements. Many thanks to Relu Patrascu and Rami Zeineh for their numerous suggestions regarding this work and also to the community at Kobra-MUD[3] for their input and guidance. Thanks to Stuart Statman for pointing in a promising direction.

References

1. Finnegan Southey, "Ossa: A Modelling System for Virtual Realities Based on Conceptual Graphs and Production Systems", MSc. thesis, supervisor Dr. James G. Linders, University of Guelph, Canada, 1998.
2. R.E. Fikes and N.J. Nilsson, "STRIPS: A new approach to the application of theorem proving to problem solving", Artificial Intelligence 2:189-208, 1971.
3. Richard Bartle, "Interactive Multi-User Computer Games", MUSE Ltd, British Telecom plc.,December 1990.

---

[3] http://kobra.et.tudelft.nl

4. T Usaka, S. Yura, K. Fujimori, H. Mori, K. Sakamuram, "A multimedia MUD system for the digital museum", Proceedings. 3rd Asia Pacific Computer Human Interaction (Cat. No.98EX110), IEEE Comput. Soc, Los Alamitos, CA, USA,1998, xviii+474 pp. 32-37.

5. T. Meyer, D. Blair, D. B. Conner, "WAXweb: toward dynamic MOO-based VRML", 1995 Symposium on the Virtual Reality Modeling Language (VRML '95), page 105-108, ACM, New York, NY, USA, 1996.

6. Paul Tarau, Veronica Dahl, Stephen Rochefort, and Koen De Bosschere, "Logi-MOO: a Multi-User Virtual World with Agents and Natural Language Programming", in S. Pemberton, editor, Proceedings of CHI'97, pages 323-324, March 1997.

7. Pavel Curtis and David A. Nichols, "MUDs Grow Up: Social Virtual Reality in the Real World",Third International Conference on Cyberspace, Xerox PARC, May 1993.

8. Sowa, John F., "Knowledge Representation: Logical, Philosophical, and Computational Foundations", Brooks/Cole, CA, 2000.

9. Jackson, Michael J., "Software Requirements and Specifications: a lexicon of practice, principles, and prejudices", Addison-Wesley, Great Britain, 1995.

10. W. R. Cyre, S. Balachandar, and A. Thakar, "Knowledge Visualization from Conceptual Structures" in "Conceptual Structures: Current Practices - 2nd International Conference on Conceptual Structures, ICCS '94", pp. 275-292, Springer-Verlag, Germany, 1994.

11. Petri, Carl Adam, "Kommunikation mit Automaten", Ph.D. dissertation, University of Bonn, English translation in technical report RADC-TR-65-377, Griffiss Air Force Base, 1966.

12. Kabbaj, Adil and Janta-Polczynski, "From PROLOG++ to PROLOG+CG: A CG Object-Oriented Logic Programming Language", Proceedings of the 8th International Conference on Conceptual Structures (ICCS 2000), Springer-Verlag, 2000.

13. Michael Chein, "The CORALI Project: From Conceptual Graphs to Conceptual Graphs via Labelled Graphs", in Proceedings of the Fifth International Conference on Conceptual Structures, ICCS'97, pages 65-77, Springer, 1997.

14. Rao, A. S. and Foo, N., "CONGRES: CONceptual Graph REasoning System", in Proceedings of the IEEE Conference on Applications of Artificial Intelligence, Orlando, Florida, pages 87-92, 1987.

15. Southey, F. and Linders, J. G., "Notio - A Java API for developing CG tools", in Proceedings of the 7th International Conference on Conceptual Structures (ICCS'99), Springer-Verlag, 1999.

16. J. McDermott and C. Forgy, "Production System Conflict Resolution Strategies", from Pattern Directed Inference Systems, edited by D. A . Waterman and F. Hayes-Roth, Academic Press, 1978.

17. John W. Esch, "Temporal Intervals", in "Conceptual Structures: Current Research and Practice", edited by Timothy E. Nagle, Janice A. Nagle, Laurie L. Gerholz and Peter W. Eklund, pp. 363-380, Ellis Horwood, 1992.

18. John W. Esch and Timothy E. Nagle, "Representing Temporal Intervals Using Conceptual Graphs" in "Proceedings of the Fifth Annual Workshop on Conceptual Structures", edited by Laurie Gerholz and Peter Eklund, pp. 43-52, Boston, USA and Stockholm, Sweden, 1990.

Uses, Improvements, and Extensions of Prolog+CG : Case Studies

Adil Kabbaj[1], Bernard Moulin[2], Jeremi Gancet[2], David Nadeau[2], and
Olivier Rouleau[2]

[1] INSEA, Rabat, Maroc, B.P. 6217
akabbaj@insea.ac.ma

[2] Université Laval, Computer Science Department
Québec, Québec, Canada G1K 7P4
moulin@ift.ulaval.ca, lapsus@videotron.ca,
david.nadeau@moncourrier.com, jgancet@yahoo.com

Abstract. Prolog+CG is a CG-Based logic programming language which
integrates Prolog, the manipulation of conceptual graphs (CGs), Java and
object-oriented constructs. It provides a powerful development environment for
the creation of knowledge-based applications and their integration on the web.
Java provides object-oriented capabilities allowing the development of multi-
platform applications. Object-oriented Prolog provides the full power of an
object-oriented logic programming language and CGs provide the expressive
power of an advanced knowledge representation language. This paper presents
the recent extensions that have been added to Prolog+CG and illustrates some
typical uses of the environment for the development of various applications.

1 Introduction

The language Prolog+CG has been developed by A. Kabbaj [2, 3] in order to extend
the Prolog language in two main directions : a *conceptual extension* allowing the
representation of goals with conceptual graphs (CGs) and the manipulation of simple
and compound CGs (type hierarchy, operations on conceptual graphs, etc.); an *object
oriented extension* allowing the manipulation of objects (object specification,
inheritance mechanisms, etc.).
Recently, Prolog+CG has been extend to include a natural and powerful interface with
Java. This integration benefits from the fact that both the Prolog+CG language and its
development environment are implemented using Java. Hence and thanks to this new
interface, Prolog+CG can be called from a Java code and inversely, Java classes can
be manipulated from a Prolog+CG program.
    The integration of Java, object-oriented Prolog and the manipulation of CGs
provides a powerful development environment for the creation of knowledge-based
applications and their integration on the web. Java allows the development of multi-
platform applications as well as the capabilities of object-oriented languages. Object-
oriented Prolog provides the full power of a logic programming language well suited
for natural language processing, inference and symbolic manipulations. CGs provide

H. Delugach and G. Stumme (Eds.): ICCS 2001, LNAI 2120, pp. 346–359, 2001.

the expressive power of an advanced knowledge representation language (advanced semantic nets, type hierarchy, schemas, notion of context, etc.). During the past six months an intensive and productive collaboration between A. Kabbaj and the other authors has enabled the development of new extensions to Prolog+CG which are strengthening the integration of its three language paradigms and providing new capabilities. This paper reports on the authors' experience using Prolog+CG language and its development environment. Some aspects of the language and its environment have been refined, some bugs have been identified and fixed and several extensions have been suggested and implemented.

The paper is organized as follows. Section 2 reports on the extension of Prolog+CG which allows calling a Prolog+CG program from a Java program. Section 3 illustrates the expert system mode available in Prolog+CG. Section 4 reports on the extension of Prolog+CG which enables the call of applications from a Prolog+CG program. Section 6 presents the extension of Prolog+CG which allows the use of Java code : creation of an instance of a class, accessing attributes and calling methods of instances or classes. It also illustrates how these extensions can be used in a natural language processing context. Section 6 and 7 present two other case studies. Section 8 discusses future orientations, and the potential use of Prolog+CG for the development of advanced applications.

Other extensions of Prolog+CG, such as types hierarchy operations and operations on simple and compound CGs with sets and co-references are described in [4].

Figure 1 shows a snapshot of the new Prolog+CG environment : the main frame is a split pane that can be used to edit a program (in the top pane) and to interact with the interpreter (in the bottom pane). Prolog+CG 2.5 provides also a visual debugger which shows the construction/deconstruction of the inference tree (Figure 1) and an "inspection" feature which allows the user to have a detailed description of the inference tree node (Figure 1, the window at the bottom) or to have the value of a variable. More details on the new Prolog+CG environment is provided through the Help menu and the site : www.insea.ac.ma/CGTools/PROLOG+CG.htm

2 Calling Prolog+CG from Java Programs

The Prolog language provides powerful reasoning and symbolic manipulation capabilities, but it is not well suited to implement the interface functionalities (windows, menus, link to data bases, etc.) that object-oriented languages provide. Hence, a good development strategy consists in using each Prolog+CG programming paradigm to implement the functionalities that it best supports. It is recommended to develop programs using Java and to call the Prolog+CG modules when it is appropriate. This section presents a set of primitives that Prolog+CG 2.5 provides in order to call Prolog+CG modules from Java programs. The use of these primitives is illustrated using a simple example.

Four primitives are used to exploit a Prolog+CG program from a Java program

✓   **void LoadFile(String fileName)** :   *"fileName"* is the name of a Prolog+CG program. The role of this primitive is to load a Prolog+CG program contained in the file *"fileName"*.

✓ **Vector Resolve(String Quest [, boolean convertResult] [, boolean ExpertSystemMode])** : the parameter *"Quest"* is the request that Prolog+CG should resolve and the two other arguments are optional :

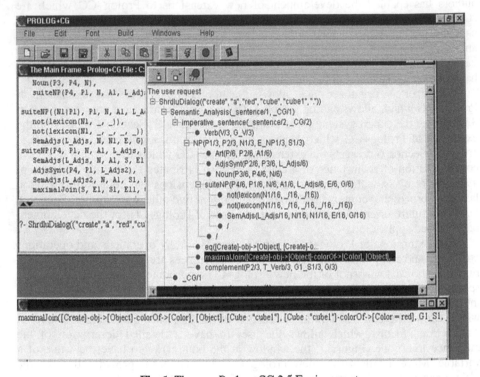

Fig. 1. The new Prolog+CG 2.5 Environment

• *"convertResult"*, if it is specified, is a boolean that indicates if the returned result should be converted to String or to be returned as Java objects. This argument is set to true (i.e. convert to String) by default.

• *"ExpertSystemMode"*, if it is specified, is a boolean that indicates if Prolog+CG should behave like an expert system shell or like a Prolog interpreter. This second argument is optional and it is set to false by default.

The primitive goal Resolve/3[1] returns all the solutions in a Vector of hashtables; one solution being represented by a hashtable. Since a solution is a set of couples <VariableIdentifier, VariableValue>, it is natural to represent it as a hashtable with VariableIdentifier as a key. To get the value of a variable in a solution, one has to use the hashtable method "get(key)" where key stands for a variable identifier.

✓ **void SaveFile(String fileName)** : saves the current program in the Prolog+CG file *"fileName"*.

✓ **void PurgeMemory()** : deletes the current Prolog+CG program from memory.

---

[1] Resolve/3 means a predicate with identifier Resolve and three arguments (i.e. arity 3).

Example :

Assume that we have a Prolog+CG program *"FamilyRelations.plgCG"* which contains, among other things, the following facts and rules :

```
[Man : John]-fatherOf->[Man : Peter].

[Woman : Deborah]-motherOf->[Man : Peter].

[Man : John]-fatherOf->[Woman : Clara].

[Person : x]-parentOf->[Man : y] :-

        [Person : x]-fatherOf->[Man : y].

[Person : x]-parentOf->[Woman : y] :-

        [Person : x]-motherOf->[Woman : y].
```

From a Java program, we can assert new facts and retract existing facts from the above Prolog+CG program and we can ask questions and get answers. Here is an example:

```
//PrologPlusCG is implemented as a Java Package
import  PrologPlusCG.PrologPlusCGFrame;
public void example() {
        ... any Java code
        // Load a Prolog+CG File
        LoadFile("FamilyRelations.plgCG");
        ... any Java code
        // Resolve a question
        Vector  solutions  =
Resolve("[Person : x]-parentOf->[Person : y].");
        // Make access to the solutions and to their
        content
        Hashtable solution;
        String valX, valY;
        For (Enumeration e = solutions.elements();
e.hasMoreElements(); ) {
        // Make access to a solution
        solution = (Hashtable) e.nextElement();
        // Make access to the variable's value for
            the solution
        valX = (String) get("x");
        valY = (String) get("y");
        // Make use of the variable's value
};
        // Assert a new fact in the Prolog+CG program
Resolve("asserta([Man : Marc]-fatherOf->[Man : Clark],
        ()).");
        // Save the current program in a new file
SaveFile("FamilyRels1.plgCG");
        // Purge the current program from the memor
PurgeMemory();
```

The four primitives "LoadFile", "Resolve", "SaveFile" and "PurgeMemory" are "external" primitives of Prolog+CG and constitute the interface from Java to Prolog+CG : they are used in a Java Program to make use of a Prolog+CG program. These simple external commands may look like details but they are a tremendous step forward in making the Prolog+CG language a concrete development tool. The interface inability to convey a sense of usability and user-friendliness has long been a major obstacle in using logic programs in real end-user developments.

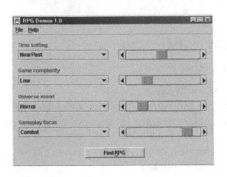
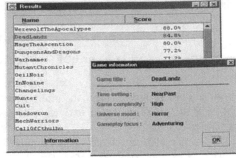

(a) an input screen   (b) output screens

Fig. 2. Front/end interface for a Prolog+CG program

For instance, it is not unusual to have many lines of Prolog code to enter before making a useful query. In addition, a lot of parameters in these lines of code are very sensible to error and require that the user know the program's internal workings fairly well to be able to get any valuable information from the system. In addition, the answer has the typical Prolog form (an output list) and it is sometimes very hard to read. The output list (i.e. the solution) is neither sorted alphabetically nor numerically. Finally, if the user wants additional information on one of the listed items, an additional query is necessary. All these considerations find an elegant solution using the *Java/Prolog+CG* interface. We have used this interface to develop a simple Prolog+CG program (Figure 2). The front/end interface has been implemented using Java and the reasoning part using Prolog+CG. The input screen (Figure 2.a) offers lists and slider bars and actually writes in the Prolog+CG program four assertions for the user, using his selections. A results-screen (Figure 2.b) then presents the output list which can be sorted using any criteria. Additional information on a given list item can be obtained with the simple press of a button (Figure 2.b). Using this example, we wanted to emphasize the importance of having interfaces between languages thus allowing each to be used for what it is best at.

3 The Expert System Mode of Prolog+CG

Prolog+CG 2.5 provides two resolution modes : a Prolog-like resolution which is the default mode and an expert system shell mode. In the first mode, when a goal is not a primitive nor a defined goal, the resolution process assumes that it is false. In the expert system mode however, the resolution process asks a question to the user in order to establish the truth of such a goal and it records it (using an assert operation).

Example : A excerpt from a simple expert system about animal classification

```
Universal > Object, Animal, Person, Action, State,
Attribute.
Object > Hair, Meat, Teeth, Claw, Eye.
Animal > Mammal, Carnivore, Cheetah.
Person > Man.
Action > Eat.
State > Belong.
Attribute > Color, Component, Tawny, Dark, Pointed,
Forward.
Man = Robert.
Animal = Yala.
[Animal : x]-is->[Carnivore] :-
[Eat] -
        -obj->[Meat],
-agnt->[Animal : x].
[Animal : x]-is->[Carnivore] :-
   [Animal : x]-poss->[Teeth]-attr->[Pointed],
   [Animal : x]-poss->[Claw].
[Belong] -
        -bnfcre->[Man : Robert],
        -pat->[Animal : Yala] -
                -colorOf->[Color]-attr->[Tawny],
                -poss->[Teeth]-attr->[Pointed] ;.
```

After the compilation of this example, the user can set the expert system mode to "on" by activating, from a Prolog+CG menu, the action "Build/Expert System Mode". Here is an excerpt from the interaction that took place :

```
?- [Animal : Yala]-is->[Cheetah], /.

==> The goal   : [Animal : Yala]-poss->[Hair] cannot be
infered from what is known, so

==> Is it true that : [Animal : Yala]-poss->[Hair] ?
tape y (for yes) or n (for no) : y.
```

As a side effect of the whole interaction, the following facts are added to the database

```
[Animal : Yala]-poss->[Hair].

[Eat] -

   -obj->[Meat],

   -agnt->[Animal : Yala].

[Animal : Yala]-partOf->[Component]-attr->[Dark].
```

The expert system mode of Prolog+CG can be exploited from a Java program by using the primitives described in Section 3. In this case, the third argument of the primitive "Resolve/3" should be used and should be set to "true".

To make the expert system mode of Prolog+CG a kernel for a complete expert system shell we are implementing the explication capability (i.e. respond to Why/How/What questions).

4 Calling Applications from a Prolog+CG Program

Prolog+CG 2.5 provides two primitives which allow calling an executable application from a Prolog+CG program. This is another way to make use of Prolog+CG 's capabilities to work as a component of a larger system.

✓   **exec(ExecFileName) :** starts the execution of the application *"ExecFileName"* and continues the current resolution.

✓   **execAndWait(ExecFileName) :** starts the execution of the application *"ExecFileName"* and waits for its termination. Then it resumes the current resolution of the Prolog+CG program.

Example :
```
ex1 :-
  eq(x, 45),
  exec("Synergy.exe"),
  write("Continue without waiting for the end of
Synergy..."), /.
ex2 :-
  eq(x, 45),
  execAndWait("Synergy.exe"),
  write("Continue after the end of Synergy ..."), /.
```

5 Using Java Code from Prolog+CG

Prolog+CG 2.5 provides several primitives allowing the use of Java code and providing access to object-oriented capabilities. Indeed, new Java objects can be created, used and destroyed from a Prolog+CG program. These objects are considered as "global objects" of a Prolog+CG program.

This section presents these primitives and briefly illustrates their use.

✓   **new(NewObjectIdent, JavaClass, ConstructorArguments) :** this primitive goal enables the creation of a new Java object. It corresponds to the following Java instruction :

```
JavaClass NewObjectIdent = new JavaClass
ConstructorArguments;
```

✓   **destroy(ObjectIdent)** : this primitive goal destroys the Java object identified by *"ObjectIdent"*.

✓   **destroyAll** : this primitive goal destroys all the Java objects that have been created in the current program.

✓   **get(Data, AttributeIdent, ObjectIdent)** : this primitive goal can be used to get the value of the attribute *"AttributeIdent"* of the object *"ObjectIdent"*. The value is then unified with the first argument *"Data"*.

✓   **set(Data, AttributeIdent, ObjectIdent)** : this primitive goal can be used to set *"Data"* as the value of the attribute *"AttributeIdent"* of the object *"ObjectIdent"*.

✓   **execMethod(Data_Or_void,   JavaClass_Or_ObjectIdent,   MethodIdent, MethodArguments)** :
this primitive goal can be used to call a method *"MethodIdent"*, of a Java class (in the case of a static method) or of a Java object with the argument list *"MethodArguments"*. If the method is of type *void* then the first argument of the primitive execMethod should be the keyword *"void"*. If the first argument of execMethod is a global variable (a new feature in Prolog+CG), then the result of the method is assigned to the variable, otherwise the returned value is unified with the first argument of execMethod.

An application that calls Java code : *SHRDLU-PCG*

To illustrate the expressive power of Prolog+CG and to show its usefulness for the development of natural language processing applications, we developed SHRDLU-PCG, a reformulation in Prolog+CG of some aspects of the classic SHRDLU program [10]. In the Prolog+CG program (see below), one can note the natural use of CG as a data structure, beside term and list, and also the very use of variables in CG : a variable can hold for a CG, a concept, a concept type, a referent, a co-referent, a concept description (or value) and a relation. This flexibility is very important from a programming perspective.

SHRDLU-PCG also illustrates how Java classes (and their attributes and methods) can be used from a Prolog+CG program.

SHRDLU-PCG simulates a very restricted natural language dialog between a user and a robot that operates in a block-world. The robot can create, move, push and pop 3D blocks. The robot is able to "understand" declarative, imperative and interrogative sentences and to react accordingly. The 3D animation that results from such a dialog is monitored thanks to the "cooperation" of the Prolog+CG program SHRDLU-PCG with a Java3D program which provides the capability to create a 3D canvas, to fill it with 3D objects (cube, cylinder, sphere, pyramid) and to do some actions on them.

First, let us consider how the semantic analysis of a sentence and especially the analysis of an imperative sentence is carried out. Examples of imperative sentences are "create a red pyramid pyramid4.", "push the red pyramid on the big cube." and "move the blue sphere at the left of cube1.".

As the components of the sentence are analyzed, CGs that represent their semantic meaning are constructed and then joined. This dynamic construction of CGs is in itself an important feature of Prolog+CG.

```
// lexicon(Word, SyntaxicCategory, TypeOrCGCanon)
lexicon("push", verb, [Push]-
              -obj->[Object],
              -on->[Object]      ).
```

```
lexicon("create", verb, [Create]-obj->[Object]-colorOf-
>[Color]).
Lexicon("sphere", noun, Sphere).
Verb(V, _CGCanon) :- lexicon(V, verb, _CGCanon).
// Syntax of imperative-sentence = Verb   NP
Complement.
// Complement = [Prep  NP].
imperative_sentence((V|P1),
[Proposition : G]-mode->[Modality : imperative]) :-
   Verb(V, G_V),
    NP(P1, P2, E_NP1, S1),
    eq([T_Verb]-obj->E_N_G1, G_V),
    maximalJoin(G_V, E_N_G1, S1, E_NP1, G1_S1, _),
    complement(P2, T_Verb, G1_S1, G).
```

Comment on Verb/2 : checks if V is a verb, if so, return the canon of the verb.
Comment on imperative_sentence/2 :
imperative_sentence(P, G) receives a sentence P, as a list of words, and produces a CG G that represents its meaning. It starts by recognizing the verb and then the noun phrase. The canon of the verb (G_V) is then joined with the CG corresponding to the noun phrase (S1). This maximal join should satisfy however the following constraint : the concept that represents the head of the noun phrase has to be joined with the concept that represents the object of the verb. So we have to locate these two concepts in the two CGs respectively and then we have to consider them as "entry concepts" for the maximal join in question; this later should start by the join of the two entry concepts.

The entry concept for the CG G_V that represents the semantic meaning of the verb is located by the following goal : eq([T_Verb]-obj->E_N_G1, G_V).

To understand the effect of the above goal, let's consider this request :

```
?- eq(G_V,[Create]-obj->[Sphere:sphere1]), eq([T_Verb]-
obj->E_N_G1, G_V).

{G_V = [Create]-obj->[Sphere:sphere1], T_Verb = Create,

E_N_G1 = [Sphere:sphere1]}
```

As a result of the above unification eq/2, the variable "E_N_G1" refers to the concept in G_V that represents the object of the verb. Note how the variable "T_Verb" stands for the type of the concept and the variable "E_N_G1" stands for the whole concept.

The defined goal NP/4 : NP(P1, P2, E_NP1, S1) returns the graph S1 that represents the meaning of the noun phrase as well as the entry concept E_NP1 in S1.

After the analysis of the verb and the noun phrase, the maximal join of their semantic representations is done, producing a CG G1_S1 and then, the semantic analysis of the complement is initiated. If the complement is specified, its semantic representation will be computed and then joined with the CG G1_S1. A detailed description of maximalJoin and other CG operations in Prolog+CG is given in [4].

After the semantic analysis of an imperative sentence, the "robot" will consider its meaning as an order to be executed. Hence and as a result of such an execution, the knowledge of the "robot" will change (for instance, the position of an object has to be modified) as well as the 3D animation that shows the visual simulation of the robot behavior. The following Prolog+CG code shows how all these aspects are related.

```
Shrdlu :-
  new(aShrdlu_Canvas3D, "PrologPlusCG.Shrdlu_Canvas3D",
      ()),
  read_sentence(_sentence),
  ShrdluDialog(_sentence), /.
ShrdluDialog(("end", ".")) :- /.
ShrdluDialog(_sentence) :-
  Semantic_Analysis(_sentence, _CG),
  _CG,
  read_sentence(_s),
  ShrdluDialog(_s), /.
```

Comment on the main goal Shrdlu : this goal initiates the 3D simulation as well as the restricted natural language dialog. In particular, it specifies a call to the primitive goal "new" in order to create an instance of the defined Java 3D class *Shrdlu_Canvas3D*. Such a creation will involve, among other actions, the display of a frame that contains a "robot" (Figure 3.a).

Comment on the goal ShrdluDialog : in the above definition of "ShrdluDialog" goal, note how the meaning of the sentence (i.e. the CG "_CG" that results from the semantic analysis of "_sentence") is put as a goal to be interpreted. Thus, in the case of an imperative sentence, the goal-variable "_CG" will be bound to the following CG

```
:    [Proposition : G]-mode->[MODALITY : imperative]
```

The proposition G (the variable G will be bound to a CG) is interpreted as an order that should be satisfied. Here is the Prolog+CG rule that naturally formulates this interpretation :

```
[Proposition : G]-mode->[MODALITY : imperative] :- G.
```

Each order is then executed according to its semantic interpretation. For instance, the order to create an object with a specific name and color is defined as follows : first assert the existence of the object in the data base, then create a physical object in the 3D canvas. Each kind of object (Cube, Sphere, Pyramid) is created by a corresponding method in the defined Java class "Shrdlu_Canvas3D".

```
[Create]-obj->[T_Obj : _IdObj]-colorOf->[Color: C] :-
  asserta(object([T_Obj : _IdObj]-colorOf->[Color :
        C]), ()),
  execMethod(void, "PrologPlusCG.Shrdlu_Canvas3D",
        T_Obj, (_IdObj, C)), /.
```

6   A Simple Natural Language Analyzer for English Sentences

Prolog is a popular language for the development of natural language processing (NLP) applications. CGs are well-suited for the semantic representation of NL sentences. Having both Prolog and CG manipulation mechanisms integrated in a single environment provides a powerful platform for NLP. We present the results of a preliminary investigation of Prolog+CG's potential for NLP. We selected a NL analyzer for English written in Prolog [1] and we reprogrammed it using Prolog+CG. The resulting program was much more concise and readable thanks to the CG manipulation mechanisms and to CGs' expressive power.

We built the lexicon as a lexical type hierarchy composed of lexical categories and elements. Here is the top-level of this hierarchy illustrated on a simple example:

> *universalGram > verb, noun, determiner.*
> *verb > transitive, intransitive, copular.*
> *noun > commonNoun, properNoun.*
> *determiner > detSing, detPlur.*
> *detSing > a, the, that, this.*
> *detPlur > the, these.*
> *commonNoun > dog, cat , ...*
> *transitive > eat, chase, ...*
> *intransitive > sleep, ...*

Such a representation is well-structured, clearer and shorter than the standard Prolog predicates used to represent lexical entries such as:

> *lexicon(dog, commonNoun).*
> *lexicon(cat, commonNoun).*
> *lexicon(eat, verb)...*

For the semantic analysis, we naturally used a concept type hierarchy which can be expressed in a Prolog+CG program in a very concise way (no need of Prolog predicates for that). Here is part of our simple example:

> *universalSem > animate, inanimate...*
> *animate > animal, human.*
> *inanimate > object, plant.*
> *animal > carnivorous, vegetarian, omnivorous.*

We also used CGs to represent conceptual schemas in a very legible and concise way compared to the equivalent Prolog entries. For example, expressing the property that "carnivorous eat meat" is a simple Prolog+CG entry:

> *mySem ([carnivorous]<-subj-[eat]-obj->[meat]).*

The program performs semantic analysis in two steps. The first step consists in building the CGs corresponding to the analysed sentence in parallel with the syntactic analysis. When the program finds a syntactic pattern which matches the NL entry, a "high level CG" is built using the syntactic concepts (*Verb, Noun, Adjective*, etc.) and relations (*Subj, Obj, Attr*, etc.). When all the parts of a sentence have been analysed, the resulting CGs can be easily joined in order to create the graph corresponding to the whole sentence, thanks to Prolog+CG's graph manipulation operations such as the *maximalJoin* primitive. For example, applying the maximalJoin primitive on the two following parts which result from the analysis of the sentence's noun-phrase and verb-phrase portions :

[verb]-obj->[meat]-attr->[adjective] and [dog]<-subj-[eat]-obj->[noun]

we get the following CG: [dog]<-subj-[eat]-obj->[meat]-attr->[adjective]

It is clear that developing the equivalent program in pure Prolog would require much more effort. Hence, thanks to Prolog+CG graph manipulation mechanisms and to few instructions integrated in the syntactic part of the NLP analyser, we get the results of the syntactic analysis in the form of CGs.

The second step of the semantic analysis consists in performing the semantic validation of the CGs obtained after the syntactic analysis. The idea is to isolate different parts of a resulting CG and to check if they match conceptual schemas contained in the system's knowledge base. For example, from the graph:

[dog]<-subj-[eat]-obj->[meat]-attr->[tasty]

the program extracts two graphs

[dog]<-subj-[eat]-obj->[meat]   and *[meat]-attr->[tasty]*

and checks if they are compatible with schemas contained in the knowledge base.

In this program we have explored only few NLP mechanisms. We think that almost all knowledge levels of a NL analyser could be encoded thanks to CGs mechanisms, from the morphological level to the semantic level, including lexical and syntactic levels. This point will be the subject of a forthcoming paper. Prolog+CG's graph manipulation mechanisms and CGs' expressive power greatly improve the construction of NLP applications, providing more concise, legible and efficient programs.

7 Evaluation of the Semantic Similarity between Two CGs

We explored the interest of using Prolog+CG to develop a program aiming at evaluating the semantic similarity between two CGs. For example such a program can be useful to search similar cases in a case base [6].

Let's consider for instance these two CGs :

```
CG1 : [Eat]-        CG2 : [Eat]-

-agnt->[Monkey]      -agnt->[Human]

-obj->[Banana]       -obj->[Vegetable]

-mnr->[Fast].        -col->[Green].
```

We can observe two main differences between CG1 and CG2. First, there is a semantic distance between corresponding nodes. For example, the agent of eating in CG1 is a "monkey" and "human" in CG2. These two concepts are linked by their common super-type "Animal" in the type hierarchy. We can also observe that the global structures of the two CG's are dissimilar.

Several authors proposed similarity algorithms which make use of these differences [7, 8, 9, 11]. The evaluation of semantic distance is called *surface similarity* while the evaluation of maximal common structure is called *structure similarity*. Maher's *Absolute Distance* (δ) and Yang *et al.'s Degree Of Inheritance*

(DOI) are useful to evaluate the semantic distance between two concepts. We have written a Prolog+CG program that compute both.

Poole and Campbell [9] presented a complete structure similarity algorithm in their publication. PROLOG+CG's advanced knowledge manipulation capabilities allow us to directly find the common generalization of two CGs using the generalize primitive : generalize(_CG1, _CG2, _CommonStructure).

Applied to our example, the program finds the following CG and the associated semantic distances.

$$[\text{Animal}] < -\text{agnt} - [\text{Eat}] - \text{obj} -> [\text{Comestible}].$$

$$\delta_{\text{Monkey,Human}} = 2 \qquad \text{DOI}_{\text{Monkey,Human}} = 0$$

$$\delta_{\text{Eat,Eat}} = 0 \qquad \text{DOI}_{\text{Eat,Eat}} = 0$$

$$\delta_{\text{Vegetal,Banana}} = 3 \qquad \text{DOI}_{\text{Vegetal,Banana}} = 1$$

An overall similarity percentage can be deduced from these information according to specific domain constraints.

In 1999, Nadeau developed this program using WinProlog. The program had 360 lines of code. The program's new version, written in Prolog+CG, has only 74 lines of code. Again, we found that Prolog+CG offers a greater expressive power than pure Prolog, greatly improving the legibility and conciseness of the programs.

8   Future Works and Conclusion

Prolog+CG is a CG-Based logic programming language; CGs play the same role as terms, i.e. they are used to represent goals and data structures. Moreover, Prolog+CG provides several primitive goals to manipulate concept type hierarchy and CGs [4].

As illustrated in this paper, Prolog+CG 2.5 integrates not only object-oriented Prolog and CG formalism, but also Java technology. It then offers a great potential for the development of a large variety of applications.

Because Prolog+CG is based on an interpreter written in Java, it is slower than commercial Prolog systems such as WinProlog. We are confident that Prolog+CG's performance will significantly improve during the coming year. However, the system's current performance is already quite reasonable for the development of prototype systems, and Prolog+CG's offers such a rich prototyping environment for the development of knowledge-based applications. More and more applications are developed using Java. Prolog+CG enables developers to integrate some kind of intelligence in such applications, taking advantage of the expressive power of CGs and the reasoning capabilities of Prolog. It will be easy to create modules that handle ontologies (using the CG type hierarchy and CGs as concept schematas), sub-systems that provide some natural language capabilities, components that provide all sorts of reasoning capabilities using structures expressed in CGs.

Many specialized libraries are available in C++. We plan to develop in the coming year a link between Prolog+CG and C++ in order to take advantage of the functionalities provided by such libraries. With this link to C++, Prolog+CG will become a very powerful platform for the development of applications integrating object-oriented components and knowledge-based systems taking advantage of the expressive power of CGs and the reasoning capabilities of Prolog.

Acknowledgments. Special thanks from A. Kabbaj to B. Moulin who suggested and pushed such an intensive and productive collaboration. Special thanks also from A. Kabbaj to the other authors who contributed strongly, by their critics and requests, to the extension of Prolog+CG. Finally, thanks to Issam Alloul who developed the Java3D program for the SHRDLU-PCG application.

References

1. Bee-gent Software agent environment (Toshiba):
   http://www2.toshiba.co.jp/beegent/index.htm
2. Gal A., Lapalme G., Saint-Dizier P., Somers H.(1991) *Prolog for Natural Language Processing*, Wiley Professional Computing editions.
3. Kabbaj A. (1996), Un système multi-paradigme pour la manipulation des connaissances utilisant la théorie des graphes conceptuels, Ph. D. Thesis, Université de Montréal, Canada.
4. Kabbaj A., Janta-Polczynski M. (2000), From PROLOG++ to PROLOG+CG : A CG Object-Oriented Logic Programming Language, in Conceptual Structures : Logical, Linguistic, and Computational Issues, Ganter B. and G. W. Mineau (eds.), Springer, LNAI 1867, 540-554.
5. Kabbaj A., The impact of sets and co-references on CG operations in Prolog+CG 2.5, Submitted to ICCS'2001
6. Kolodner J. (1993) *Case-Based Reasoning*, Morgan Kaufmann.
7. Maher P.E. (1993) A similarity measure for conceptual graphs, *International journal of intelligent systems*, Vol 8, No. 8, 1993, 819-837.
8. Myaeng H. , Lopez-Lopez A. (1992) Conceptual graph matching : a flexible algorithm and experiments, *Journal of experimental and Theoretical Artificial Intelligence*, Vol 4, April-June, 107-126.
9. Poole J., Campbell J.A. (1995) A novel algorithm for matching conceptual and related graphs, Conceptual structures: applications, implementation and theory, In G. Ellis et al. (eds.), Springer Verlag Lecture notes in Artificial Intelligence 954, 293-307.
10. Winograd T. (1972), *Understanding Natural Language,* Academic Press, NY.
11. Yang G-C., Choi Y.B. and Oh J.C. (1993) CGMA : A novel conceptual graph matching algorithm, Conceptual structures : theory and implementation, In H.D. Pfeiffer, T.E. Nagle (eds.), Springer Verlag Lecture notes in Artificial Intelligence754, 252-261.

An Application of the Process Mechanism to a Room Allocation Problem Using the pCG Language

David Benn[1] and Dan Corbett[2]

[1] School of Computer and Information Science, University of South Australia, Mawson Lakes, 5095
David.Benn@motorola.com
[2] School of Computer and Information Science, University of South Australia, Mawson Lakes, 5095
corbett@cs.unisa.edu.au

Abstract. The Sisyphus-I initiative consists of a constraint satisfaction problem in which a group of people in a research environment must be allocated rooms. Numerous constraints are detailed by the problem description which together impose a partial ordering on any solution. A solution to Sisyphus-I is presented in order to illustrate pCG, a CG-oriented programming language which embodies Mineau's (1998) state-transition based process mechanism. We consider pCG to be an experimental language and believe that feedback from the CG community would be useful at this stage of development. A non-trivial application with which the community is already familiar is an effective means by which to accomplish this. The solution involves automatic extraction of most of the information required to represent the problem from the Sisyphus-I web page, and a pCG program which produces suitable room allocations via a process. The means by which the presented solution could be further constrained to increase its robustness is briefly discussed, as is the likely future development of pCG.

1 Sisyphus-I

Sisyphus-I was used at ICCS'99 as an example of a non-trivial problem that could be represented using conceptual graphs (CGs) [9]. In this paper, we show a solution to the problem using Mineau's process mechanism [3], [13]. The problem statement [9] calls for "...traces of the problem-solving processes...This means that...we want descriptions of operational problem solvers."

The Sisyphus-I room allocation problem is a constraint satisfaction problem in which a group of people in a research department each with particular needs must be allocated rooms. It is a means by which to test knowledge acquisition and reasoning strategies. Numerous constraints are detailed by the problem description and are intended to impose a partial ordering on any solution. One constraint is that the head of the YQT research department must be allocated alone to a large room. Another is that smokers and non-smokers may not share rooms. Yet another is that researchers should share rooms with people from different projects to prevent insularity. There are two slightly different problem statements given in [9]. The difference between them is that Katharina, the head of a project and a smoker is replaced by Christian, a researcher and smoker. Katharina is immediately eligible for a single room by virtue of her lofty status as a project lead. However, Christian ought to share a room with

H. Delugach and G. Stumme (Eds.): ICCS 2001, LNAI 2120, pp. 360–376, 2001.

another person due to his more lowly status, unless doing so would place him with a non-smoker, or unless there are only single rooms remaining.

Many techniques were used in the 1999 solutions of the original Sisyphus-I initiative. Agents were used to represent the knowledge in [19] while other authors used a rule base, either for the meta-knowledge [5] and [1] or for the domain facts [12]. Mineau's solution in [15] was to represent the constraints and goals on graphs, and then use subsumption to compare graphs and find falsehoods, which represent invalid states. A similar method for finding invalid states, using projection, was employed by [1].

While our solution, on the surface, seems only to represent yet another rule-based solution, the rules are represented by dynamic CGs with processes. The significance of our solution is that the graphs are used to represent both the declarative facts of the current context and the process knowledge of the domain. Projection is the basis for graph matching in the solution presented here, with type subsumption employed in one of the rules, and restriction on concepts used extensively elsewhere in the process definition. Matched graphs representing rooms and people are retracted from a knowledge base (KB) as constraints are satisfied, with room allocation graphs being asserted into the KB. The frame problem[1] [18] is avoided since graphs are retracted, not negated, decreasing the burden for future KB searches.

2 Processes

Mineau proposed the notion of *processes* to overcome the fact that Conceptual Structures Theory (CST) does not cater for the dynamic retraction and assertion of graphs in a CG-based KB [13]. Mineau's processes are one kind of executable conceptual graph formalism. Other work has been carried out in this domain by the CG community, for example [2], [4], [6], [7], [8], [10], [11], [16], [17]. Some authors have focussed upon the execution of a graph by special status being given to particular nodes, while others have taken a state transition approach, or both. Some mechanisms manipulate only concept referents, while others utilise concepts or graphs.

A number of authors use the term *process* either to describe what their formalism is simulating, or in the case of [13], as the name of a formalism. In [18] Sowa remarks that processes can be classified as continuous and discrete. An example of the former is a physical process such as the world weather system. Continuous processes are best simulated on analog computers, although they can be approximated by sufficiently fine-grained discrete processes. A discrete process consists of a series of events and states over time, which is the kind of process that is catered for by Mineau's formalism.

A recurring theme is the apparent utility of state transition based systems, of which [13] is an example. Some authors use state-based systems to solve particular problems, such as [2], [11], and [16]. Sowa remarks that State Transition Diagrams are a common way of representing discrete processes. Sowa discusses Petri nets as a powerful and generic state transition mechanism. [18]

Mineau suggests that actors and demons are specialisations of processes [13]. Actors take concepts of a predetermined type as input and yield a specialisation (by referent) of a predetermined concept type as output, by means of some computation

---

[1] Increasingly more is asserted about what is *not* the case, eg that a room is no longer vacant.

based upon the input concept referent [17]. Demons consume input concepts and assert other concepts as output [6]. Processes generalise demons by taking CGs as input and asserting or retracting CGs as the result of their processing. The relationship is not really this simple however since, for example, Mineau's 1998 factorial example uses actors within preconditions. So, on the one hand we have the suggestion that there exists a generalisation hierarchy between actors, demons and processes, ie `actor < demon < process`. On the other hand, there is at least one example of a process using an actor as part of its definition, essentially a *uses* or *has-a* relationship. It is arguable that without actors, process preconditions of even moderate complexity would not be possible since arbitrary referent values could not be easily computed or manipulated. A concept is a singleton graph, so a case can be made for a process being a generalisation of a demon as Mineau has suggested, especially since assertion and retraction is the primary mechanism for both.

Mineau's claim is that his process formalism minimally extends CST. Essentially, [13] calls for the following: CGs are grouped into rules consisting of pre and post conditions, an attempt is made to specialise a rule's preconditions to graphs in a KB, and if this succeeds, the KB is updated using post-condition graphs. This is repeated as many times as there are precondition matches. The first rule's preconditions and the final rule's post-conditions may be added to via parameters.

Lukose describes Executable Conceptual Structures (ECS) in [10], an executable conceptual modelling language. Components of the system include: concept types (differentia graphs defined in terms of a lambda expression); actor types (scripts with a main method); precondition and post-condition graph lists; lists of graphs to be deleted; a notion of short and long term working memories. ECS also provides special concepts representing predicates (eg `[LTEQ]`) and statements (eg `[ASSERT]`), along with constructs to express temporal relationships, and others to package problem solutions. When taken together, complex systems can be built for problem solving. A complete description of ECS cannot be provided here. What can be said however is that ECS provides a similar capability to Mineau's process formalism, at the expense of a greater number of parts.

Cyre asks the question: given an arbitrary CG, what would be required to execute it? He points out that complex, possibly hierarchical graphs may be difficult to reason about, so may require simulation to understand, and believes that simulation can be useful for predicting the consequences of behaviour described by a CG. A ball-throwing example is given in [4] involving a CG containing a `Throw` action concept. Cyre laments that most researchers have extended CST with special nodes (eg actors or demons), and suggests that instead, `Throw` be associated with some procedure which "knows" what kinds of concepts it can work with, and how `Throw` affects them. Here, concepts rather than relations are being used as the active elements in a graph. What this amounts to, however, is that certain types must be marked as having special roles by some interpreting system, instead of explicitly using actor nodes. Moreover, special relations are added to express inclusive and exclusive logical disjunction [4].

These and other executable CG mechanisms are reviewed in detail in [3]. Mineau's executable CG mechanism strikes a balance between simplicity and expressive power. Correspondence with Mineau at the time [3] was embarked upon confirmed that there had been no implementation of the process mechanism, "...only a theory which needs to be refined, implemented and applied."[2] That is exactly what the work described herein entails. A major motivation for any implementation of the process mechanism is to determine whether it works, and if so, in what ways it requires improvement.

3   The pCG Language

We believe that someone wishing to use processes should be able to do so by expressing them directly in a suitable source language, so as to be able to work closely with the fundamental entities. Mineau provides the beginnings of such a language in [13], and that paper's stated long term goal of the development of a Conceptual Programming Environment (CPE) combining logic and imperative programming suggests a direction. Such a language would also provide a means to embody the process engine. An alternative is to develop an Application Programming Interface (API) in a popular language, such as C++ or Java. One problem with this approach is that the complex details of using an API can quickly obscure the problem domain. For the purpose of understanding the likely use of processes and for ease of their application to a particular problem, it was decided that a language, pCG, and an interpreter for it be developed.

The primary motivation for pCG is to make available a concrete implementation of Guy Mineau's process mechanism [3], [13] for empirical investigation of its strengths and weaknesses. A secondary motivation is to provide an easily extensible general-purpose language with CG processing functionality which makes the primary entities of interest in the domain (eg CGs, actors, and processes) first class values, and permits object-based, functional, and declarative programming styles, which is consistent with Mineau's idea for a CPE.

The pCG distribution is available on the web,[3] and is available under the Gnu Public Licence. The complete pCG solution to the Sisyphus-I problem described in this paper is also provided in the distribution, along with many other example programs. We consider pCG to be an experimental language and believe that feedback from the CG community will be useful at this point in time. A non-trivial application seems to be an effective means by which to accomplish this.

3.1   Design

CG tools such as [8] have been developed to support purely visual programming. We have taken the approach of acknowledging that traditional programming languages have a role to play. In short, CGs are used where knowledge representation is required, while traditional constructs are used where imperative or functional programming is most appropriate. Languages such as Perl (lists, `foreach`), Lisp (dynamic typing, lambda), Java (objects), and Prolog (assert, retract, pattern matching) have all influenced pCG. The choice of a new language is a deliberate one, to clearly state that something different is being represented, and to avoid confusing similarities with existing languages. Given that major new constructs are required (eg actor type definition, process statement) which will take the form of new syntax, and that significant new types must be introduced (eg concept, graph), an existing

---

[2] E-mail correspondence from Guy Mineau to David Benn, December 1ˢᵗ, 1999.

[3] See http://www.adelaide.net.au/~dbenn/Masters/ for the archive and other information about pCG.

language would require significant extension. CG languages and platforms other than Synergy are considered in [3]. One drawback of creating a new language is that no pre-existing utility source code in that language exists, but since pCG has an extensible type system, utility code written in Java can be used.

3.2 Key Features

The pCG language is interpreted for rapid development. It is *dynamically typed* since values not variables determine type. pCG is *lexically scoped* since functions and processes introduce a new lexical scope when invoked. pCG can also be characterised as an *object-based* language since its fundamental types have state and operations on that state. However, pCG does not support the creation of arbitrary object types within the language, but it is possible to easily add new types or modify existing ones using Java without change to the interpreter itself. All values in pCG are objects with associated *attributes* (eg length), *operations* (like methods in Java), and *operators* (eg +). For example, CG operations such as *join* and *project* are supported as operations on graph objects. Some attributes may appear on the left side of an assignment statement (eg myConcept.designator=2). All attributes may appear in expressions. Objects have a type attribute indicating the name of a value's type, and the *is* operator can be used for run-time type identification. Objects in pCG are garbage collected when they can longer be reached.

The pCG language is *multi-paradigm*, since apart from its object-based nature, pCG supports *imperative* (variables, assignment, operators, selection, iteration), *functional* (higher order functions, values, recursion), and *declarative* styles of programming. The latter is supported in the form of processes, since one specifies rules containing pre and post conditions representing knowledge. The details of testing preconditions against the KB, and the assertion/retraction of post-conditions in the KB are left to the process execution engine. Apart from the essential iteration and selection constructs found in any modern language, three control constructs are central to pCG: functions, actors, and processes. There is currently no provision for concurrent execution in pCG.

3.2.1 Actors

Consider the following function and actor definitions in pCG:

```
function sqr(n,m)
  nVal = n.designator;
  if not (nVal is number) then
    exit "Input operand to " + me.name + " is not a number!";
  end
  m.designator = nVal*nVal; // example of setting an attribute
end

actor SQR(x) is `[Num:*a'*x'][Num:*b'*y']<sqr?a|?b>`;
```

The actor defines a dataflow graph which contains a single actor node or *executor*. This executor is defined in terms of a pCG function, sqr, which takes a source and a sink concept as parameters. The sink concept's designator is mutated according to the square of the source concept's designator, which is first tested to ensure it is a number. In the actor definition, *a and *b are defining variables, ?a and ?b are bound variables for identifying concepts, while *x and *y are variables representing designator values. Such variables derive from the original actor notation of [17].

One way to invoke this actor is to write: g = SQR(4) which results in a mutated copy of the defining graph being assigned to g thus: [Num:*a 4] [Num:*b 16]<sqr?a|?b>.

A point of departure in pCG's implementation of actors compared to [17] is that only the input parameter is specified in the actor definition's parameter list for the purpose of binding. The actor mechanism can otherwise determine the correct order in which to pass concepts to a particular executor, so long as the arcs are correctly ordered. Assuming the executor (a function or other dataflow graph) performs correctly, the returned mutated graph copy will have appropriately mutated sink concepts. The above invocation method is not useful when an actor appears in a process rule's precondition. An invocation graph may be constructed and the graph activated directly, for example:

```
n = 4;
mySqrGrStr = "[Num:*a " + n + "] [Num:*b'*y']<sqr?a|?b>";
g = activate mySqrGrStr.toGraph();
```

Notice that no actor definition is required here. The graph is constructed as a string, the source concept's referent is bound in that string, and the string is converted to a graph which is then activated, returning the mutated graph copy. This is akin to actor invocation that occurs during process execution, except that the process engine implicitly activates precondition graphs containing actors. The process definition of section 4 contains no actors, but [3] gives an implementation of Mineau's iterative factorial process [13] which does use actors in preconditions.

3.2.2 Processes

An abstract process is defined in [13] as: *process p(in g1, out g2,...) is { r_i = <pre, post>, $\forall_i \in [1, n]$ }* . In pCG, the syntax for a process is:

```
'process' name '(' in | out parameter [, …] ')'
  ['initial' block]
  ('rule' ident
    [option-list]
    'pre'
      ['action' block] (['~'] pre-condition)*
    'end'
    'post'
      ['action' block] (post-condition [option export])*
    'end'
  'end')*
'end'
```

A process has a name and a list of in and out parameters, followed optionally by a block of code for miscellaneous initialisation purposes, followed by a set of zero or more rules. Each rule consists of an arbitrary identifier, optionally followed by a list of one or more options for the rule, and a pre and post condition section. A precondition section consists of an optional action block and zero or more possibly negated graph expressions. A post-condition section consists of an optional action block and zero or more graphs (in contexts — see below) to be asserted or retracted.

A block consists of arbitrary pCG code. The intent of such blocks is to aid in the debugging of processes, to provide useful output during process execution, or to otherwise combine procedural and declarative programming styles. Such code may also be used to construct graphs which are subsequently used in precondition graph matching, or post-condition graph assertion or retraction. Only retraction of whole graphs from a KB, not partial graphs is possible in pCG.

Contexts with concept types PROPOSITION, ERASURE, and CONDITION are essentially used as a packaging mechanism for passing parameters to a process. PROPOSITION and ERASURE context types are also used in the body of post-conditions to distinguish between assertions and retractions. An alternative to the latter is discussed in section 6. A precondition is an arbitrary graph expression. If preceded by a '~' character, the sense of the match for that graph is reversed.

An input parameter must be a context and have a type of PROPOSITION or CONDITION. The descriptor of a PROPOSITION context is asserted before process execution starts, the corresponding graph acting as a trigger which (along with other asserted graphs from the caller's KB) causes a suitable rule to fire. A CONDITION context's descriptor graph is added to the precondition list of the first rule of a process. Output parameters — contexts of type PROPOSITION or ERASURE — are appended to the post-condition list of the last rule. When this rule is reached, the specified graphs are asserted or retracted in the caller's KB.

When a process is invoked, the first rule is retrieved, and each graph of the rule's precondition is tested in turn. If one does not match, matching for that rule is discontinued and the next rule is retrieved. If no matching rule is found, the process terminates. If one is found, the post-conditions of the matched rule are retrieved. For each post-condition, its descriptor graph is either asserted or retracted — depending upon the type of its context — from the local KB, or the caller's KB if an export option (see section 4.3) is active. The process engine then starts again at the top of the ordered rule collection, and attempts to find a matching rule during the next cycle.

The projection operation is central to the matching algorithm of pCG's process execution engine, and the algorithm used by pCG is similar to that given in Theorem 3.5.4 of [17] except that relation type subsumption is also permitted. A match may be partial. For example, a process rule precondition such as `[Person:*a'*name'] [Role:*b'Researcher'] (Chrc?a?b)` would be sufficient to retrieve the complete graph of Michael T. shown in section 4.2 if present in a KB, and if it were encountered first during search.

A process invocation introduces a new KB scope. So, there are two run-time stacks in pCG: one for variables and another for KBs. There is one of each at the top-level for global code execution. KBs in pCG store concept and relation type hierarchies, and a set of graphs. The content of a process caller's KB is copied to the invoked process's KB. Look-up and, by default, assertion and retraction are confined to the KB which is top-most on the stack. The alternative is a proliferation of graphs in the caller's KB, meaningless outside of a given process, which would need to be retracted by some other means upon exit from that process.

4   A Sisyphus-I Solution in pCG

What follows is a detailed description of how the Sisyphus-I problem is represented and solved with a pCG program, the bulk of which consists of a process definition.

4.1   Assumptions and Interpretations

This section briefly documents assumptions made and interpretations of [9] for the purpose of problem representation and solution.

The criterion to be used for determining room centrality is unclear from [9]. The criterion used for the current solution is physical connection to room C5-118 (The Tower), implying that rooms C5-113 to C5-119 are central. An alternative interpretation is that any room which is *near* C5-119 could be considered to be central. For example, C5-120 could be argued to be central by virtue of physical proximity to C5-119. Another reasonable interpretation is given by [1] which suggests that C5-116, C5-117, and C5-119 are central. It is assumed that the black square on the room graphic in [9] corresponds to room C5-123.

In section 2.1.1 of [9], "true" and "yes" are both used to indicate boolean true values. The problem representation converts all occurrences of "true" to yes".

4.2   Representing the Problem

All the information required to solve the problem is contained in section 2.1.1 of the HTML found at the URL of [9]. A Perl script was applied to the HTML to generate simple CGs each of which represents some aspect of each YQT member. One such graph set is shown below in CGIF:[4]

```
[Person:*a'Michael T.'] [Role:*b'Researcher'] (Chrc?a?b)
[Person:*a'Michael T.'] [Project:*b'BABYLON Product'] (Member?a?b)
[Person:*a'Michael T.'] [Smoker:*b'No'] (Chrc?a?b)
[Person:*a'Michael T.'] [Hacker:*b'Yes'] (Chrc?a?b)
[Person:*a'Michael T.'] [Person:*b'Hans W.'] (Coworker?a?b)
```

The following graphs for Katharina N. apply to the first problem statement, with almost identical graphs for Christian I. in the second.

```
[Person:*a'Katharina N.'] [Role:*b'Researcher'] (Chrc?a?b)
[Person:*a'Katharina N.'] [Project:*b'MLT'] (Member?a?b)
[Person:*a'Katharina N.'] [Smoker:*b'Yes'] (Chrc?a?b)
[Person:*a'Katharina N.'] [Hacker:*b'Yes'] (Chrc?a?b)
```

It is inefficient to process many small graphs, so the graphs for each person are joined into a single graph by pCG code. These multi-relation graphs are then asserted into the top-level knowledge base (KB). As an example, Michael T's graphs are joined into the single graph shown below in LF and CGIF, respectively:

```
[Person:'Michael T.'] -              [Person:*a'Michael T.']
  (Chrc) -> [Role: 'Researcher']     [Role:*b'Researcher']
  (Member) -> [Project: 'BABYLON     [Project:*c'BABYLON
                    Product']                  Product']
  (Chrc) -> [Smoker: 'No']           [Smoker:*d'No']
```

---

[4] The Conceptual Graph Interchange Format is used in pCG examples throughout this paper.

```
(Chrc) -> [Hacker: 'Yes']          [Hacker:*e'Yes']
(Coworker) -> [Person: 'Hans W.']  [Person:*f'Hans W.']
                                     (Chrc?a?b)
                                     (Member?a?c)
                                     (Chrc?a?d)
                                     (Chrc?a?e)
                                     (Coworker?a?f)
```

The following pCG code joins and asserts the graphs into the KB:

```
yqt = file inFilePath;
lines = yqt.readall();
yqt.close();
g = "";
foreach line in lines do
  if line.length > 0 then
    if g is graph then
      g = g.join(line.toGraph()); // add to current graph
    end else
      g = line.toGraph(); // new person
    end
  end else
    assert g; // add current person's graph to KB
    g = ""; // prepare for next person
  end
end
```

Although most information can be automatically extracted, a few facts can not, so these are joined to existing graphs and reasserted, thus:

```
function addTo(addition)
  newConcepts = addition.concepts;
  foreach g in _KB.graphs do
    gConcepts = g.concepts; // are the head concepts identical?
    if newConcepts[1] == gConcepts[1] then
      retract g; assert g.joinAtHead(addition); return;
    end
  end
end
```

```
addTo('[Person:*a'ThomasD.'][Head:*b'YQT'](Chrc?a?b)');
addTo('[Person:*a'Hans W.'][Head:*b'BABYLON
Product'](Chrc?a?b)');
addTo('[Person:*a'Joachim I.'][Head:*b'Other'](Chrc?a?b)');
addTo('[Person:*a'Katharina N.'][Head:*b'Other'](Chrc?a?b)');
```

The additional information above concerning Katharina N. ensures that her room allocation will have a higher priority than it would otherwise, which will become apparent in the discussion of process rules to come.

Information about the available rooms is next asserted in the top-level KB with large central rooms being C5-117 and C5-119, small central rooms being C5-113..116, and large non-central rooms being C5-120..123.

```
rooms={"[LargeRoom:*b'C5-117'][Location:*c'Central']
         [Integer:*d 2](Chrc?b?c)(Vacancy?b?d)",
       ...
       "[LargeRoom:*b'C5-120'][Location:*c'NonCentral']
         [Integer:*d 2](Chrc?b?c)(Vacancy?b?d)",
       ...
       "[SingleRoom:*b'C5-
116'][Location:*c'Central'](Chrc?b?c)"};
foreach room in rooms do
  assert room.toGraph();
end
```

The top-level KB now contains graphs representing information such as that there
exists a large room whose location is central and whose vacancy count is 2, which
corresponds to the first graph above. While the characteristic of largeness and the
vacancy count may seem to be redundant information, the latter is used in the process
of the next section to differentiate between full and partially tenanted large rooms.

4.3 The Rules

This section details a pCG process definition which yields a solution to the first and
second statements of the Sisyphus-I problem. Action blocks are shown for the first
rule, but omitted thereafter to save space. In each rule, the precondition action block
prints a question that asks whether a particular kind of person needs to be allocated to
a room. The post-condition action block calls functions which direct information to
standard output and to a graphical room display regarding a room allocation if the rule
of which it is a part fires.

The process definition that follows is punctuated with commentary regarding
particular rules where deemed necessary. The process is dependent upon other user-
defined functions. For the complete code listing see the pCG distribution.

The Sisyphus process is first named for invocation purposes, and in this case, no
input or output parameters are passed to it, but [3] and [13] provide examples of their
use. The first rule simply allocates any large central room to the head of department.

```
process Sisyphus()
  rule allocateHeadOfYQT
    pre
      action
        println "Need to allocate room for the head of YQT?";
      end
      `[Person:*a'*name'][Head:*b'YQT'](Chrc?a?b)`;
      `[LargeRoom:*a'*roomLabel'][Location:*b'Central']
        (Chrc?a?b)`;
    end
    post
      action
        label = getRoomLabel(); // uses _MATCHES array
        name = getPersonName(); // also uses _MATCHES array
        showAllocation("Head of YQT", name, label);
        plotName(name, label, 1, headColor);
      end
      `[PROPOSITION:[Person:*a'*name'][Room:*b'*roomLabel']
```

```
          (Occupant?a?b)]`; option export;
      mkErasureGraph(_MATCHES[1]); // erase person
      mkErasureGraph(_MATCHES[2]); // erase room
   end
end // rule allocateHeadOfYQT
```

The PROPOSITION context in the post-condition block directs pCG to assert its descriptor graph into the KB. The caller's KB is targeted rather than the local KB of the process by the `export` directive. Any modifications to the process's local KB will not affect the caller's KB, except when "`option export`" is applied to a post-condition graph, or when the final rule fires. The former applies here with the result that the selected post-condition graphs are asserted in the top-level KB.

The _MATCHES list contains each complete graph that was matched by the rule's precondition graphs via projection onto the graphs in the current KB. The list is created by the pCG interpreter during rule invocation. The getter functions in the action block are invoked to iterate over another special list, `_KB.corefvars`, which stores name-value pairs of concept designator variable bindings (such as `*name` and `*roomLabel`) from the matching operations. The only reason for invoking `mkErasureGraph()` is to yield an ERASURE context.[5] The fact that this is necessary at all suggests that the generic `post..end` syntax ought to be replaced by something more specific. See also section 6.

The next rule *must* come before `allocateFirstSecretary`, otherwise the latter will fire twice, yielding allocation into two different large rooms. The rule `allocateFirstSecretary` allocates one secretary to an empty large central room and decrements the vacancy count of that room by the retraction and assertion of room graphs. The `allocateSecondSecretary` rule looks for a large central room with one vacancy. Both secretaries are thus allocated to the same room, due to a particular rule sequencing.

```
   rule allocateSecondSecretary
      pre
         `[Person:*a'*name'][Role:*b'Secretary'](Chrc?a?b)`;
         `[LargeRoom:*a'*roomLabel'][Location:*b'Central']
            [Integer:*c 1](Chrc?a?b)(Vacancy?a?c)`;
      end
      post
         `[PROPOSITION:[Person:*a'*name'][Room:*b'*roomLabel']
            (Occupant?a?b)]`; option export;
         mkErasureGraph(_MATCHES[1]); // erase person
         mkErasureGraph(_MATCHES[2]); // erase room
      end
   end // allocateSecondSecretary

   rule allocateFirstSecretary
      pre
         `[Person:*a'*name'][Role:*b'Secretary'](Chrc?a?b)`;
         `[LargeRoom:*a'*roomLabel'][Location:*b'Central']
            [Integer:*c 2](Chrc?a?b)(Vacancy?a?c)`;
```

---

[5] Code such as `("[ERASURE:" + _MATCHES[1] + "]").toGraph()` is used.

```
end
post
   `[PROPOSITION:[Person:*a'*name'][Room:*b'*roomLabel']
      (Occupant?a?b)]`; option export;

   mkErasureGraph(_MATCHES[1]); // erase person

`[ERASURE:[LargeRoom:*a'*roomLabel'][Location:*b'Central']
      [Integer:*c 2](Chrc?a?b)(Vacancy?a?c)]`;

      `[PROPOSITION:[LargeRoom:*a'*roomLabel'][Location:*b
      'Central'][Integer:*c 1](Chrc?a?b)(Vacancy?a?c)]`;
   end
end // allocateFirstSecretary
```

The third post-condition expression in `allocateFirstSecretary` could be replaced by `"mkErasureGraph(_MATCHES[2])"` in order to erase the corresponding room graph, but the current code displays a symmetry between erasure of the old room graph and assertion of the new room graph.

The next two rules (`allocateManager` and `allocateAHead`) are similar to `allocateHeadOfYQT` except that different constraints are applied in the person and room preconditions.

Rules for allocating first and second researcher pairs are similar to those for secretary allocation shown above. In a similar manner, `allocateSecondResearcher` must come before `allocateFirstResearcher`, otherwise the latter will fire twice, yielding allocation into two different large rooms, making the final 3 pairs of allocations (see actions 8, 9, and 10 in [9]) impossible without a smoker/non-smoker conflict. This ordering is akin to a degenerate recursive step. The lesson here is that the more constrained rule should appear first. We allocate non-smoking researchers with this and the next rule, handling smokers later.

Once again, `allocateSecondSmoker` must precede `allocateFirstSmoker`, otherwise the latter will fire twice, yielding allocation into two different large rooms. Here we are trying to pair off smokers into large rooms. If we have an *odd* number of smokers or non-smokers, we may have no choice but to allocate smokers with non-smokers, or to allocate a smoker to a single room.

The final catch-all rule is designed to cope with the Second Problem Statement (section 2.2 of [9]). It simply says that any remaining researcher should be allocated to *any* remaining room. There should in fact be three rules: two to handle a large room in the manner above, and one to handle a single room (eg allocate{First, Second}RemainingResearcher, and allocateRemainingResearcher). It just so happens that we know from the problem statement that there is only one small room remaining by the time this rule applies, and that the process will terminate on the following cycle since no room or people graphs will be left in the local KB of the process.

```
rule allocateRemainingResearcher
   pre
      `[Person:*a'*name'][Role:*b'Researcher'](Chrc?a?b)`;
      `[Room:*a'*roomLabel'][Location:*b'*somewhere']
```

```
            (Chrc?a?b)`;
    end
    post
        `[PROPOSITION:[Person:*a'*name'][Room:*b'*roomLabel']
            (Occupant?a?b)]`; option export;
        mkErasureGraph(_MATCHES[1]); // erase person
        mkErasureGraph(_MATCHES[2]); // erase room
    end
  end // allocateRemainingResearcher
end // process Sisyphus
```

4.4 The Results

The pCG program described above is invoked via the following code: "Sisyphus()". The result of this invocation is text on the standard output and a window with a room graphic, both of which are updated during the course of the program run.[6] A sample of the text output is shown for the first problem statement below.

```
Asserting YQT member graphs.
Asserting more information about certain YQT members.
Need to allocate room for the head of YQT?
--> Head of YQT 'Thomas D.' allocated to C5-117
Need to allocate room for the head of YQT?
Need to allocate room for second secretary?
Need to allocate room for first secretary?
--> First secretary 'Monika X.' allocated to C5-119
Need to allocate room for the head of YQT?
Need to allocate room for second secretary?
Second secretary 'Ulrike U.' allocated to C5-119
...
Need to allocate room for the head of YQT?
Need to allocate room for second secretary?
Need to allocate room for first secretary?
Need to allocate room for the manager?
Need to allocate room for a head?
Need to allocate room for a second researcher?
Need to allocate room for a first researcher?
--> First researcher 'Michael T.' allocated to C5-122
Need to allocate room for the head of YQT?
Need to allocate room for second secretary?
Need to allocate room for first secretary?
Need to allocate room for the manager?
Need to allocate room for a head?
Need to allocate room for a second researcher?
Second researcher 'Harry C.' allocated to C5-122
...
```

The text output reflects the pattern of rule firings. The final several lines of code iterate over the top-level KB once the process has terminated, sending to standard

---

[6] This takes approximately 10 seconds on a 550 MHz Pentium III Linux workstation.

output the final room allocation graphs. This code and the output for the first problem statement is shown below, in that order.

```
println "Room allocation graphs:";
filter = `[Person:*a'*x'][Room:*b'*y'](Occupant?a?b)`;
foreach g in _KB.graphs do
  h = g.project(filter);
  if not (h is undefined) then println " " + h; end
end

[Person:*a"Thomas D."][Room:*b"C5-117"](Occupant?a?b)
[Person:*a"Monika X."][Room:*b"C5-119"](Occupant?a?b)
[Person:*a"Ulrike U."][Room:*b"C5-119"](Occupant?a?b)
[Person:*a"Eva I."][Room:*b"C5-113"](Occupant?a?b)
[Person:*a"Hans W."][Room:*b"C5-114"](Occupant?a?b)
[Person:*a"Joachim I."][Room:*b"C5-115"](Occupant?a?b)
[Person:*a"Katharina N."][Room:*b"C5-116"](Occupant?a?b)
[Person:*a"Werner L."][Room:*b"C5-120"](Occupant?a?b)
[Person:*a"Jurgen L."][Room:*b"C5-120"](Occupant?a?b)
[Person:*a"Marc M."][Room:*b"C5-121"](Occupant?a?b)
[Person:*a"Angi W."][Room:*b"C5-121"](Occupant?a?b)
[Person:*a"Michael T."][Room:*b"C5-122"](Occupant?a?b)
[Person:*a"Harry C."][Room:*b"C5-122"](Occupant?a?b)
[Person:*a"Andy L."][Room:*b"C5-123"](Occupant?a?b)
[Person:*a"Uwe T."][Room:*b"C5-123"](Occupant?a?b)
```

5 Discussion

The key difference in the second problem statement is that Christian is allocated last after all higher priority constraints have been satisfied, whereas in the first problem statement Katharina is allocated much earlier, due to her higher status. This is reflected in the static rule orderings of the process. [9] suggests the general approach that should be followed, such as allocating the head of the YQT department, followed by secretaries, then heads of projects, then smokers, and other researchers. Examples of who should be allocated to which rooms are also given. The current solution departs from the one suggested in that smokers are allocated after all non-smoking researchers, which still produces the desired outcome. The rationale is that the non-smoking mandate is a high-priority constraint.

In a number of cases, two rules are required for correct allocation to large rooms that can hold two occupants. The ordering of these rules is important, as others have found (eg [1], [5]). For example, the allocateSecondResearcher rule looks for a large room in which there is already an occupant. The immediately following rule is allocateFirstResearcher which will fire first in the absence of such half-tenanted rooms, assigning a non-smoking researcher and decrementing the room's vacancy count. Such rule pairs have already been noted. An earlier version of the program used an actor to obtain a decremented vacancy count in a rule's precondition block, which was utilised in a post-condition graph of that rule. Since the vacancy count is first 2 then 1, actors are overkill in this context, hence the current solution which simply retracts a room graph with a vacancy count of 2 and asserts an otherwise identical room graph with a vacancy count of 1. This is also faster since actor execution involves additional execution steps.

The final rule in the Sisyphus process (`allocateRemainingResearcher`) uses type subsumption on the `Room` concept. Concept and relation type hierarchies are declared before the process definition, but only two actually matter:

```
concept Room > SingleRoom;
concept Room > LargeRoom;
```

This permits the final rule to ask for any room, rather than a single or large room. These types were inspired by [15]. An earlier version of the pCG program instead had a size relation in each room graph. While using a size concept such as [`Size: *dontCareHowBig`] and an associated relation in precondition graphs works, having one less node in each room graph permits more efficient precondition matching, and a more elegant solution based upon type subsumption rather than referent specialisation.

As already mentioned, the program solves both problem statements, however *no rule* imposes a constraint which ensures that people on the same project are not allocated to the same room. The process works without such a constraint given the supplied ordering of YQT members in [9], but if the order of graphs originally asserted in the KB were to change, this constraint could be violated. One approach is for graphs representing two different individuals to be compared in terms of the coworker or project relations, perhaps via a `NotSame` actor. However, identical precondition graphs in a single pCG rule will yield identical matches, for example:

```
[Person: *a '*name1'][Project: *b '*proj1'](Member?a?b)
[Person: *a '*name2'][Project: *b '*proj2'](Member?a?b)
```

This suggests that process rules ought to employ backtracking, essentially treating the conjunction of precondition graphs as a goal, to ensure that different `*proj1` and `*proj2` combinations are tried. It is not currently possible in a single precondition block to find all combinations of graphs in the KB which match the graphs in that precondition. One can only do this in pCG via explicit iteration and projection over a KB's graphs. [14] and [15] also point to a possible solution in terms of the application of constraints. This problem requires further investigation.

Information that is not used in the current solution is the project, hacker, and coworker concepts and their associated relations. An interesting question is whether the current solution could be made more robust by making use of some or all of these relations. However, coworker information sometimes does not correspond to project membership as obtained from the information provided in [9].

6 Conclusions and Future Work

We believe that pCG shows promise as a general CG language, and in particular as a means by which to explore Mineau's process mechanism, but improvements are of course possible. An operational solution to the Sisyphus-I room allocation problem was presented. The pCG language was used as the basis for this, and while successful in solving the problem, the attempt revealed shortcomings in both the solution and the language. The requirement that two researchers on the same project should not share an office might be solved by an application of constraints (eg [14], [15]), or by treating the conjunction of preconditions in a rule as a goal and employing backtracking if necessary. If such functionality were implemented in pCG, the language's problem solving power would be increased.

The generic `post..end` syntax in rules ought to be replaced by `assert..end` and `retract..end`, each of which would contain a sequence of simple graphs instead of ERASURE and PROPOSITION contexts. The content of `_KB.corefvars` could be made available to process action block code on the local stack frame of the executing code.

Where complex pre or post-condition graphs are required, a CG graph editor capable of generating CGIF can be used, and later read from a file for use in a process definition. Perhaps a graphical front end could be used to draw and generate complete process definitions.

The pCG language has the advantage of simplicity, and being interpreted, "compile" is removed from the edit-compile-run cycle, speeding up development of pCG programs. It is also extensible by the addition of Java code, as has been noted. However, it is not a refinement of a widely used existing language.[7] Given that the implementation of pCG cleanly separates the parser (around 500 lines of code) from the core interpreter, the source language for pCG could just as easily be Lisp, if so desired. Feedback from the CG community would be of benefit with respect to the usability of pCG.

Other future work relating specifically to pCG is detailed in [3]. Such work includes the reconciliation of the graph assertion and retraction mechanism with the rules of inference [17], [18], improvements to graph matching efficiency, implementation of the conformity relation, and the addition of further attributes and operations to pCG objects to make the intrinsic types more useful.

References

1. Baget, J.F., Genest, D., and Mugnier, M.L., A Pure Graph-Based Solution to the SCG-1 Initiative. In *Proceedings of the 7th International Conference on Conceptual Structures*, Blacksburg, VA, USA, pp. 355–376, July 1999.
2. Bos, C., Botella, B., and Vanheeghe, P., Modelling and Simulating Human Behaviours with Conceptual Graphs. In *Proceedings of the 5th International Conference on Conceptual Structures*, Seattle, Washington, USA, pp. 275–289, August 1997.
3. Benn, D., Implementing Conceptual Graph Processes. Master of Computer and Information Science Minor Thesis, April 2001, University of South Australia. URL: http://www.adelaide.net.au/~dbenn/Masters/
4. Cyre, W., Executing Conceptual Graphs, In *Proceedings of the 6th International Conference on Conceptual Structures*, Montpellier, France, pp. 51–64, August 1998.
5. Damianova, S. and Toutanova, K., Using Conceptual Graphs to Solve a Resource Allocation Task. In *Proceedings of the 7th International Conference on Conceptual Structures*, Blacksburg, VA, USA, pp. 355–376, July 1999.
6. Delugach, H., Dynamic Assertion and Retraction of Conceptual Graphs. In *Proceedings of the 6th Annual Workshop on Conceptual Graphs*, Eileen C. Way, editor, SUNY Binghampton, New York, pp. 15–26, July 1991.
7. Ellis, G., Object-Oriented Conceptual Graphs. In *Proceedings of the 3rd International Conference on Conceptual Structures*, Santa Cruz, CA, USA, pp. 144–157, August 1995.

---

[7] For example, Prolog+CG (http://www.insea.ac.ma/CGTools/CGTools.htm).

8. Kabbaj, A., Synergy: A Conceptual Graph Activation-Based Language. In *Proceedings of the 7ᵗʰ International Conference on Conceptual Structures*, Blacksburg, VA, USA, pp.198–213, July 1999.

9. Linster, M., Documentation for the Sisyphus-I Room Allocation Task, [Accessed Online: November 2000], URL: http://ksi.cpsc.ucalgary.ca/KAW/Sisyphus/Sisyphus1/

10. Lukose, D., Executable Conceptual Structures. In *Proceedings of the 1ˢᵗ International Conference on Conceptual Structures*, Quebec City, Canada, pp. 223–237, August 1993.

11. Mann, G.A., A Rational Goal-Seeking Agent using Conceptual Graphs. In *Proceedings of the 2ⁿᵈ International Conference on Conceptual Structures*, College Park, Maryland, USA, pp. 113–126, August 1994.

12. Martin, P. and Eklund, P., WebKB and the Sisyphus-I Problem. In *Proceedings of the 7ᵗʰ International Conference on Conceptual Structures*, Blacksburg, VA, USA, July 1999.

13. Mineau, G., From Actors to Processes: The Representation of Dynamic Knowledge Using Conceptual Graphs. In *Proceedings of the 6ᵗʰ International Conference on Conceptual Structures*, Montpellier, France, pp. 65–79, August 1998.

14. Mineau, G., Constraints on Processes: Essential Elements for the Validation and Execution of Processes. In *Proceedings of the 7ᵗʰ International Conference on Conceptual Structures*, Blacksburg, VA, USA, pp. 66–82, July 1999.

15. Mineau, G., Constraints and Goals under the Conceptual Graph Formalism: One Way to Solve the SCG-1 Problem. In *Proceedings of the 7ᵗʰ International Conference on Conceptual Structures*, Blacksburg, VA, USA, pp. 334–354, July 1999.

16. Nenkova, A. and Angelova, G., User Modelling as an Application of Actors. In *Proceedings of the 7ᵗʰ International Conference on Conceptual Structures*, Blacksburg, VA, USA, pp. 83–89, July 1999.

17. Sowa, J.F., Conceptual Structures: Information Processing in Mind and Machine, Addison-Wesley, 1984.

18. Sowa, J.F., Knowledge Representation: Logical, Philosophical, and Computational Foundations, Brooks/Cole, 2000.

19. Thanitsukkarn, T. and Finkelstein, A., Multiperspective Analysis of the Sisyphus-I Room Allocation Task Modeled in a CG Meta-Representation Language. In *Proceedings of the 7ᵗʰ International Conference on Conceptual Structures*, Blacksburg, VA, USA, July 1999.

Author Index

Lecture Notes in Artificial Intelligence (LNAI)

Lecture Notes in Computer Science